Crop Pest Management
At a Glance

(for Agricultural Competitive Exams)

The Authors

Mr. Rakesh Kumar Sharma is Assistant Director of Agriculture at Department of Agriculture, Bilaspur (C.G.). He has also worked as Seed Inspector, Senior Surveillance Inspector and SRF in Indira Gandhi Krishi Vishwavidyalaya, Raipur. He received his M.Sc (Ag.) degree in 2008 from Indira Gandhi Krishi Vishwavidyalaya, Raipur (C.G.). He has qualified JRF (Agronomy) during 2006 and National Eligibility Test (ASRB, ICAR). He has published 7 books namely *"Guide for Agricultural Entrance Examinations"*, *"Agriculture at a Glance"* (in Hindi and English), *"Question Bank for Agricultural Competitions"*, *"Basic of Agriculture for Engineers"*, *"Agricultural Competition Explorer"*, *"Agricultural Extension Explorer"* many research papers and technical articles. He has vast experience of teaching of under-graduate programme,, research and extension activities.

Mr. Hemkant Chandravanshi is a Ph.D. Scholar in the Department of Entomology at College of Agriculture, IGKV, Raipur, (C.G.). Presently he is the State President of AIASA Chhattisgarh-2016 (All India Agricultural Students Association) and Former University President of Indira Gandhi Krishi Vishwavidyalaya Students Union 2014-15. He received his B.Sc. (Ag.) in 2013 and M.Sc. (Ag.) degree in 2015 in the field of Entomology from Indira Gandhi Krishi Vishwavidyalaya, Raipur, Chhattisgarh. He was awarded "Best NSS Volunteer Award" in 2011-12.

Mr. Chandramani Sahu is a Ph.D. Scholar in the Department of Entomology at College of Agriculture, IGKV, Raipur, (C.G.). He received his M.Sc. (Ag.) degree in 2014 from Indira Gandhi Krishi Vishwavidyalaya, Raipur, Chhattisgarh. He has qualified National Eligibility Test (NET) during 2015. He is now continuing his research on biological control of insect pests. He has published many research papers and technical articles and has a lot of experience of teaching of under-graduate programme and research activities.

Ms. Mohanisha Janghel is a Ph.D. Scholar in the Department of Entomology at Orissa University of Agriculture and Technology, Bhubaneswar. She received his B.Sc. (Ag.) in 2012 from IGKV, Raipur, (C.G.) and M.Sc. (Ag.) degree in 2014 at O.U.A.T Bhubaneswar. She has qualified JRF and SRF, 3 gold medals in post graduation, awarded inspire DST fellowship during Ph.D programme. She has published many research papers and technical articles and has experience of teaching of under-graduate programme.

Crop Pest Management
At a Glance

(for Agricultural Competitive Exams)

R.K. Sharma
H.Chandravanshi
C.Sahu
M. Janghel

2017
Daya Publishing House®
A Division of
Astral International Pvt. Ltd.
New Delhi – 110 002

ISBN 9789386071019 (Paperback)

Published by : **Daya Publishing House®**
A Division of
Astral International Pvt. Ltd.
– ISO 9001:2008 Certified Company –
4760-61/23, Ansari Road, Darya Ganj
New Delhi-110 002
Ph. 011-43549197, 23278134
E-mail: info@astralint.com
Website: www.astralint.com

Preface

The book entitled *"Crop Pest Management: At a Glance"* is a compilation and a fairly competitive work of Entomology subject covering overall Crop pest management (*Kharif* and *Rabi*) syllabus of B.Sc. (Ag.) course. This book consists of topic on Entomology dealing with both basic and applied aspects of Entomology. The book would be of important for the graduate students and to those students appearing for various competitive examinations. All these examinations are mostly objective based and students always look for study material that is ready to use and easy to grasp.

The present book *"**Crop Pest Management: At a Glance**"* has been prepared in most simple, clear and appropriate manner. While preparing the book, a wide discussion has been made with the students, teachers and scientists and as per their desire, this book is brought out.

Pests damage crops and have serious impact on the economic output of a farm. In the present book, relevant information and modern way to crop pest management have been discussed. It comprises pest management of cereal crops, pulses, oilseed, fiber crops, cash crops, fruit crops and stored grains. It also provides detailed information on crop losses due to pests and their distribution, host range, life history and nature of damage are presented systematically. Available management practices like preventive, cultural, mechanical and biological measures are discussed. It is hoped that the book will be of interest and benefit to the students, entomologists, plant pathologists, extension workers, farmers, researchers and all others interested in crop pest management.

We humbly thanks to Dr. Narendra Pandey and Dr. K.L. Nandeha, Department of Agronomy, Dr. R.B.Tiwari, Dean COA, Bhatapara and Dr. Vinay Pandey, Dean COAE, Raipur, Mr. Anil Koushik, Assistant Director Agriculture and Mr. Nitesh Gupta, Dept. of Agriculture, Bilaspur to encourage us to compile a best Pest Managment book.

We hope that this book impart basic and competitive knowledge of Entomology fields of agriculture. We will be grateful to readers, if errors are pointed out so that necessary corrections can be incorporated in the book. We humbly welcome critical suggestions for further improvement of this book.

R.K. Sharma

Contents

1

Introduction to Crop Pest

Pest

Any living organism that causes harm to man, his crops or animals or possession or simply cause annoyance to human being, qualifies to refer as pest.

Pest Management

It is the system in the context of associated environment and population dynamics of the pest, utilizes all possible techniques or practices to maintain the pest population that will not cause economic damage or losses. Pests will be dealt in respect with the following points.

Nomenclature: (Taxonomic Position)

Every living organisms are known by common name and scientific name. Particular insects are known by common name in certain area/locality and not all over the world. They are recognized in scientific community by scientific names which consist of two names, *viz. Earias vitella* the first name indicates genera and the second specify the species name. This system of nomenclature is called as binomial system of nomenclature. Similarly, trinomial system of naming is in existence for some insects where in three names are given *e.g. Pyrilla perpusilla coimbatorensis, Pyrilla perpusilla pusana etc.*

Marks of Identification

Description of different developmental stages for *e.g.* shoot fly egg, larva, pupa and adult is important for correct identification of pest.

Hosts

These are the plants on which insect use to feed upon for completion of its life cycle. When main host is not available insect can feed on other hosts for survival is called alternate hosts.

Life History

Means the development of insect for instance in most of the insects development take place from egg to adult stages, *e.g.* Jowar shootfly. The object to study the life history is to find out certain weak points of the insects *viz.* site of population, carry over from one season to next, habit and habitat of pest. These are to be pointed out for deciding the control strategies of the pests.

Nature of Damage

There is hardly any plant which is not infested by the pest. Pests injure to host plants. They damage one or the other parts of the plant *viz.* roots (root feeders), Stem/shoot (stem borers) leaves (leaf feeders), buds, flowers, fruits (fruit borers) and grains also. Depending upon feeding habit, pests are categorized as sucking pests and chewing pests. Accordingly the symptoms are produced on damaged plant parts.

Economic Threshold Level (ETL)

It is the pest density at which control measures should be applied to prevent an increasing pest population from reaching the economic injury level.

Management of the Pest

While managing the pest, there should be an integrated pest management (IPM) approach in order to keep pest population below a level of economically acceptable damage (ETL).

I - Integration that is harmonious use of multiple methods to control the impact of single pest as well as multiple pests.

P - Pest- any organism that is detrimental to humans including vertebrates and invertebrate or weed or pathogens.

M - Management refers to a set of decisions or rules based on ecological principles, economic and social consideration.

Different methods of pest control/components or tools of IPM are:

1. Cultural methods
2. Mechanical methods
3. Biological methods, and
4. Chemical methods

1. Cultural Methods of Pest Control

The manipulation of cultural practices at an appropriate time for reducing or avoiding pest damage to crops is known as cultural control.

1. Crop rotation
2. Trap crops
3. Tillage
4. Altered timings
5. Clean cultures
6. Pruning and thinning
7. Crop refuse and destruction
8. Soil manuring and fertilization

2. Mechanical Methods of Pest Control

Reduction or suppression of insect pest population by means of manual devices or labour.

1. Hand picking and collection with hand nets and killing insects
2. Provision of preventive barriers – Deep trenches, Bagging/wrapping, Tin bands *etc.*
3. Construction of rat proof godowns
4. Sticky bands around tree trunks
5. Use of an alkathene band around the tree trunks
6. Shaking the trees and bushes by which the insects fall to the ground and they can be collected.
7. Sieving and winnowing against stored grain pests

3. Biological Methods of Pest Control

The successful management of a pest by means of another living organism (parasites, parasitoids, predators and pathogens) that is encouraged and disseminated by man is called biological control. In such programme the natural enemies are introduced, encouraged, multiplied by artificial means and disseminated by man with his own efforts instead of leaving it to nature.

1. **Introduction or classical biological control:** It is the deliberate introduction and establishment of natural enemies to a new locality where they did not occur or originate.
2. **Augmentation:** It is the rearing and releasing of natural enemies to supplement the numbers of naturally occurring natural enemies.
3. **Conservation:** Conservation is defined as the actions to preserve and release of natural enemies by environmental manipulations or alter production practices to protect natural enemies that are already present

in an area or non use of those pest control measures that destroy natural enemies.

5. **Parasite:** A parasite is an organism which is usually much smaller than its host and a single individual usually doesn't kill the host. Parasite may complete their entire life cycle (*e.g.* Lice) or may involve several host species. Or Parasite is one, which attaches itself to the body of the other living organism either externally or internally and gets nourishment and shelter at least for a shorter period if not for the entire life cycle. The organism, which is attacked by the parasites, is called hosts.
6. **Parasitoid:** It is an insect parasite of an arthopod, parasitic only in immature stages, destroys its host in the process of development and free living as an adult. *e.g.*: Braconid wasps.
7. **Predator:** A predator is one which catches and devours smaller or more helpless creatures by killing them in getting a single meal. It is a free living organism throughout its life, normally larger than prey and requires more than one prey to develop.
8. **Pathogens:** Exploitation of disease causing organism *i.e.* Bacteria (B.t.), Viruses (NPV), Fungi, Protozoa and Entomopathogenic nematodes (EPNs) to reduce the population of insect pest below the damaging levels.

2. Chemical Method of Pest Control

Control of insects with chemicals is known is chemical control. The term pesticide is used to those chemicals which kill pests and these pests may include insects, animals, mites, diseases or even weeds. Chemicals which kill insects are called as insecticides.

2

Pest Management of Cereal Crops

(A) PESTS OF RICE

1. Rice Gall Midge (RGM)

Also known as: Asian rice gall midge

Scientific Name, Order and Family

Orseolia oryzae (Wood-Mason)

(Diptera, Cecidomyiidae)

Taxonomical Position/Scientific Classification

Kingdom:	Animalia
Phylum:	Arthropoda
Class:	Insecta
Order:	Diptera
Family:	Cecidomyiidae
Genus:	Orseolia
Species:	*O. oryzae*

Biology and Life Cycle

1. **Egg**: Eggs are laid singly or in small groups of 2-6 usually on the underside of leaf blade or leaf sheath. Single female lays 100-300 eggs. They are Long, tubular, and shiny white, sometimes as pink, red or yellow, elongate, eggs hatch in 3-4 days.

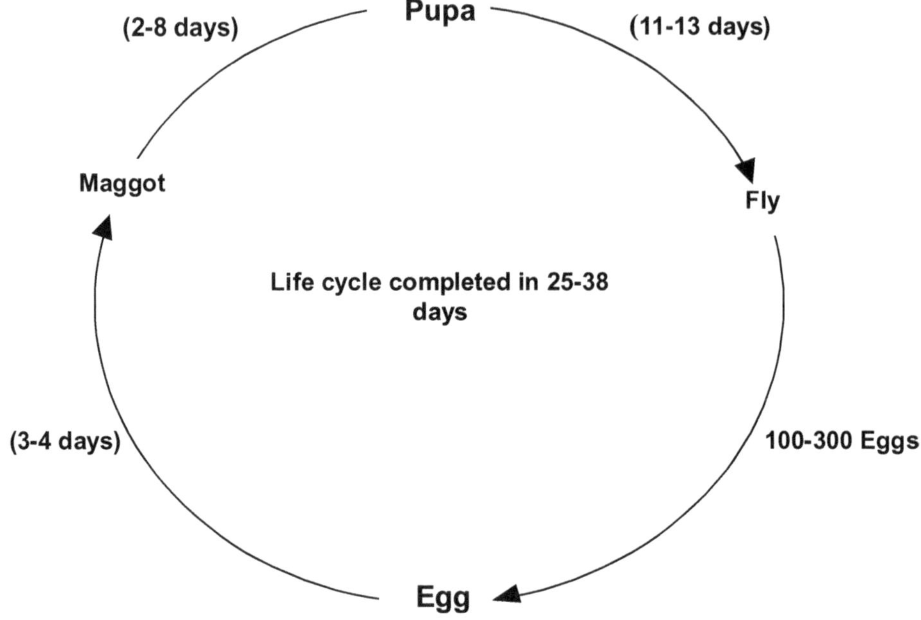

Figure 1: Life Cycle of *Orseolia oryzae*.

2. **Maggot**: Maggot is pale reddish, apodous moves down to the shoot apex without boring into plant tissue. Throughout its development it feeds at the base of the apical meristem leading to suppression of apical meristem, formation of radial ridges from inner most leaf primordium and elongation of leaf sheath. Three to four larval stages are observed in 15-20 days.
3. **Pupa**
   - Pupation is 2-8 days.
   - Before adult emergence, the pupa uses its abdominal spines to reach the top of the Gall.
   - It makes exit holes for adult emergence.
4. **Adult/Fly**: Fly is mosquito like and is 3-3.5 mm long. Female has bright orange red abdomen, swifter with a reddish telescopic body. Male is darker and smaller. Adult longevity is 1-3 days.
   - The Gall (modified leaf sheath) enlarges at the base of the tiller as a result of the continuous feeding of the maggot.
   - Larval and pupal development is completed inside the Gall.
   - One maggot occupies one tiller.
   - The entire life cycle is completed in 25-38 days.

Marks of Identification

Adult is a dipteran fly with long slender legs. Female is bright red while male is darker in colour. Maggot is pale red, apodous, tapering anteriorly.

Host Crops

Rice (Monophagous).

Nature and Symptoms of Damage

- ☆ Newly hatched maggot moves down from the leaf blade to the shoot apex and feeds on the growing point. It results in the stim form stimulation of leaf sheath to form a hallow pale green cylindrical tube similar to onion leaf. It is known as Gall/Onion shoot/silver shoot.
- ☆ The feeding by the maggot and the larval secretion, which contains an active substance called cecidogen, is responsible for cell proliferation of the meristematic cells and gall formation.
- ☆ Affected tillers do not bear panicles. Infestation in early growth period of crop induces vigorous subsidiary tillerings.

Economic Threshold Level (ETL)

One silver shoot/m^2 or 5 per cent affected tillers.

Management

(A) Preventive Measures

1. Avoid late transplanting in endemic areas. Early planted kharif crop escapes pest.

(B) Cultural Measures

1. Ploughing under the ratoon of previous crops can reduce infestation.
2. Control of grassy weeds and wild rice (alternate hosts) from surrounding areas can reduce gall midge incidence.
3. Draining of rice fields for 5-7 days affects midge populations.
4. Planting of early and using early maturing varieties may help to avoid high infestations.
5. Grow gall midge resistant varieties *i.e. Ruchi, Abhya, Mahamaya, Danteshwari, Indira sugandhit dhan-1, Karma masuri, Indira sona* (hybrid), *Samleshwari, Chandrahasni, Jaldubi etc.*
6. Using only moderate amounts of nitrogen and potassium fertilizers and adopting split applications to reduce population growth rates.
7. Avoiding staggered planting (complete planting in an area within 3 weeks) to reduce infestation.
8. Seed treatment with Chlorpyriphos 0.2 per cent emulsion for 3 hours or seed mixing with either Chlorpyriphos (0.75 kg a.i./100 kg seeds) or

Imidacloprid (0.5 kg a i./100 kg seeds) provide protection for 30 days in the nursery.

(C) Mechanical Measures

1. Install light traps.

(D) Chemical Measures

1. Seedling root dip in 0.02 per cent chlorpyriphos emulsion before transplanting for 12 -14 hours gives protection for 30 days.
2. Granule application of Phorate 10 G @ 10 kg/ha or carbofuran 3G @ 25 kg/ha or Quinolphos 5 G @ 10 kg/ha after 10 and 30 days after transplanting in presence of thin film of water in the field.

(E) Biological Measures

1. **Parasitoid:** *Platygaster oryzae* – Larval parasitoides control RGM at early phase.
2. **Predator:** Wolf Spider ***Pardosa psuedoannulata*** **and Lynx spider** ***Oxyopes javanus*** **feeds on maggots.**

2. Rice Stem Borer

Also known as - Yellow Stem Borer

Scientific Name, Order and Family

Scirpophaga incertulas Walker

(Lepidoptera, Pyralidae)

Taxonomical Position/Scientific Classification

Kingdom:	Animalia
Phylum:	Arthropoda
Class:	Insecta
Order:	Lepidoptera
Family:	Crambidae
Genus:	Scirpophaga
Species:	***S. incertulas***

Biology and Life Cycle

1. **Eggs**: Eggs are round and are covered with buff-coloured hairs and laid on upper leaf surface in masses of 15-80. The number of eggs laid by a single female moth is upto 150. Eggs hatch in about 5-10 days.
2. **Larvae**: Fully grown larva is formed in about 20- 40 days depending upon the climatic conditions. A developed larva measures about 20 mm in length and is white or yellowish-white in colour. Larva makes an exit

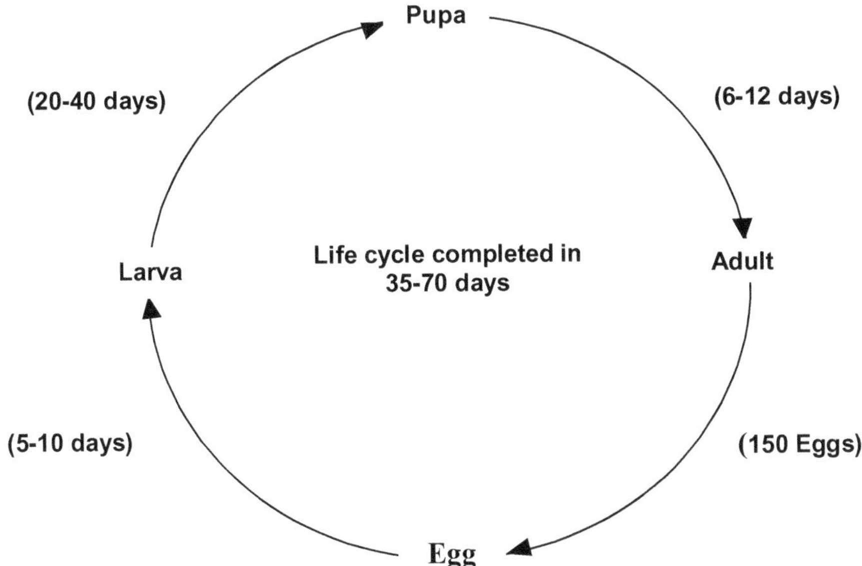

Figure 2: Life Cycle of *Scirpophaga incertulus*.

hole and pupates within the larval tunnel, usually at the base of the plant. The cocoon is silky white in appearance. It remains in pupal stage for 6-12 days (may prolong upto a month in certain season).

3. **Pupa**: Pupa is yellowish white with greenish tinge and measures about 16 mm long and 2.5 mm wide.
4. **Adult**: Adult is medium sized, 12 mm long, straw coloured moth with wing expanse of 30 mm. Forewings are yellowish, hindwings are whitish. However, one black spot is present at the centre of each of the forewings of female moth.

Marks of Identification

1. Female is bright yellowish brown and male is pale yellow with pointed head. A prominent black spot present on each of the forewing of the female but absent in male. Turf of anal hairs present in female. Caterpillar is white or yellowish white with dark brown head and prothoracic shield.

Host Crops

Rice (Monophagous).

Nature and Symptoms of Damage

- Newly hatched larva moves on the leaf blade for 1-2 hours and later reaches the leaf sheath and bores into the stem near the nodal region at ground level.
- When this type of damage occurs during vegetative phase of the crop, the central leaf whorl does not unfold, turn brownish and finally dry off. The

lower leaves on the same plant remain green and healthy. This condition is commonly known as 'dead heart'. Affected tillers dry out without producing panicles.

☆ Larval feeding inside the stem results in the formation of "dead heart" and 'white ear head" during vegetative and reproductive stages, respectively which can be easily pulled out.

☆ The pest is active from seedling stage till the maturity of crop. In kharif and rabi crops, the dead hearts are formed in months of August-September and February-March, respectively while the white heads are formed during October-November and April-May, respectively.

Economic Threshold Level (ETL)

1. 1 egg mass per square metre area.
2. 5-10 per cent dead hearts appear in the field.

Management

(A) Preventive Measures

1. Destruction of stubbles after harvest by ploughing in or burning helps in reducing carry over of the borer larvae.

(B) Cultural Measures

1. Clipping the tips of seedlings before transplanting greatly reduces the carryover of eggs from the seedbed to the transplanted fields.
2. Rice varieties with short stature and shorter growth duration periods suffer less damage than long growth duration varieties.
3. Rice with shorter growth duration varieties suffer less damage than long duration varieties. This may be because of stem-borer mortality due to harvests occurring twice in the double cropping system.
4. Destruction of stubbles through tillage after harvest, followed by flooding, reduces stem borer populations resulting in low incidence in the next crop.
5. Planting or seeding times may be delayed to avoid the peak emergence of moths from the diapausing populations.
6. Rice seedbeds may be used as a trap crop for moths emerging from diapause.

(C) Mechanical Measures

1. Setting up of light traps to trap and destroy the moth.
2. Install sex pheromone traps to monitor and mass trap.

(D) Chemical Measures

(I) Nursery treatment: Apply any one of the following granules at 15 days after sowing: Phorate 10 G @ 20 kg/ha or Carbofuran 3 G @ 25 kg/ha or Quinalphos 5 G @ 15 kg/ha or Sevidol® (carbaryl + lindane) 4:4 G @ 25 kg/ha. Soil must be saturated with water at the time of application of granules. The water level of 3 to 4 inches should be maintained in the field atleast for four days after treatment. The water should not be drained in or drained out during above period. If sufficient moisture is not there in the field, apply following insecticides in the nursery 15 days after sowing: Fenitrothion 50 EC @ 0.08 per cent or Quinalphos 25 EC @ 0.08 per cent or Phenthoate 50 EC @ 0.08 per cent.

(II) Field treatment: Apply carbofuran 3 G @ 25 kg or Benfuracarb 3G @ 33 kg or or chlorantraniliprole 0.4G @ 10 kg or Cartaphydrochloride 4G @ 18.75 kg or spray acephate 75 SP @ 600-1000 gm or cartap hydrochloride 50 SP @ 1 kg or Quinalphos 25 EC @ 1.0 lt or Azadirachtin 0.15 W/W @ 1.5-2.5 lt or Azadirachtin 5 per cent 400 ml or Fipronil 5 SC @ 1-1.5 lt or fipronil 80 WG @ 50-62.5 gm or Phosphamidon 40 SL @ 1.25 lt or Thiacloprid 21.7 SC @ 500 ml or Thiamethoxam 25 WG 100 gm per ha using water @ 500 Litre/ha.

(E) Biological Measures

1. Releasing of Egg parasitoids – *Tetrastichus schoenobii* and *Trichogramma japonicum* (released 3 to 4 times at weekly interval @ 50,000 parasitoides/ha), *Trichogramma australicum*.
2. Releasing of Egg larval parasitoids – *Platygaster oryzae*.
3. Apply *Bacillus thuringiensis* var *kurstaki* and neem seed kernel extract in the combination of 2.5 g/l and 1 per cent to reduce the oviposition by the stem borer.

3. Green Leaf Hopper (GLH)

Also known as: Paddy leaf hopper.

Scientific Name, Order and Family

Nephotettix virescens (Dist.),

Nephotettix nigropictus (Stal.)

(Hemiptera; Cicadellidae)

Taxonomical Position/Scientific Classification

Kingdom:	Animalia
Phylum:	Arthropoda
Class:	Insecta
Order:	Hemiptera
Family:	Cicadellidae
Genus:	Nephotettix

Species: ***N. virescens***

N. nigropictus

Biology and Life Cycle

1. **Egg**: Female lays 50-75 eggs in mosses or groups in the midrib of leaf blade or leaf sheath. The pre-oviposition period usually ranges between 3 to 6 days. Eggs hatch out in 4 to 7 days.
2. **Nymph**: Nymphs of most species moult 5 times. Nymphal period varies from 14-25 days.
3. **Adult**: The active period of insect is August to October. The total life cycle of the insect is completed in 15-28 days.

Marks of Identification

Adult of both the species of green leaf hoppers have grass green forewings with a terminal dark patches. They are greenish, wedge shaped insect measuring about 4 to 5 mm in length. The presence of a distinct submarginal black band in the vertex of *Nephotettix nigropictus* distinguishes it from *N. virescens*. Nymphs of both the species are smaller than adults, wingless and pale green in colour.

Host Crops

Rice, millets, grasses

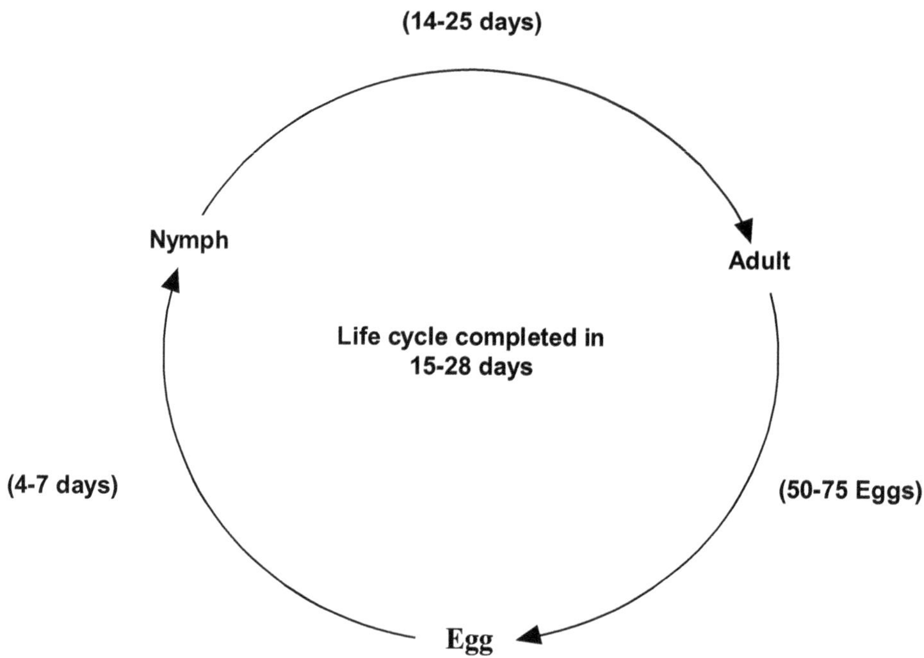

Figure 3: Life Cycle of *Nephotettix virescens*.

Nature and Symptoms of Damage

- Both nymphs and adults congregate in large number on leaf sheaths and leaf blades and suck the cell sap from them.
- As a result of pest infestation, leaves turn pale yellow initially, later on become brownish and finally dry up producing the symptom of "***hopper burn'***.
- They secret honeydew on the leaves and develop shooty mould. It interferes with the photosynthesis activity of plant.
- The vigour of the plant is reduced. Both the species of leaf hoppers are known to act as vector of some viral diseases *i.e.* Yellow dwarf, Tungro, Yellow orange leaf.
- The peak activity of leaf hopper population is in October or November.

Economic Threshold Level (ETL)

Nursery – 60 hoppers/25 sweeping

Flowering stage - 10 hoppers/hill

Vegetative stage - 5 hoppers/hill

Tungro endemic area - 2 hoppers/hill

Management

(A) Cultural Measures

1. Reducing the number of rice crops to two per year and synchronized establishment across farms reduces leafhoppers and other insect vectors of rice virus or phytoplasma diseases.
2. Transplanting older seedlings (>3 weeks) also reduces viral disease susceptibility transmitted by leafhoppers.
3. Avoid planting at peak activity (shown by historical records) period to avoid infestation.
4. Early planting within a given planting period, particularly in the dry season, reduces the risk of insect-vector disease.
5. Nitrogen should be applied at an optimal level to discourage population buildup and influence plant recovery.
6. Good weed control in the field and on the bunds removes the preferred grassy hosts and promotes crop vigor.
7. Crop rotation with a non-rice crop during the dry season decreases disease reservoirs.

(B) Mechanical Measures

1. Install light traps, only for 6-10 light hours.

(C) Chemical Measures

1. Spraying of Chloropyriphos 20 EC @ 2-2.5 lt/ha or Quinolphos 25 EC @ 1.5-2.0 lt/ha or Imidachloprid 17.8 SL @ 100 ml/ha or dust with Carboryl 50 WP @ 2 kg/ha.
2. Spray application of phosalone or etofenprox or cartap hydrochloride or acephate or chlorpyriphos or carbaryl, at 0.5 kg a.i./ha or fipronil at 50 g a.i./ha or application of granular insecticides such as phorate or sevidol or cartap hydrochloride or carbofuran at 1 kg a.i./ha or fipronil 0.3 per cent G @ 25 kg/ha.

(D) Biological Measures

1. Release of egg parasitoides *i.e. Oligosita* sp., *Tetrastichus* sp. and *Anargus* sp.
2. Release of nymphal and adult parasitoides *i.e. Haplogonatopus* sp.
3. Release of predators *i.e. Coccinellid beetle,* Mirid bug, Spiders, *Lycosa* sp. *etc.*

4. Brown Plant Hopper (BPH)

Scientific Name, Order and Family

Nilaparvata lugens Stal.

(Hemiptera; Delphacidae)

Taxonomical Position/Scientific Classification

Kingdom:	Animalia
Phylum:	Arthropoda
Class:	Insecta
Order:	Hemiptera
Family:	Delphacidae
Genus:	Nilaparvata
Species:	*N. lugens*

Biology and Life Cycle

1. **Egg**: Eggs are laid in a group of 2 to 12 in leaf sheath (near the plant base or in the ventral midribs of leaf blades). White, transparent, slender cylindrical and curved eggs are thrust in straight-line in two rows. They are covered with a dome - shaped egg plug secreted by the female. A single female lays about 100 to 250 eggs. Eggs hatch out in 4 to 8 days.
2. **Nymph**: Freshly hatched nymph is cottony white, 0.6 mm long and it turns purple-brown, 3.0 mm long in the fifth instar. Nymphs of most species moult 5 times. Nymphal period varies from 13-15 days.

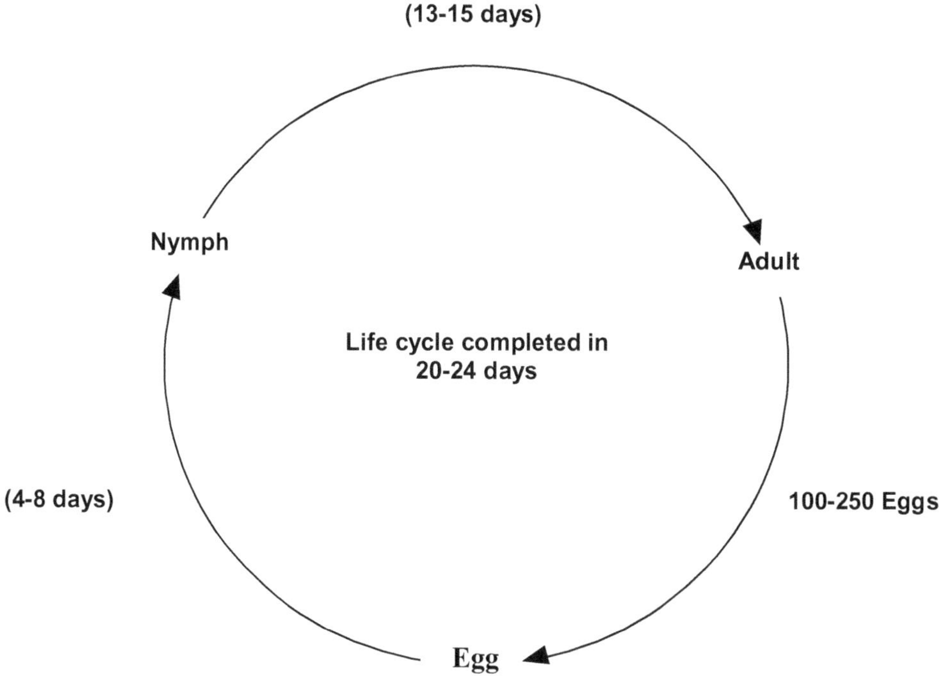

Figure 4: Life Cycle of *Nilaparvata lugens*.

3. **Adult**: Adult hopper is 4.5-5.0 mm long and has a yellowish brown to dark brown body. The wings are sub hyaline with a dull yellowish tint. It has two characteristic wing morphs: macropterous (long winged) and brachypterous (short winged).
4. The total life cycle of the insect is completed in 20-24 days.

Marks of Identification

1. Nymph is pale white with brownish tinge having grayish blue eyes.
2. Adult is brown to brownish black. Both short and long winged adults are produced. Males are comparatively smaller and darker in colour. Abdominal tip of female is rounded.

Host Crops

Rice, sugarcane and some grasses.

Nature and Symptoms of Damage

- ☆ Both nymphs and adults suck the sap from basal parts of the plant causing yellowing of leaves followed by complete drying up of whole plant upto the base and lodging of the crop.
- ☆ Infestation starts in patches and later spreads to entire area. There is no grain formation if attack appears before flowering.

- ☆ In post flowering attack, yield is reduced by 20 to 30 per cent.
- ☆ These plant hoppers show negative phototaxis and prefer high humidity. They congregate in areas of luxuriant plant growth and multiply near the basal parts of the plants.
- ☆ The brown plant hopper occurs in association with other species of leaf and plant hoppers. Their larger population results into characteristic drying up of crop in patches, which is commonly known as "hopper burn".
- ☆ The plant hoppers are known to transmit a viral disease called 'grassy stunt' but is has not been reported so far from India. Pest is invariably noticed in ill-drained soils.

Economic Threshold Level (ETL)

8-10 hoppers/hill or 20 hoppers/hill when spider is present at 1 hopper/hill.

Management

(A) Cultural Measures

1. Avoiding high dosages of nitrogenous fertilizers, close spacing, and high relative humidity which increases planthopper populations.
2. Sensible use of fertilizer by splitting nitrogen applications can also reduce chances of plant hopper outbreaks.
3. Draining rice fields can be effective in reducing initial infestation levels. The field should be drained for 3 - 4 days when heavy infestations occur.
4. Growing no more than two crops per year and using early-maturing varieties reduces planthopper abundance and damage.
5. Synchronous planting (planting neighboring fields within 3 weeks) and maintaining a rice-free period may be effective.
6. Use BPH resistant varieties *i.e.* IET-7568, IET-7575, IET-6315, IET-7943, IET-8115 *etc.*

(B) Mechanical Measures

1. Set up light traps to monitor and control pest population.

(C) Chemical Measures

1. Spray neem seed kernel extract 5 per cent (25 kg/ha) (or) neem oil 2 per cent (10 L/ha).
2. Spray imidacloprid 17.8 SL 125 ml or buprofezin 25 SC 325 ml or acephate 75 SP 625 per ha.
3. Spraying carbaryl (0.75 kg a.i./ha) or etofenprox or moncrotophos or phosalone or chlorpyrifos @ 0.5 kg a.i./ha or lindane 20 EC at 1 litre/ha in the early stages of the crop.

(D) Biological Measures

Release of natural enemies like wolf spider (*Lycosa pseudoannulata)* and green mirid bug (*Cyrtorrhinus lividipennis).*

1. Egg parasite – *Anagrus* spp.
2. Nymphal parasite – *Gonatopus* spp.
3. Egg and nymphal predator – *Cyrtorhinus lividipennis*
4. Coccinelid beetles – *Coccinella arcuata*

5. White Backed Plant Hopper (WBPH)

Scientific Name, Order and Family

Sogatella furcifera Horvath

(Hemiptera : Delphacidae)

Taxonomical Position/Scientific Classification

Kingdom:	Animalia
Phylum:	Arthropoda
Class:	Insecta
Order:	Hemiptera
Family:	Delphacidae
Genus:	Sogatella
Species:	***S. furcifera***

Biology and Life Cycle

1. **Egg**: Cylindrical eggs are laid upto 600-750 eggs in as many as 112 egg masses with 1-24 eggs in each in leaf sheath and in the mid rib of leaves. The eggs are dirty white in colour and shaped of *Dolichus lablab*. Egg period lasts from 3-6 days.
2. **Nymph**: White to a strongly mottled dark grey or black and white in colour and 0.6 mm size when young. Nymphal period lasts from 12-17 days with five instars.
3. **Adult**: The adult hopper is 3.5 - 4.0 mm long. The forewings are uniformly hyaline with dark veins. There is a prominent white band between the junction of the wings. Macropterous males and females and brachypterous females are commonly found in the field. One life cycle is completed in 18-24 days.

Marks of Identification

1. **Nymph**: Nymphs are white to dark gray or black and white in colour. In white nymphs, vertex characteristically gives a narrow face to the hoppers. Forewings hyaline with dark veins and a dark spot in the middle

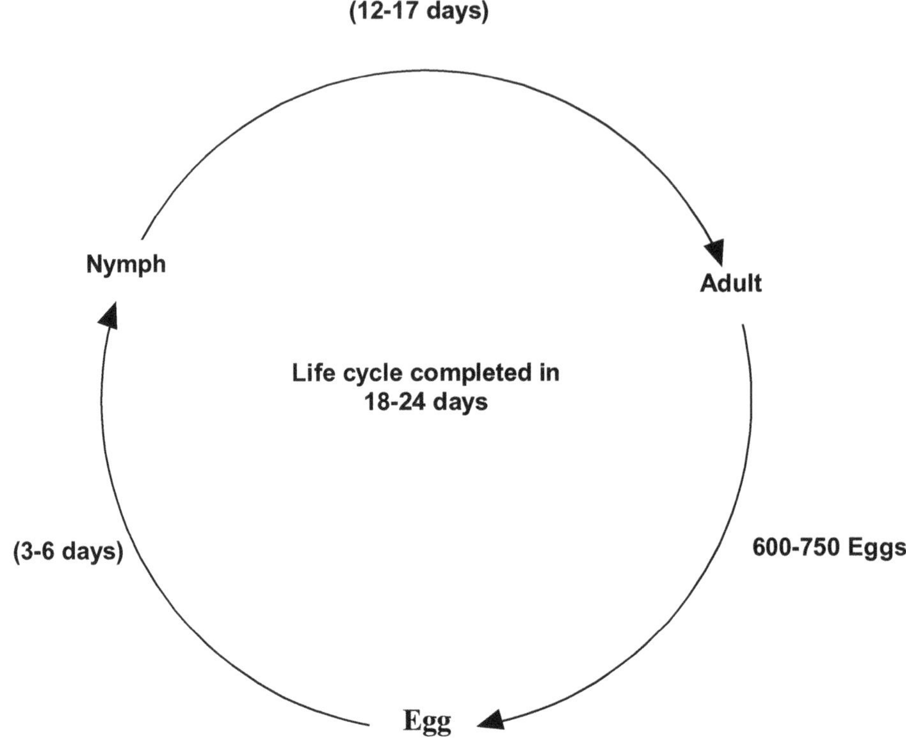

Figure 5: Life Cycle of *Sogatella furcifera*.

of posterior edge. Pronotum pale yellow and adults possess a diamond like marking on the thorax.

2. **Adult**: 3-4 mm long, straw coloured, slightly smaller than green leaf hopper. Very active insect with white strip on their back.

Host Crops

Rice, maize, millets, sugarcane, some grasses.

Nature and Symptoms of Damage

- ✰ Damage is caused by both nymph and adult through feeding and oviposition.
- ✰ Both nymphs and adults suck phloem sap from leaves and stem in general at the base of the plant causing reduced vigour, stunting, yellowing of leaves and delayed tillering and grain formation.
- ✰ Feeding puncture and lacerations caused by ovipositor predispose the plants to pathogenic organisms and honey dew excretion encourages the growth of sooty mould.

- ☆ White backed plant hopper is more abundant during the early stage of the growth of rice crop, especially in nurseries. Young seedlings die due to attack of WBPH.
- ☆ Rice is more sensitive to attack at the tillering phase than at the boot and heading stages.
- ☆ Rice crop fails to produce complete grains and this condition is known as red disease.

Economic Threshold Level (ETL)

10-20 hoppers/hill.

Management

Same as given for BPH.

(A) Cultural Measures

1. Avoid use of excessive nitrogenous fertilizers.
2. Control irrigation by intermittent draining.
3. Synchronous planting within 3 weeks of staggering and maintaining a free-rice period could also decrease the build-up of Brown plant hopper.
4. There are varieties released by IRRI, which contain genes for White backed plant hopper resistance, like IR26, IR64, IR36, IR56, and IR72.
5. Avoid close planting and provide 30 cm rogue spacing at every 2.5 to 3.0 m to reduce the pest incidence.

(B) Mechanical Measures

1. Set up light traps during night. Installation of light traps with incandescent light at 1-2 m height @ 1/acre to monitor the population.

(C) Chemical Measures

1. Spray any one of the following: Phorate 10G @ 10 kg/ha (or) Carbofuran 3 G 17.5 kg/ha (or) Dichlorvos 76 WSC 350 ml/ha.
2. When WBPH population reaches threshold level of 10 insect/hill, apply chlorpyriphos 20 EC @ 2.5 lit./ha or Imidacloprid 17.8 SL @ 80 ml/ha or carbaryl 50 WP @ 2.5 kg/ha, Acephat 75 SP @ 0.625kg/ha or Fibronyl 5 EC@ 600 ml/ha.

(D) Biological Measures

1. Release egg parasitoid *Anagrus* sp. and adults and nymphs of *Pachygonatopus* sp.
2. Effective predators are *Coccinella arcuata, Cyrtorrhinus lividipennis* and *Tytthus parviceps.*

3. Nymphs and adults are eaten by general predators, particularly spiders and coccinellid beetles.
4. Hydrophilid and dytiscid beetles, dragonflies, damselflies, and bugs such as nepid, microveliid, and mesoveliid eat adults and nymphs that fall into the water surface.
5. Spiders, stapphylinid beetles, carabid beetles and lygaeid bugs search the plant for White backed plant hopper nymphs and adults.

6. Leaf Folder of Rice

Also known as: Rice Leaf roller/Patti Modak/Chhitri.

Scientific Name, Order and Family

Cnaphalocrocis medinalis Guenée

(Lepidoptera; Pyralidae)

Taxonomical Position/Scientific Classification

Phylum:	Arthropoda
Class:	Insecta
Order:	Lepidoptera
Family:	Pyralidae
Genus:	*Cnaphalocrocis*
Species:	***C. medinalis***

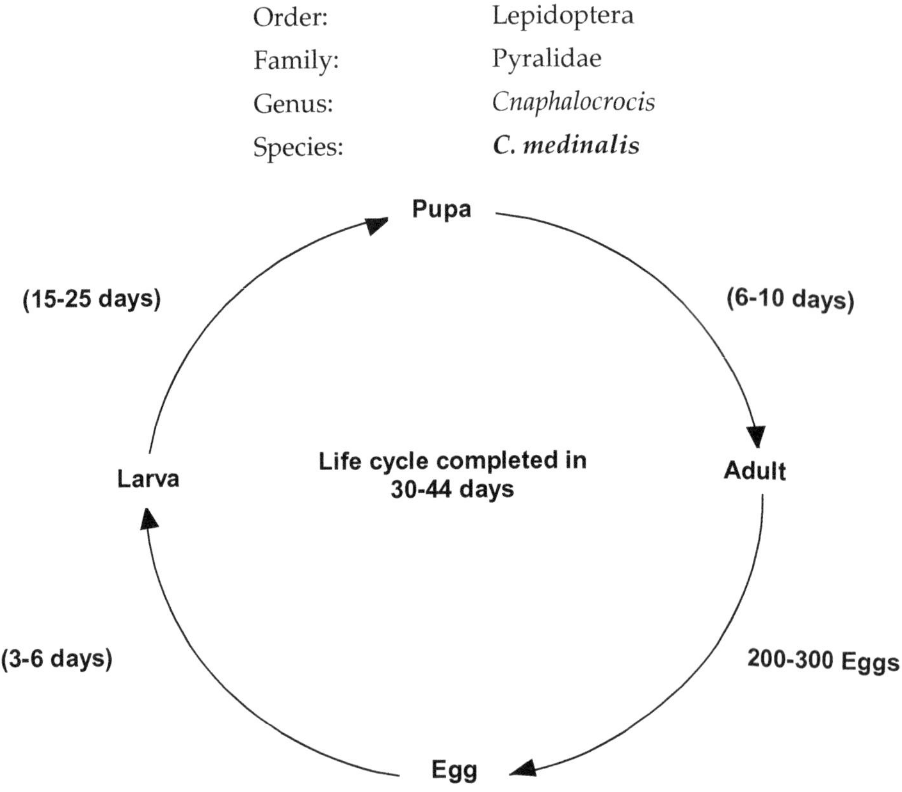

Figure 6: Life Cycle of *Cnaphalocrocis medinalis*.

Biology and Life Cycle

1. **Egg**: Eggs are laid singly or in clusters of 8-12 on leaf blades and leaf sheath. Eggs are flat or oval in shape and yellowish white in colour. The fecundity is 200-300 eggs and incubation period is of 3-6 days.
2. **Larva**: Greenish translucent, prothoracic shield straight apically and rounded laterally. Larvae hatch in 3 to 5 days. The first and second instar larvae are gregarious and generally feed the slightly folded basal regions of theyoung leaves in a tiller. Generally only one larva is found per leaf roll and after feeding on one fold for about 2 to 3 days it moves on to another leaf. The larval period lasts 15 to 25 days, during which time; the larvae may invade on more than three leaves.
3. **Pupa**: Pupation takes place in the leaf rolls and the pupal period lasts for 6 to 10 days.
4. **Adult**: Moth is golden or yellowish brown and with brownish wings with many dark wavy lines in centre and dark band on margin of wings. Total life cycle is completed in 30-44 days and pest completed 4-5 generations in a year.

Marks of Identification

Adult is a small moth with golden yellowish brown wings. Two distinct dark wavy lines present on forewings while only one wavy line on hindwings. Caterpillar is dull white, but it turns to greenish yellow, when fully grown.

Host Crops

Rice, Grasses, Maize, Sorghum.

Nature and Symptoms of Damage

The destructive stage of this insect is the larval stage. The larva/caterpillar folds the leaves longitudinally and remains inside. They tie the leaf margins with the thread like silk. They feed the green mesophyll of leaf inside the folded leaf. It scrapes the green tissues of the leaves and makes them white and dry.

There is one larva in one leaf. After feeding one leaf, larva moves to another leaf. In this way, one larva can damage a number of leaves. Feeding reduce the photosynthesis area of leaves. So vegetative growth and finally yield is hampered. During severe infestation, the whole field exhibits scorched appearance. Yield loss is maximum, when flag leaf is damaged.

Economic Threshold Level (ETL)

One damaged leaf/hill or 10 per cent damaged leaves in vegetative stage, 5 per cent damaged leaves (flag leaf) in flowering stage.

Management

(A) Cultural Measures

1. Remove grass weeds from the bunds around the paddy fields.

2. Early planting may help to avoid greater degrees of leaf damage.
3. Wider spacing (22.5 x 20 cm and 30 x 20 cm) and low usage of nitrogenous fertilizers decreases leaf damage.
4. Highly fertilized plots seem to attract females for oviposition. Therefore, it is advisable to avoid over-fertilization.
5. Egg predators (crickets) inhabit surrounding grass habitats and move to the field at night for predation. Maintenance of non-rice habitats might be worthwhile.
6. Higher damages will occur in shaded areas. Therefore, remove the causes of shading within the field.

(B) Mechanical Measures

1. Set up light traps to attract and kill the moths. Set up light trap one for at least 5 ha.
2. Set up bird perches (40 to 50/ha) of Insectivorous Birds at vegetative phase of crop.
3. Monitor through pheromone traps @ 10 to 12/ha for timely control measures. Change the lure at 15-20 days intervals.

(C) Chemical Measures

1. Spraying the crop with 0.05 per cent Fenitrothion or 0.15 per cent carbaryl taking into consideration 1- 2 leaf rolls/hill ETL.
2. Need based spraying of phosalone or carbaryl or monocrotophos or etofenprox or cartap hydrochloride or quinalphos or spray of fipronil 5 SC at 1 litre/ha.
3. Spraying of any following insecticides at economic threshold level of 10 per cent damage leaves: Fenitrothion at the dose of 454 ml per acre area, Malathion at the rate of 404 ml per acre area, Carbaryl at the rate of 688 g per acre area, Cypermethrin at the rate of 202 ml per acre area.

(D) Biological Measures

1. Release of *Trichogramma chilonis* (Egg Parasitoids) thrice on 37, 44 and 51 DAT @ 5 cc (1 lakh egg parasitoids)/ha/release.
2. Release the egg cards in field during morning hours.
3. Tie the egg cards under the leaf surface facing outside.
4. Avoid spraying of chemicals three days before and upto seven days after field release of egg cards.

Natural Enemies

1. *Itoplectis narangae* - Ichnuemonidae Wasp

Predators

1. Release of mirid bug (*Cytorhinus lividipennis*) @ 50 – 75 egg/m^2.

7. Rice Case Worm

Also known as: Dhan ka Banki

Scientific Name, Order and Family

Nymphula depunctalis Guenée

(Lepidoptera; Pyralidae)

Taxonomical Position/Scientific Classification

Phylum:	Arthropoda
Class:	Insecta
Order:	Lepidoptera
Family:	Pyralidae
Genus:	Nymphula
Species:	***N. depunctalis***

Biology and Life Cycle

1. **Egg**: Eggs are laid singly on the underside of leaves. Individual egg is circular, flattened, and measures 0.5 mm in diameter. It is light yellow and has a smooth surface. Mature eggs are darker and develop two purplish

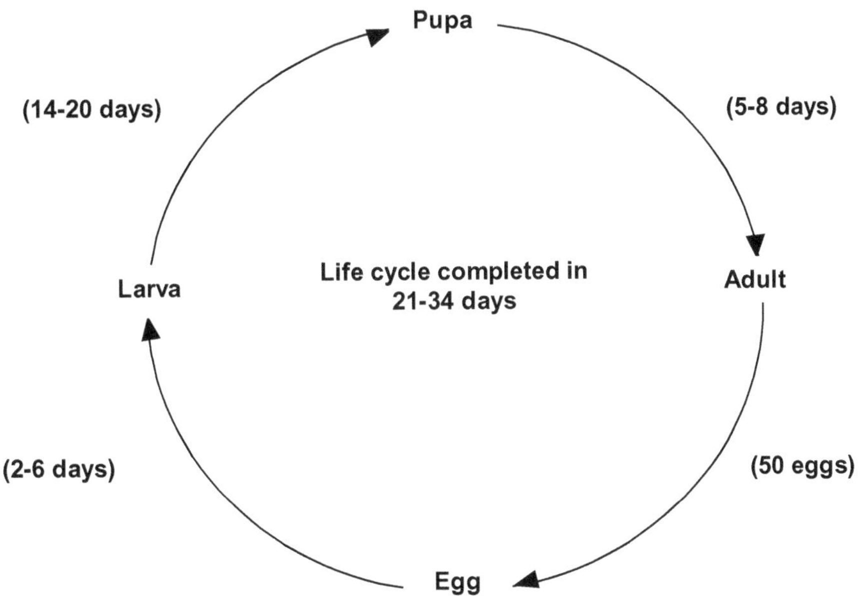

Figure 7: Life Cycle of *Nymphula depunctalis*.

dots. Fecundity is about 50 eggs. The incubation period is completed in 2-6 days.

2. **Larva**: Pale translucent green with orange head. It has filamentous gills on the sides of the body. The larvae are found hanging from the leaf and measures upto 15mm long. The larval period is completed in 14-20 days.
3. **Pupa**: The pupa is cream in colour and about 5.5 mm long. Mature pupa is silvery white. Leaf case attach to the base of tillers. The pupation period is completed in 5-8 days.
4. **Adult**: Moth is small, delicate white with pale brown wavy markings. The adult moth is about 5 mm long. It is bright white with light brown and black spots. The total life cycle occupies 21-34 days.

Marks of Identification

1. **Adults** are small fragile moth with a wing expans of 16 mm. The wings have white speckled.
2. **Full grown larvae** are light green in colour with a light brownish orange head. Larvae are semi-aquatic in habit with filamentous gills on the side of body.

Host Crops

Rice (Monophagous).

Nature and Symptoms of Damage

The early stages of the crop are damaged by the caterpillars of this pest. The leaf blades are eaten away completely leaving the mid rib only. They also construct tubular cases inside leaves and remain inside these leave rolls and feeds upon the foliage. The weeds in rice field serves as alternate host for this insect.

1. Larvae cut small pieces of the leaf tips, form tubular cases, remain inside the case, which is attached to the plant or seen floating on water surface.
2. Larvae feed on foliage by scraping chlorophyll leaving horizontal rows of green material, giving a ladder like appearance.

Economic Threshold Level (ETL)

1-2 case/hill.

Management

(A) Cultural Measures

1. Rice fields with wider hill spacing (30 x 20 cm) usually suffer less damage from caseworm.
2. Early planting may escape the peak caseworm moth activity period.
3. Dislodging the leaf cases from the plant by passing a rope and draining the water later will be helpful.

4. Addition of small quantity of kerosene in the water form a thin film on the water kills the larvae.
5. Draining of fields for 5-7 days kills caseworm larvae.
6. Use of older seedlings reduces the duration of the susceptible stage of the crop.
7. Nitrogen fertilizer use at optimal dosages and split applications reduce the rice caseworm's abundance.

(B) Mechanical Measures

1. Install light traps.

(C) Chemical Measures

1. Avoid the use of granular insecticides where the crop is attacked by case worm every year.
2. Spraying the crop with Quinolphos 25 EC @ 2 litre/ha or Chloropyriphos 20 EC @ 1.4 litre/ha or monocrotophos 36EC @ 850 ml/ha or carbaryl @ 0.15 per cent.

(D) Biological Measures

1. Introduction of *Elasmus* sps; *Apalteles* sps; *Bracon* sps; *Hormius* sps. is effective in controlling the destruction caused by the larvae because they paracitises the larval stages.
2. Introduction of *Pediobius* sp. and *Apsilops* sp. are effective in controlling this pest because they parasitises the pupal stage.

8. Rice Army Worm

Also known as: Oriental armyworm/Rice earhead cutting caterpillar/Dhan ka Surhi.

Scientific Name, Order and Family

Mythimna separate

(Noctuidae: Lepidoptera)

Taxonomical Position/Scientific Classification

Kingdom:	Animalia
Phylum:	Arthropoda
Class:	Insecta
Order:	Lepidoptera
Family:	Noctuidae
Genus:	Mythimna
Species:	***M. separata***

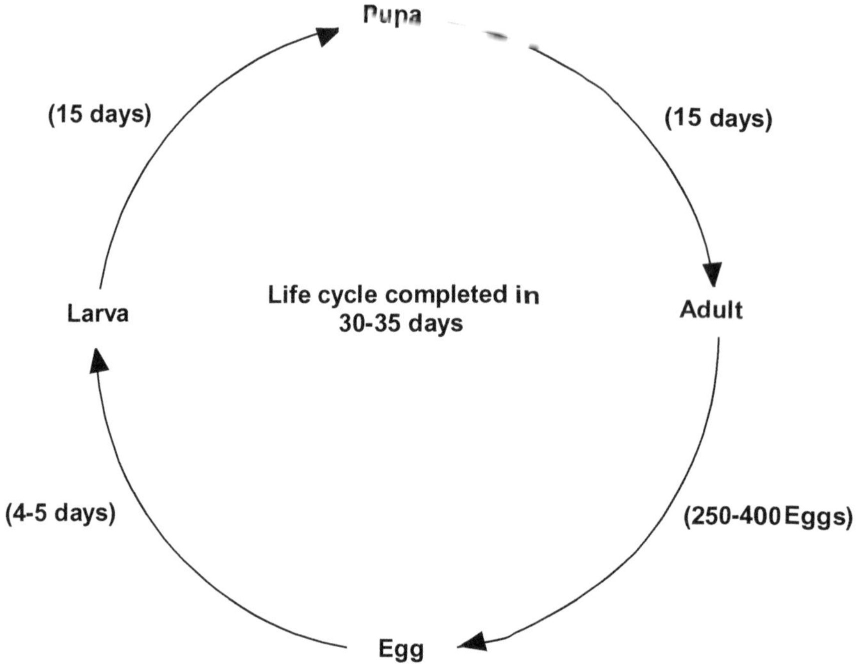

Figure 8: Life Cycle of *Mythimna separata*.

Biology and Life Cycle

1. **Egg**: Eggs are laid singly or in cluster of 20-75 on the inner side of old partially dry leaf sheath. fecundity consisting of approximately 250-400 eggs. The hatching period varies from 4-5 days.
2. **Larva**: After hatching the caterpillars starts feeding on the leaves of the seedlings. Generally the caterpillars move in swarm from one field to the other. The caterpillars are fully grown in about 15 days and measures 3-5 cm in length.
3. **Pupa**: After attainment of full size the larva pupates inside the soil and remains in this condition for about 15 days.
4. **Adult**: Thus, the life cycle is completed in about 30-35 days which is repeated several times each year.

Marks of Identification

1. **Larva**: Full grown larvae are dirty pale brown or dark. Larva always lives in group. The head is gray coloured. The larvae are gregarious in habit.
2. **Adult**: The adult moths are stoutly build and pale brown in colour with dark spots in hind wings and have brownish tinged.

Host Crops

Rice, Jwar, Maize, Wheat, Barley, Pearl millet, Sugarcane, Chickpea, Linseed, Tobacco, Sorghum, Pea, Grasses *etc.*

Nature and Symptoms of Damage

- ☆ Caterpillars cause severe damage to rice plants in nursery beds. They appear suddenly in masses and move like an army from field to field so that seed beds or the direct seeded fields look as if grazed by cattle.
- ☆ The caterpillar feed at night and hides during the day. Larvae cut the seedlings in large scale and present an appearance to the field, which is grazed by cattle by its nocturnal feeding.
- ☆ Peduncles of ears are bitten through in maturing crop. They feed gregariously and march from field to field.
- ☆ The damage is severe in July -September. It breeds on a variety of grasses. Yield loss ranges from 25-50 per cent.

Economic Threshold Level (ETL)

One larva/hills.

Management

(A) Preventive Measures

1. Keeping the bunds clean *i.e.* free off weeds in the beginning of the season.
2. Deep ploughing after harvesting to expose the hibernating stages (Larvae and Pupa) to the action of sun and predatory birds.
3. Digging a trench and flooding it with water for preventing the migration of larvae from one field to another.

(B) Cultural Measures

1. Collection and destruction of egg masses.
2. Use Crop rotation.
3. Flood the fields with water for 4-5 days.
4. If the infestation is lesser in magnitude, the infected seedlings should be uprooted and destroyed.

(C) Mechanical Measures

1. Use of light trap.

(D) Chemical Measures

Spraying with cypermethrin 25 EC @ 0.06 per cent or carbaryl 50 WP @ 2 kg/ha, or Quinolphos 25 EC @ 2 lt/ha or Nuvan 76 EC @ 200 ml/ha or dusting with carbaryl 10 per cent dust @ 20kg/ha or Dichlorvos 80 EC @ 600 ml/ha.

(E) Biological Measures

1. Introduction of *Apanteles ruficrus, Sarcophaga orientaloides* and *Exorista fallax* is quite helpful in controlling the destruction caused by the caterpillars since, these insects parasitise the caterpillars.
2. Sparrows and crows act as predators of this pest.

9. Rice Swarming Caterpillar

Also known as: Grass Army worm

Scientific Name, Order and Family

Spodoptera mauritia Boisduval

(Noctuidae: Lepidoptera)

Taxonomical Position/Scientific Classification

Phylum:	Arthropoda
Class:	Insecta
Order:	Lepidoptera
Family:	Noctuidae
Genus:	Spodoptera
Species:	***S. mauritia***

Biology and Life Cycle

The pest is active throughout the kharif season but its damage is noticeable in the early part, especially after good rains when it appears in the form of a true army worm.

1. **Egg**: Eggs are spherical and creamy white in colour. Eggs are laid in masses/group on underside of leaves and in grasses. Eggs are covered with buff coloured hairs. The incubation period is of about 5-10 days. The fecundity is 150-200 eggs.
2. **Larva:** Caterpillars are light green with yellowish white lateral and dorsal stripes in the early stages and later become dark brown or grayish green in colour with a crescent (semi-circular) shaped black spot on the side of each segment. The larval period is completed in 15-23 days.
3. **Pupa**: It pupates in an earthen cocoon in soil. Pupa is dark brown and measures 16-17 mm long. The pupal period is completed in 7-14 days.
4. **Adult:** Moth is medium sized, stout built dark brown with a conspicuous triangular black spot on the forewings. Hind wings are brownish white with thin black margins. The total life cycle of the pest is completed in 27-47 days.

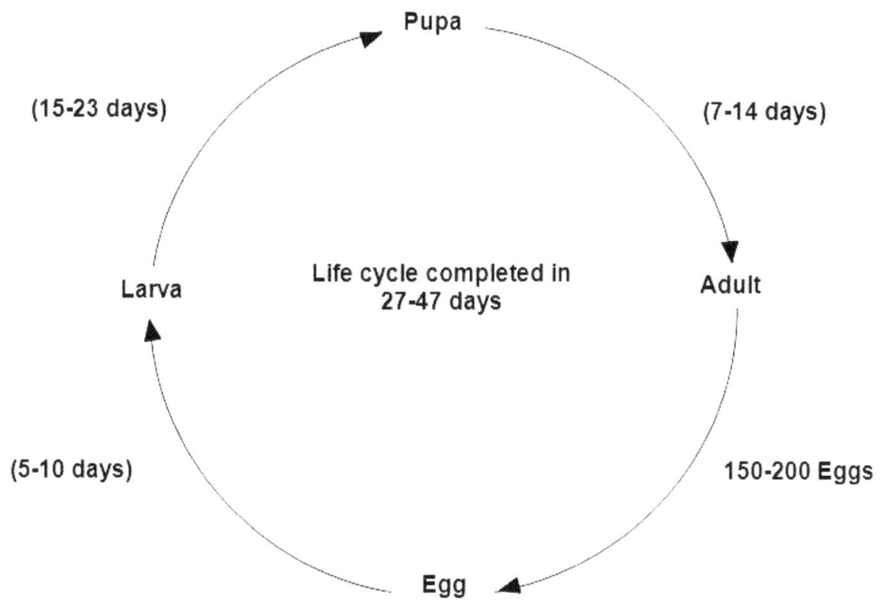

Figure 9: Life Cycle of *Spodoptera mauritia*.

Marks of Identification

1. **Adult**: The adult is stoutly build dark-brown moth with a conspicuous black spot on the fore wing. The males are characterised by immense tuft of hairs on the fore legs. The adult moth measures 15 mm in length and 35 mm in wing span.
2. **Larva**: Caterpillar is small dark to pale green with dull strips laterally. Early instar of larvae is green with yellowish tinge, whereas older ones are dark brown or grayish green in colour with dorsal and sub-dorsal stripes. Its head is black in colour.

Host Crops

Rice, Maize, Jowar, Wheat, Barley, Sugarcane and a variety of grasses.

Nature and Symptoms of Damage

- ✰ The young caterpillars damage paddy crop by cutting off leaf tips, leaf margins, leaves and even the plants at the base, more severely on the seedlings in nursery and direct seeded crops and early tillering stage in transplanted rice.
- ✰ Newly hatched larvae cause the plants to look sick with withered tips and cut leaves. They appear suddenly in masses and move like an army from field to field so that the seeds or direct seeded fields look as if grazed by cattle, warranting to re-sowing or replanting.

- ✰ Generally a transplanted crop is not seriously affected. Rice plants older than 6-7 weeks are usually not attacked by this pest.
- ✰ Loss to the paddy yield caused by this pest varies from 10 to 20 per cent.

Management

(A) Cultural Measures

1. Crop rotation in endemic pockets helps.
2. Deep ploughing the field in summer exposes the larvae and pupae for predation by birds.
3. Remove excess nurseries and weeds from the field and bunds.
4. Flooding the nurseries and small fields brings out the larvae to the surface, which get predated by the birds. Ducks if let into field, will feed on the caterpillars. A herd of ducks can easily destroy these caterpillars if let in to the fields.
5. Use of bamboo perches facilitates predation by birds.
6. In case of severe infestation, small plots can be isolated and the movement of the caterpillars can be prevented by digging a trench around the infested field wherever possible.

(B) Mechanical Measures

1. Collection of caterpillars with a hand net or sweeping basket and their destruction.
2. The inability of *S. mauritia* larvae to swim in water is a weakness and in flooded fields they are forced to stay on the plants which they defoliate. Therefore, kerosene oil may be poured into the stagnant water in the bunded fields (2 lt. kerosene per hectare). With the use of a long rope stretched across the field (two persons walk through the field) the paddy plants are shaken rigorously. The larvae fall into the kerosenized water and ultimately die.
3. Before the paddy season, grasses near the fields earmarked for paddy, may be destroyed mechanically which provide shelter and act as reservoir for migration of larvae.
4. Light traps can also be utilized for mass trapping of the moths.

(C) Chemical Measures

1. Spray of Acephate 50 WP @ 1200 g/ha or Carboryl 50 WP @ 2 kg/ha or Chlorpyriphos 20 per cent EC 1.25 litres/ha or quinalphos 25 EC @2.0 litres/ha or triazophos 40 EC @ 1.0 litres/ha or dichlorvos 76 SL 600 ml/ha on the paddy crop during evening hours. Spraying at early stage is very effective

(D) Biological Measures

Armyworms are held in check by egg and larval parasites. When these parasites fail, usually because of drought, armyworms become epidemic:

1. Parasitoids such as Trichogrammatids, Scelionids, tachinids, ichneumonids, eulophids, chalcids, and braconid wasps parasitize this pest.
2. Ants, birds and toads feed on the pest. Fungal diseases and a nuclear polyhedrosis virus also infect the larvae.

10. Gundhi Bug of Rice

Common Name

Earhead bug, Leaf forted bug

Scientific Name, Order and Family

Leptocoria varicornis Thunberg

Leptocoria acuta

(Hemiptera, Coreidae)

Taxonomical Position/Scientific Classification

Phylum:	Arthropoda
Class:	Insecta
Order:	Hemiptera
Genus:	Leptocorisa
Species:	*L. varicornis*

Biology and Life Cycle

1. **Egg:** The female lays eggs in cluster of 10-20 usually in two rows along the midrib on the upper surface of the leaf blade. Eggs are circular, brownish seed like 2 mm long. The eggs period lasts from 5-7 days. The fecundity is about 200-300 eggs.
2. **Nymph:** First instar is small, 2mm long, pale green in colour which grows to deepen green through different instars. Grown up nymphs are similar to adult in colour. The nymphal period lasts from 14-21 days.
3. **Adult:** Adults are greenish yellow, long and slender, above 20 mm in length with a characteristics buggy odour. They have long leg and antennae with 4 joints. The pests give fowl smell. The total life cycle of the pest is completed in 25-30 days.

Marks of Identification

It is an active greenish-yellow and or brownish insect with long legs and characteristic buggy odour. The adult measures about an inch in length. Early instar nymphs are pale green while older ones brownish.

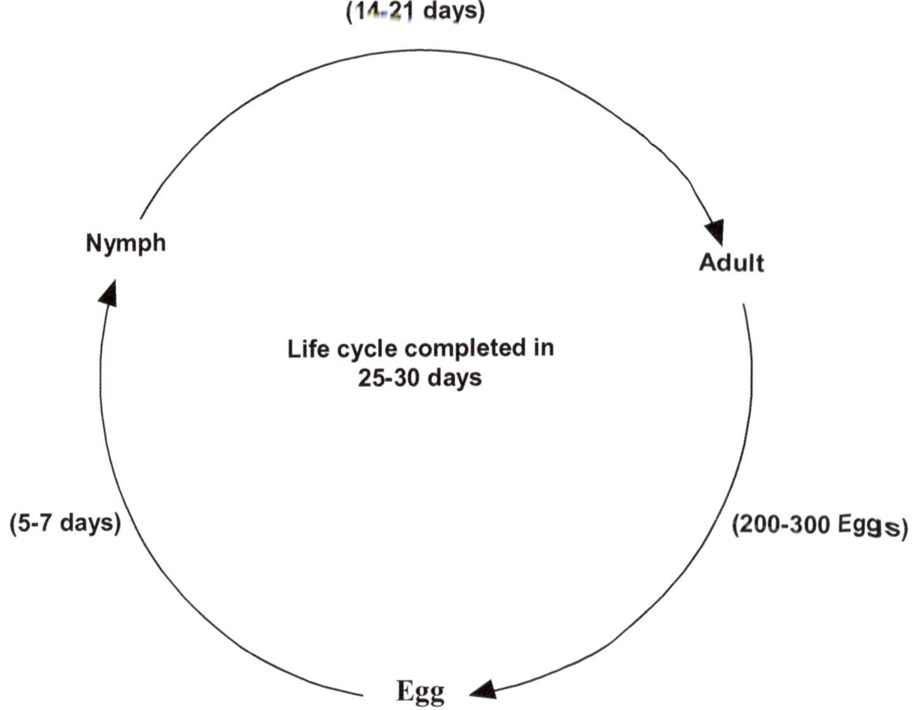

Figure 10: Life Cycle of *Leptocorisa varicornis*.

Host Crops

Rice, millets, maize, sugarcane and some grasses.

Nature and Symptoms of Damage

- ☆ Both nymphs and adults suck the sap from individual grains at milky stage.
- ☆ Affected grains become chaffy with black or brown spots at the site of injury (feeding puncture).
- ☆ Infested paddy straw contains foul smell.
- ☆ Obnoxious odour emanates on disturbing the bugs in the field.
- ☆ In case of severe infestation, ear head becomes chaffy, giving the appearance of a white ear.
- ☆ Yield loss may be up to 10- 40 per cent.

Economic Threshold Level (ETL)

5 bugs/100 panicles or 1 bug/hill - flowering stage,

16 bugs/100 panicles or 3 bug/hill - milky stage.

Management

(A) Preventive Measures

1. Remove weeds from fields and surrounding areas.

(B) Cultural Measures

1. As the bugs feed and breed on various types of grasses, especially during the offseason, removal of grasses from field and field bundhs help in reducing the pest population.
2. Draining out the water from infested field for three to four days is also helpful.
3. Crop rotation is advisable.

(C) Mechanical Measures

1. Collection of the bugs with a hand net and their destruction is a useful mechanical method.

(D) Chemical Measures

1. Dust any one of the following at 25 kg/ha twice, the first during flowering and second a week later: Quinalphos 1.5 D or Methyl parathion 2 per cent DP.
2. Spray twice as Malathion 50 EC @ 500 ml/ha or Carboryl 50 WP @ 1 kg/ha.

(E) Biological Measures

1. *Cicendalasix punctata pentatumid* prey upon the nymph and adults of *Leptocorisa*.
2. Small wasps parasitize the eggs and the meadow grasshoppers prey on them.
3. Both the adults and nymphs are preys to spiders, coccinellid beetles and dragonflies.

(B) PESTS OF WHEAT

1. Stem Borer of Wheat

Also known as: Pink Stem Borer

Scientific Name, Order and Family

Sesamia inferens Walker.

(Lepidoptera; Noctuidae)

Taxonomical Position/Scientific Classification

Phylum:	Arthropoda
Class:	Insecta

Order:	Lepidoptere
Family:	Noctuidae
Genus:	Sesamia
Species:	***S. Inferens***

Biology and Life Cycle

1. **Egg**: The eggs are laid in clusters in 2-3 rows within the cover of the leaf sheath. The eggs are found in creamy white with hemi-spherical shaped. The egg stage lasts for about 7-10 days.
2. **Larva**: The larvae are pinkish brown and have a smooth cylindrical body. The head is reddish brown. The larvae after hatching bore into the stem and feed upon the tissues of the stem. Larvae stage for about 20-30 days.
3. **Pupa**: A pupal stage for about 8-10 days. The pupation takes place into the stem.
4. **Adult**: Moth is medium sized, straw coloured with forewings having marginal black streaks. Hind wings and thorax are white. The female lays about 100 yellowish pearl like eggs between the stem and the leaf sheath in 1-3 rows. Life cycle is completed in about 40-70 days, depending upon the climatic conditions.

Marks of Identification

The adult moth is small, stout and straw coloured, measuring about 28 mm in wing expands. The caterpillars are pinkish in colour without any stripes.

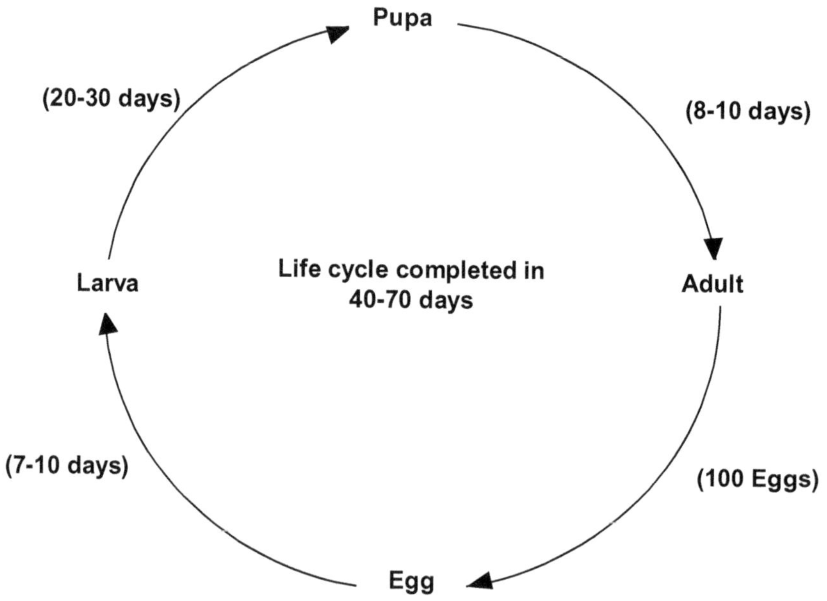

Figure 11: Life Cycle of *Sesamia inferens.*

Adults of pink borer are stout and straw coloured moths. The forewings are straw in colour while hind wings are whitish. Adults are nocturnal in habit. Full grown larvae are cylindrical, pinkish and measure about 25 mm in length. Body possesses black spots with one hair each. Eggs are creamy white and spherical. Pupae are light brown and robust.

Host Crops

Wheat, Maize, sorghum, rice, sugarcane, pearl millet and finger millet.

Nature and Symptoms of Damage

- ☆ The damage to the wheat crop is mainly caused by the caterpillars.
- ☆ The caterpillars bores into the stem and feed upon the tissues of the central shoot. The infected plant produces 'dead heart'.
- ☆ The young larvae after hatching congregate inside the leaf whorls and feed on the folded central leaves causing typical 'pin hole" symptoms.
- ☆ Severe feeding results in killing of the central shoot and consequent dead heart formation.
- ☆ Usually the second instar larvae migrate to neighbouring plants by coming out from the whorls and suspending themselves from the plants by silken threads, these are then easily blown off by wind to other plants. Infested plants become weak and bear very small carheads.
- ☆ Generally, the pest attacks only on young plants. The older plants are not killed but they produce few grains.

Economic Threshold Level (ETL)

10 per cent dead hearts

Management

(A) Cultural Measures

1. The infected wheat plant should be uprooted and destroyed.
2. Destruction of stubbles at the time of ploughing after harvesting, the crop decreases the carry over to the next crop..
3. Rotation of crop.
4. Remove dead hearts along with larvae and destroy them.

(B) Chemical Measures

1. Spraying of Quinolphos 25 EC @ 1.5-2 litre or Chloropyriphos 20 EC @ 2-2.5 litre/ha.

(C) Biological Measures

1. The egg is parasitized by *Telenomeus* spp. and *Trichogramma minutum.*

2. The larva is parasitized by *Apanteles flavipes, Bracon chinensis* and *Sturmiopsis inferens.*
3. The pupa is parasitized by *Xanthopinpla* sps and *Tetrastichus aygari.*

Release parasite *Trichogramma chilonis* @ 1.5 lakh/ha, 3 to 4 times in 10 days interval, starting from two weeks after germination.

2. Wheat Termites

Also known as: White Ant

Scientific Name, Order and Family

Odontotermes obesus Rambur

Microtermes obesi

Isoptera, Termitidae

Taxonomical Position/Scientific Classification

Kingdom:	Animalia
Phylum:	Arthropoda
Class:	Insecta
Order:	Isoptera
Family:	Termitidae
Genus:	*Odontotermes*
Species:	***O. obesus***

Biology and Life Cycle

Soon after first monsoon showers, the sexual forms leave their colony for nuptial flight during evening. After a short flight mating takes place. They shed their wings and the queen and king settle down in the soil. The female burrows in the soil, lays eggs and establish new colony. The queen gradually grows in size and start egg laying very rapidly at the rate of one egg per second or 70,000 to 80,000 eggs in 24 hours. It lives for 5 to 10 years. It can live for several years also. There is only one queen in a colony. Incubation period is about 24-90 days and within 6 months larvae develop to form soldiers or workers. The reproductive castes when produced mature in 1 to 2 years.

- ☆ Eggs: Dull, kidney shaped and hatches in 30-90 days
- ☆ Nymphs: Moult 8-9 times and are full grown in 6-12 months
- ☆ Adult: Creamy coloured tiny insects resembling ants with dark coloured head

Marks of Identification

These are social insects living in a colony. Polymorphic forms are noticed.

(A) Reproductive Caste (Winged)

They live in royal chambers.

1. **Queen:** This is the only perfectly developed inviduals, develop from fertilized eggs from the female. It is much larger in size and has creamy white abdomen which is marked with transverse dark brown strips. The queen is housed in a special area reffered as "Royal chember", which is situated in the centre of the nest at a depth of about 0.5 m. below the ground surface. It lives for 5 to 10 years and lays thousands of eggs.
2. **King:** It is develop from unfertilized eggs. It is much smaller than queen and slightly bigger than workers. It is secondarily wingless insect.

(B) Sterile Caste (Wingless)

1. **Workers:** Develop from fertilized eggs. They are whitish yellow. Head wider than reproductive castes. Mandibles are stronger, meant for feeding on. They avoid light and need high humidity for their survival.
2. **Soldiers:** Develop from unfertilized eggs. They have large head and strongly chitinized sickle shaped mandibles. They defend the colony by fighting (mandibulate type soldiers).

Host Crops

Wheat, sugarcane, groundnut, cotton, chillies, brinjal, fruit trees *etc.*

Nature and Symptoms of Damage

Termites damage the wheat crop soon after sowing and near maturity. Workers of termites feed on the roots and stem parts of the plants. The damaged plants dry up completely and are easily pulled out. The plants damaged at later stages give rise to white ears. Other forms do not cause any direct damage to the crops. Attack in earhead stage results in loss in 25 per cent of the grain production.

Economic Threshold Level (ETL)

No action threshold has been set for termites.

Integrated Pest Management

1. Destroy termite nest by deep ploughing is the only permanent remedy.
2. Stubbles of the prevoius crop and other dead decaying matter should be removed since they attract termite and form sources of infestation.
3. Keep the crop healthy and vigorous. Shortage of water leading to initial drying of the plant, may lead to termite infestation. Hence, it is very necessary to keep the crop healthy and vigorous.
4. Use of well rotten organic manure.
5. Treat the seeds @ 4 ml of chlorpyriphos 20 EC per kg of seed.
6. Locate termitoria (mounds) and destroy queen by digging out termitoria or fumigating with fumigants like CS_2/CS_2 + chloroform mixture @ 250 ml/mound.

7. Termites damage in standing crop can be minimized by dilute 1 liter of Chloropyriphos in 2 litre of water and mix with 50 kg of soil and broadcast evenly in 1 hactare in rainfed acrop.
8. In irrigated crop after application, provide light irrigation.
9. Flood irrigation at the time of planting.
10. Apply 125 kg of heptachlor 3 per cent D per ha in the furrows at time of planting.
11. Damage in standing crop can also be minimize by application of 2.5 litre of Chloropyriphos per ha. in irrigation water.

(C) PESTS OF SORGHUM and MAIZE

1. Stem Borer

Also known as: Maize Stalk borer

Scientific Name, Order and Family

Chilo partellus Zell.

Lepidoptera; Pyralidae

Taxonomical Position/Scientific Classification

Kingdom:	Animalia
Phylum:	Arthropoda
Class:	Insecta
Order:	Lepidoptera
Family:	Crambidae
Genus:	Chilo (moth)
Species:	***C. partellus***

Biology and Life Cycle

1. **Egg:** Creamy white colored, scale like and 1.5 mm wide. Typically they are deposited in overlapping cluster on the underside of leaves close to midribs, with an incubation period of 4-5 days in the warm season.
2. **Larva:** Immature larvae are yellowish, spotted and 1-2 mm in length. Mature larvae are crystalline ans spotted with black warts on each body segment. They have 4 purple/brown longitudinal stripes on the back of the body, 20-25 mm in length, and have a prominent reddish-brown head. Larval stages last 14-28 days.
3. **Pupa:** pupae are light yellow/brown to dark red/brown. Pupae have belts of small spines on the dorsal anterior margins from the 5th to 7th abdominal segments, 6 dorsal spines, and 2 large flattened ventral spines on the last abdominal segment. Pupation occurs within the tunneled stem. Pupae are about 15 mm long.

4. **Adult:** Adult moths are straw-coloured and small (approx. 15 mm). Hindwings are white, while forewings are straw-coloured with darker scale patterns forming longitudinal strips close to the wing margins. The adult stage lats from 2-12 days. Adult females can lay between 200-600 eggs during a life time, in separate batches of 10-80 eggs.

Marks of Identification

- ✰ **Egg:** Scale-like flat oval eggs in batches on the under surface of leaves near the midribs.
- ✰ **Larva:** Yellowish brown with a brown head and prothoracic shield and having many dot spots on the body.
- ✰ **Adult:** Moth is medium size, straw coloured with pale yellowish grey forewings, having minute dots on the apical margin. The hindwings are whitish.

Host Crops

Sorghum, maize, bajra, rice, sugarcane, ragi and other grasses.

Nature and Symptoms of Damage

Newly hatched larva initially feeds on the tender leaves particularly in the central leaf whorl causing numerous shot holes in the leaf lamina. Thereafter, larva bores into the stem portion thereby causing death of central shoot commonly known as 'dead hearts'. These dead hearts can be easily pulled out. The borer attacks all parts of sorghum plant except roots. Bore holes are visible on the stem near the nodes. Young larva crawls and feeds on tender folded leaves causing typical "shot hole" symptom. Affected parts of stem may show internally tunneling caterpillars. Pest is active from June to November.

- ✰ Withering and drying of central shoot -"dead heart"
- ✰ Red mining in the midrib.
- ✰ Bore holes visible on the stem near the nodes.
- ✰ Tender folded leaves have parallel "shot hole"
- ✰ Affected parts of stem may show internally tunneling of caterpillars.

Economic Threshold Level (ETL)

10 per cent dead heart

Management

(A) Cultural Measures

1. Remove and destroy infested plant parts *i.e.*dead hearts along with larvae.
2. Plough the field after harvest for destroying stubbles and hibernated larvae.

3. Sowing cowpea as an intercrop to minimise stemborer damage (Sorghum: cowpea 4:1).

(B) Mechanical Measures

Set up of light traps till mid night to monitor, attract and kill adults of stem borer, grain midge and earhead caterpillars.

(C) Chemical Measures

Mix any one of the following insecticides with sand to make up a total quantity of 50 kg/ha and apply in the leaf whorls:

1. Apply Phorate 10 G 8 kg/ha or Carbaryl 4 G @ 10 kg/ha in leaf whorl of sorghum at 10 days interval commencing from 30 days after germination
2. Spray Carbaryl 50 WP @ 0.2 per cent at 10 days interval starting from 30 days after germination.

(D) Biological Measures

The use of two parasitic wasps (*Cotesia flavipes*) and (*Xanthopimpla stemmator*) can attack and subsequently kill *C. partellus* pests. These parasitic wasps can lay eggs into *C. partellus* (*C. flavipes* on adult and *X. stemmator* on the pupae) and upon hatching, these eggs feed internally into the pest. They then exit and spin cocoons. Therefore, management of habitats that conserve these parasitic wasps could also result in the decline of *C. partellus* populations.

2. Sorghum Shoot Fly

Also known as: Seedling pest of Jwar.

Scientific Name, Order and Family

Atherigona soccata Rond.

(Diptera; Muscidae)

Taxonomical Position/Scientific Classification

Kingdom:	Animalia
Phylum:	Arthropoda
Class:	Insecta
Order:	Diptera
Family:	Muscidae
Genus:	Atherigona
Species:	***A. soccata***

Biology and Life Cycle

1. **Egg**: Eggs are laid on the underside of the leaves of 7-8 days old seedlings, or on young tillers. 1-3 eggs are laid per leaf. The milky white eggs have

an elongated shape, and measure 0.8 x 0.2 mm. They hatch after 2-3 days. Fecundity is about 20-40 eggs.

2. **Larva**: The young larvae crawl down inside the sheath. Then they bore into the base of the young shoot, killing the growing point and the youngest leaf. This leaf turns brown and withers (dead heart). The full grown (third instar) larvae are 8-10 mm long and has a white or yellowish colour. The larval period takes 7-14 days.
3. **Pupa**: Pupation usually takes place in the base of the dead shoot, but sometimes in the soil. The pupal period takes about 7 days.
4. **Adult**: The adult fly is about 4 mm long. It looks like a small house fly. Head and thorax of the female are pale grey. The abdomen is yellowish with paired brown patches. The male is more blackish. The pupal period lasts from 7-8 days.
5. This pest is found throughout the year when hosts are available at the right stage. Highest population levels are observed in August and September. The life cycle is completed in 17-25 days.

Marks of Identification

Adult fly is dark grey, like the common house fly but much smaller in size, 6 and 4 dark spots on abdominal segments of female and male respectively (arranged in rows of two). Adults love sunlight and have comparatively longer life span of about a month or more. They feed on honeydew and other sweet material available in nature.

Host Crops

Jowar and grasses like *Andropogan sorghum*, *Cynodon dactylon* and *Panicum* spp.

Nature and Symptoms of Damage

Damage to the plant is **caused by maggot** begins on 7-10 days crop. Maggots after hatching from the eggs creep down within the leaf sheath till they reach to the base of seedling. They bore into the shoot and feed internally. As a result of maggot feeding, central leaf wilts and later dries up giving the typical symptom of 'dead heart'. These 'dead hearts' can be easily pulled out. Infested shoots emit bad odour. Maggots feed on decaying tissues. Normally, damage occurs from first week to about one month after emergence. If the attack occurs little later, plants may produce side tillers that may also be attacked. Late sowing during the rainy season increases the likelihood of attack.

The shoot fly is reported to be a serious pest on crop sown late in the kharif season or early in rabi season. Pest attacks the sorghum crop only in early stage of growth. The hybrid varieties are comparatively more susceptible to the attack. The total loss in yield is sometimes as high as 60 per cent.

- ☆ The maggot bores inside the stem and cuts the growing point.
- ☆ Central shoots dried and produce "dead heart" symptom.
- ☆ The infested plant produces side tillers.

Economic Threshold Level (ETL)

1 egg/plant in 10 per cent of plant in the first two weeks of sowing or 10 per cent dead hearts

Management

(A) Cultural Measures

1. Sowing should be completed as early as possible after the onset of monsoon or within 15 days after receiving of rains (upto 1st week of july) because the infestation of shoot fly is more in late sowing crop (sowing on late July to August).
2. Use high seed rate from 10 kg to 12.5 kg/ha to maintain optimum plant population while destruction of affected plant. High seed rate helps to compensate the losses caused by pest.
3. Use of resistant and less susceptible varieties *i.e.* IS-5450, IS-2130 *etc.*
4. Water logging condition should be avoided.

(B) Mechanical Measures

1. Uprooted and destroy the infested seedlings because larval or pupal stage present in the stem.

(C) Chemical Measures

1. Use seeds treated with carbofuran 50 SP. Prepare solution of gum arabic by adding 25 g of gum arabic in 85 ml of water. Pour 60 ml of this solution in 1 kg of seed. Shake it thoroughly. Add 100 g of carbofuran 50 SP to this seed and again shake it thoroughly. Dry this seed in shade for atleast 12 hours before sowing. 200 g of carbosulfan may be rubbed to 1 kg of seed, mixed thoroughly and used for sowing.
2. Application of 10 per cent Phorate granules in furrows before sowing @15 kg/ha should be done and the insecticide should be covered with thin layer of soil after which seed should be sown.
3. Spraying with 0.01 per cent Metasystox after one week of germination.
4. Two spraying of Carboryl 35 EC @ 0.05 per cent at 7 and 10 days stage of plant.

(D) Biological Measures

1. Among the parasitoids, *Trichogramma chilonis* and *Trichogrammatoidea simmondsi* on the eggs, and *Neotrichoporoide snyemitawus* on the larvae are most important.
2. Predators: *Brumoides suturalis, Cheilomenes sexmaculata, Coccinella septempunctata, Coccinella undecimpunctata* L., and *Scymnus trepidulus.*

3
Pest Management of Pulses

(A) PESTS OF URD and MOONG

1. Pod Borer

Common Name

Spotted pod borer/Bean or legume pod borer/Mung Moth.

Scientific Name, Order and Family

Maruca vitrata

(Lepidoptera; Crambidae)

Taxonomical Position/Scientific Classification

Kingdom:	Animalia
Phylum:	Arthropoda
Class:	Insecta
Order:	Lepidoptera
Family:	Crambidae
Genus:	Maruca
Species:	***M. vitrata***

Biology and Life Cycle

1. **Egg**: The female lays 6-189 pale cream or light yellow, translucent eggs, singly or in batch of 2-16 eggs on the stems, young leaves, flowers and pods. Eggs are measured 0.45-0.65 mm in size. Eggs hatch in about 4-5 days.
2. **Larva**: The caterpillars feed inside the flowers for about a week; then they move to the pods. They are pale cream, with two rows of markings on their backs. They grow to 18 mm before they exist the pods and pupate in the soil. There are five larval instars. Total larval development is completed in 8-14 days. The head is dark brown. The larvae are translucent and shining, and have six rows of black spots running from thorax to abdomen. Because of the prominent black spots on the larva, it is also called a spotted caterpillar.
3. **Pupa**: Pupation occurs in a silken cocoon amongst webbed leaves/pods or in soil. The prepupal period lasts for two days.
4. **Adult**: The moths have brown front wings, with white patches. The hind wings are mostly white with a brown border. The life cycle is completed in 18 to 35 days. In their normal resting posture, the moths hold the wings in a horizontal position, unlike other moths which rest with folded wings. There is no diapause in this insect, and the populations during the off-season are maintained on wild hosts such as *Vigna triloba*.

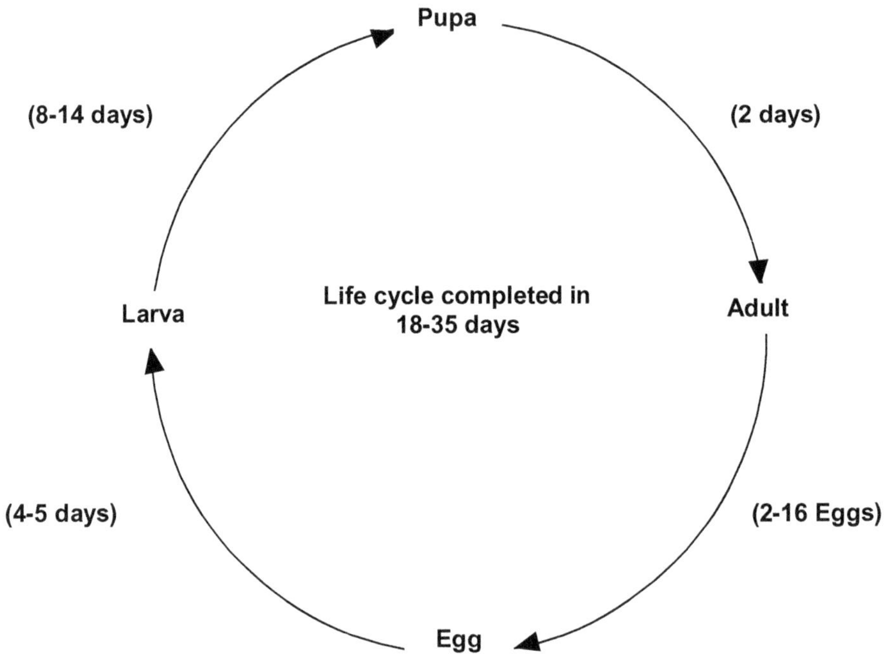

Figure 12: Life Cycle of *Maruca vitrata*.

Marks of Identification

- ☆ **Larvae**: They are pale with two rows of black markings on their backs.
- ☆ **Moth**: The moth is brown with a white patch on the front wings.

Host Crops

Cowpea, pigeonpea, mung bean, soybean, groundnut, field pea, kidney bean *etc.*

Nature and Symptoms of Damage

- ☆ The larvae web the leaves and inflorescence, and feed inside on flowers, flower buds, and pods.
- ☆ The flower bud stage is preferred most for oviposition, and it is at this stage that the young larvae cause substantial damage, and reduce the crop potential for flowering and fruit setting.
- ☆ The young larvae bore into the flower buds, and cause flower shedding by destroying the young flower parts enclosed in the sepals. Young larvae feed on the style, stigma, anther filaments, and ovary; besides a limited feeding on the internal components of the corolla.
- ☆ The larvae move from one flower to another as they are consumed, and a larva may consume 4-6 flowers before larval development is completed. Third to fifth-instar larvae were capable of boring into the pods and consuming the developing grains.
- ☆ The moths and larvae of *M. vitrata* are nocturnal. The larvae, which are photo-negative, emerge early in the evening and feed on the plant throughout the night.
- ☆ Losses in grain yield of 20 to 60 per cent due to *Maruca* damage in grain legumes have been estimated.

Economic Threshold Level (ETL)

10 larvae per 100 flowers.

Management

(A) Cultural Measures

1. Timely sowing of the crop decrease the pod borer infestation.
2. Sowing of cowpea 12 weeks after maize reduced the legume pod borer damage.
3. Inter/Mix cropping of maize - cowpea – sorghum. Pod borer incidence was significantly lower in intercropped, and at higher plant populations than in pure stands.
4. Growing of resistant varieties (Mungbean: JRUM 1, JRUM 11, JRUM 33, DPI 703, LAN 14-2, UPM 83-6, UPM 83-10, Pusa 115, PDM 116 and ML 353.

(B) Mechanical Measures

1. Use of pheromone traps @ 5/ha for monitoring of pest population.
2. Collection and destruction of infested pods.

(C) Chemical Measures

1. Seed treatment with imidacloprid 17.8 SL @ 3 ml/kg + foliar spray of profenophos 50 EC @ 2.0 ml/l.
2. Foliar application of Cypermethrin (0.008 per cent) or Dimethoate (0.07 per cent) at flowering, and then repeated at 10-15 days interval provided effective protection against *M. vitrata*.
3. Spray of profenophos 50 EC @ 2.0 ml/l +DDVP @ 0.5 ml/l found to highly effective against *M. vitrata*.
4. Spray chlorpyriphos 20 EC @ 2.5 ml/l and lambdacyhalothrin 10 EC @ 0.5 ml/l.
5. Foliar spray of lambdacyhalothrin 10 EC @ 0.5 ml/l + NSKE 5 per cent is effective against *M. vitrata*.
6. Spray of spinosad 45 SC @ 0.2 ml/l is most effective in controlling this pest.

(D) Biological Measures

1. Spray *Bacillus thuringiensis* 5 WG @ 1.0 g/l.
2. Rlelease of larval/pupal parasitoids *i.e. Aplomya metallica* (Wiedemann), *Exorista xanthaspis, Palexorista solennis, Peirbaea orbata, Zygobothria atropivora, Zygobothria ciliate, Thelairosoma* sp., *Pseudoperichaeta laevis, Apanteles* sp., *Bracon greeni, Tetrastichus sesamiae, Tetrastichus* sp.
3. Release of Predators include Dermapterans *[Diaperastichus erythrocephala* Olivier], mantids *[Polyspilota* sp. and *Spodromantis* sp.], carabids *[Chlaenius* sp. and *Cicindela lacrymosa*], coccinellids *[Coccinella repanda*] *etc.*
4. *Bacillus thuringiensis* (Bt) and neem seed powder and neem kernel extract are effective against legume pod borer.

2. Red Hairy Caterpillar

Scientific Name, Order and Family

Amsacta moorei

(Lepidoptera; Arctiidae)

Taxonomical Position/Scientific Classification

Kingdom:	Animalia
Phylum:	Arthropoda
Class:	Insecta

Order:	Lepidoptera
Family:	Arctiidae
Genus:	Amsacta
Species:	*A. moorei*

Biology and Life Cycle

1. **Egg**: The female lays about 600-700 eggs in masses on the underside of leaf, sometimes on weeds and grasses. The eggs are light yellow in colour. The egg period is 4-5 days. Fecundity is 146-615 eggs.
2. **Larva**: Tiny greenish caterpillar feeds on the leaves gregariously. A full grown larva measures 5 cm in length, reddish brown hairs all over the body arising on warts.The larval period is of 23-40 days.
3. **Pupa**: With the receipt of showers, the grown up larva pupates in earthern cells (in soil) at a depth of 10-20 cm. They pupate mostly along the field bunds and in moist shady areas under the trees in the field and undergo pupal diapause till the next year. The pupation period is about 6-7 days.
4. **Adult**: Adults are medium sized moths. All markings are red in white wings. On receipt of heavy rains, about a month after sowing in kharif season, white moths with black markings on the hind wings emerge out from the soil in the evening hours. The total life cycle is completed on 33-52 days.

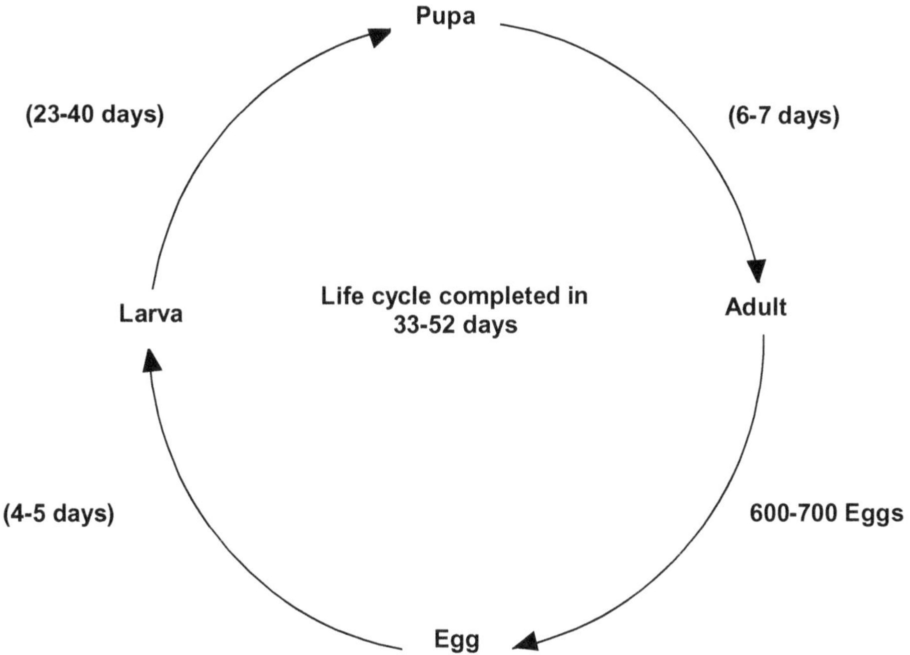

Figure 13: Life Cycle of *Amsacta moorei.*

Marks of Identification

- ☆ **Full grown caterpillar** is 27-30 mm long, reddish to greenish in colour. Body covered with dense brown coloured hairs and red colour anal turf, hence the name called Red hairy caterpillar.
- ☆ **Adult**: Both the wings are white in colour with black spots. The ant margin of thorax and the entire abdomen are scarlet red. There are black bands and dots on the abdomen. There is a red line in the outer margin of forewings.

Host Crops

It is a polyphagous pest, majorly infested to maize, jwar, bajra, ragi, paddy, sugarcane, groundnut, castor, mung, soybean, til, sunflower, urd and cruciferous crops.

Nature and Symptoms of Damage

- ☆ The newly emerged larvae/caterpillar feed on the green matter of leaves gregariously by scraping the under surface of tender leaflets leaving the upper epidermal layer intact in early stages. Later, they feed voraciously on the leaves and main stem of plants.
- ☆ The larvae from 2nd instar onwards spread to the entire field and fed on the leaves from edges onwards the midrib. In severe infestation, the crop is completely infested.
- ☆ They march from field to field gregariously. Severely affected field looks as though they are grazed by cattle. Sometimes it results in the total loss of pods.
- ☆ They also feed on sorghum, cotton, finger millet, castor, pulses and cowpea, *etc.*
- ☆ Resowing becomes necessary, if the attack of this pest occurs in early stage of the crop.

Economic Threshold Level (ETL)

8 egg masses/100 meter

Management

(A) Cultural Measures

1. Deep ploughing after harvest of the crop.
2. Organize campaign to collect and destroy the pupae after summer ploughing on receipt of showers.
3. Grow cowpea or red gram as an intercrop to attract adult moths to lay more eggs.

(B) Mechanical Measures

1. Set up 3-4 light traps and bonfires immediately at the onset of rains at 4 weeks after sowing in the rainfed season to attract and kill the moths and to know brood emergence.
2. Collect and destroy egg masses in the groundnut, cowpea and redgram.
3. Collect and destroy gregarious early instar larvae on lace like leaves of inter crops *viz.*, red gram and cowpea.
4. Dig out a trench around the field to avoid the migration of caterpillars, trap larvae and kill them.

(C) Chemical Measures

1. Spraying of Quinolphos 25 EC @ 2 lt/ha or Cypermethrin 25 EC @ 250 ml/ha or Fenvelrate 20 EC @ 400 ml/ha.
2. For young caterpillars – apply Carbaryl 10 D @ 25 kg/ha or Fenvelrate 0.4 per cent @ 15 kg/ha.
3. For grown up caterpillars - spray dichlorvos 625 ml/ha (or) chlorpyriphos 1250 ml/ha in 375 litres of water.

(D) Biological Measures

1. Use nuclear polyhedrosis virus @ 250 LE/ha and *Bacillus thuringiensis* (Bt).
2. Use of Predatory bugs – *Canthonidia furcellate.*
3. Release of *Bracon hebetor* @ 5000/ha. two times at 7-10 days interval.
4. Conserve dominant predators like *Coccinella* sp. and *Minochilus sexmaculata* and parasitoids like *Chelonus* spp.
5. Conserve the bio control population of spiders, long horned grasshoppers, praying mantis, robar fly, ants, greenlace wing, damsel flies/dragon flies, flower bugs, shield bugs, lady bird beetles, ground beetle, predatory cricket, earwig, braconids, trichogrammatids, NPV, green muscular fungus.
6. Use 5 per cent neem seed kernel extract on need basis.
7. Inter cropping with pigeon pea, mung bean and soybean provides increase in population of spiders.
8. Population of coccinellids is higher on groundnut with maize, mung bean and soybean and *Chrysoperla* spp. is higher with maize and soybean intercrops.

(B) PESTS OF SOYBEAN

1. Girdle Beetle

Scientific Name, Order and Family

Oberea brevis

(Coleoptera; Cerambycidae)

Taxonomical Position/Scientific Classification

Kingdom:	Animalia
Phylum:	Uniramia
Class:	Insecta
Order:	Coleoptera
Family:	Cerambycidae
Genus:	Oberea
Species:	***O. brevis***

Biology and Life Cycle

1. **Egg**: With the help of mouth, the female makes 2 rings on the stem or leaf petiole at a distance of 6-15 mm, then it makes 3 punctures near the basal ring, later on it lays eggs in the middle puncture. The fecundity is about 100-150 eggs. The egg period is of 4-8 days.

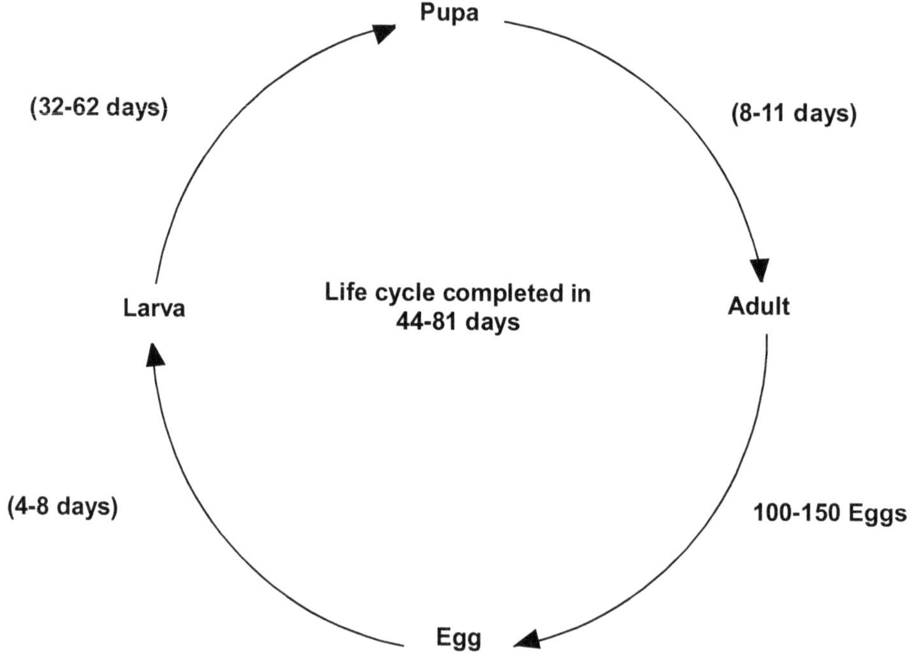

Figure 14: Life Cycle of *Oberea brevis.*

2. **Larva**: The larval period is 32-62 days having 4-5 instar.
3. **Pupa**: The pupation is takes place on stem tunnel. The pupal period is completed in 8-11 days.
4. **Adult**: The total life cycle is completed on 44-81 days.

Marks of Identification

1. The adult beetle has a hard shell-like exterior and rather long antennae.
2. In newly emerged beetle – head, thorax and forewings are yellowish in colour and turn black later on. The antennae is larger than the body.
3. The larva is a white, soft-bodied worm with a dark head. The full grown grub is yellowish in colour with spines on each segment.

Host Crops

Soybean, Urd, Moong and other leguminous crops.

Nature and Symptoms of Damage

- ☆ Female beetle makes a ring on branches making three holes, where it lays eggs. After hatching, grub enters into main branch, there after into main stem.
- ☆ The grub tunnel into the stem upto bottom and may be filled with excreta. As a result of feeding, the leaves and growing point dries up.
- ☆ The cropwith broken stem is an indication of attack of girdle beetle. This pest active on soybean from july to harvest of the crop.
- ☆ Early symptoms include the drying of the edges of trifoliate leaves.
- ☆ Seedlings and young plants are wilted or dead.
- ☆ On older plants, all or part of the leaves are wilted and brown.
- ☆ Plants may fall over (lodge).
- ☆ Petioles or the main stem have two parallel girdles.
- ☆ The crop is badly affected to extent of 65-85 per cent in terms of yield of grains.

Economic Threshold Level (ETL)

10 per cent infestation at flowering stage.

Management

(A) Preventive Measures

1. Grow resistant/tolerant varieties *i.e.* JS-71-05, NRC 7, JS 97-52, MAUS 32 and Indira Soya 9.

(B) Cultural Measures

1. Destruction of infested branches along with the grubs inside.

2. Seed treatment with imidacloprid 3 g/kg seed gives protection upto 30 days.
3. Treat the seed with Thiamethoxam 70WS.

(C) Mechanical Measures

1. Use of light traps.
2. Erection of bird perches @ 10-12/ha.
3. Use of *Dhaincha* as a trap crop for Girdle beetle.

(D) Chemical Measures

1. Apply Acephate 75 SP @ 750 g/ha.
2. Spraying of Dimethoate 30 EC @ 1 litre/ha or Quinolphos 25 EC @ 1.5 litre/ha.
3. Spray Profenophos 50 EC @ 1 lt/ha or Fenvalrate 20 EC @ 5 ml in 10 litres of water.
4. Application of Triazophos 40 EC @ 1 lt/ha or Chlorantraniliprole 18.5 SC @ 150 ml/ha.

(E) Biological Measures

1. Use of predator – *Chrysoperla carnea.*
2. Conserve spiders, *Coccinellid beetle*, tachinid fly, praying mantids, dragon fly, damsel fly, *Chrysoperla* and meadow grass hoppers.

2. Stem Fly

Scientific Name, Order and Family

Melanagromyza sojae

(Diptera; Agromyzidae)

Taxonomical Position/Scientific Classification

Kingdom:	Animalia
Phylum:	Arthropoda
Class:	Insecta
Order:	Diptera
Family:	Agromyzidae
Genus:	Melanagromyza
Species:	***M. sojae***

Biology and Life Cycle

1. **Egg**: Shiny bluish to black fly deposit eggs in punctures made by fly on young leaves. Generally, eggs are light green owing to the colour of leaf tissues in which they are inserted. Fecundity is about 40-64 eggs. The incubation period is 2-7 days.
2. **Maggot**: Barrel shaped and white in colour. Female is slightly bigger in size than male. The maggot stage is completed in 10-15 days.
3. **Pupa**: Pupation takes place at ground level within the stem. Pupation period is 7-10 days.
4. **Adult**: Adult fly exits through a thin semi transparent window. The total life cycle is completed in 16-25 days.

Marks of Identification

Adult flies comparatively smaller in size than housefly and with mettalic black in colour. Maggot sare without legs, pale yellow and 3.4 mm in length.

Host Crops

Soybean, urd, moong, tur, til, pea and jwar.

Nature and Symptoms of Damage

- ✰ Fly is active throughout the year. But it is very destructive at Kharif season especially at 2 leaf stage of the crop.

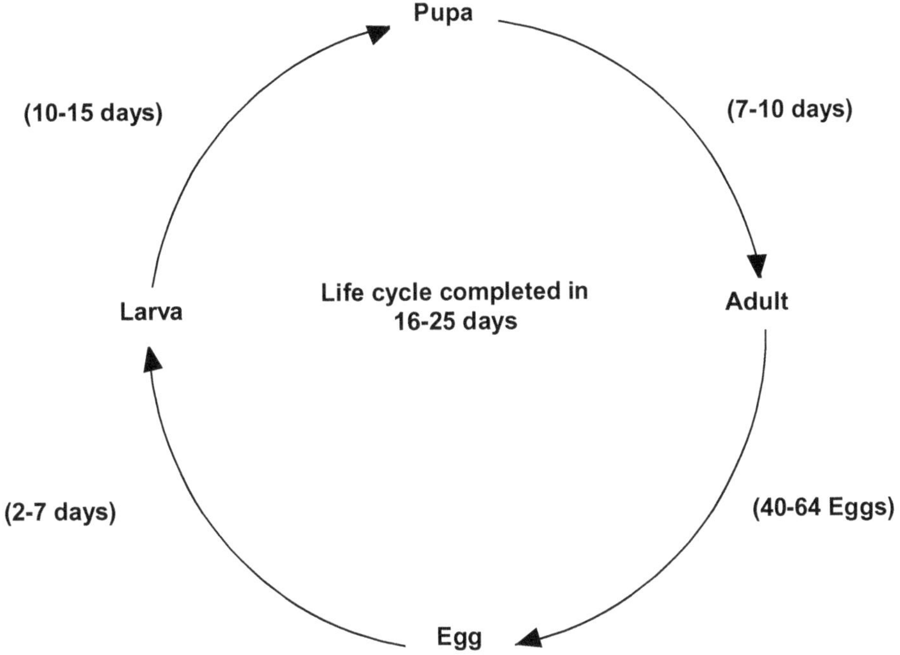

Figure 15: Life Cycle of *Melanagromyza sojae.*

- Newly hatched larvae bores into the midrib of leaf and feed on the green matter. Ultimately, it bore into the stem and makes tunnel downwards. Infested tunnel turns deep red in colour usually and maggot completes its life cycle.
- Yellowish maggots bore into nearest vein, reach the stem through petiole, bore down the stem and feed on cortical layers and may extend to tap root resulting in the following symptoms :
  - Distinct tunnel of stem split open.
  - Death of plant or branches.

Economic Threshold Level (ETL)

Economic threshold level is 5 per cent plant infestation.

Management

(A) Preventive Measures

1. Deep summer ploughing.
2. Proper crop rotation with dissimilar crops.
3. Removing and destroying the damaged plant parts can reduce the population.
4. Grow resistant/tolerant varieties *i.e.* JS-335, PK262, NRC 12, NRC 37, MACS 124, and MAUS 2, MAUS 47.

(B) Cultural Measures

1. Destruction of infested branches along with the grubs inside.
2. Seed treatment with imidacloprid 3 g/kg seed gives protection upto 30 days.
3. Treat the seed with Thiamethoxam 70WS.

(C) Mechanical Measures

1. Use of light traps.
2. Erection of bird perches @ 10-12/ha.

(D) Chemical Measures

1. Foliar spray with Trizophos 40 EC @ 10 ml or Chlorpyriphos 20 EC @ 20 ml or Acephate 75 WP @ 10 g or Dimethoate 20 ml or per 10 litre of water.
2. Apply Phorate 10G @10 kg/ha.
3. Spray with Thiamethoxam 25 WG @ 100 g/ha or Quinolphos 40 EC @ 1 lt/ha.

(E) Biological Measures

1. Use of predator – *Chrysoparila carnea.*

2. Conserve spiders, *Coccinellid beetle*, tachinid fly, praying mantids, dragon fly, damsel fly, *Chrysoperla* and meadow grass hoppers.

(C) PESTS OF PIGEONPEA

1. Pod Borer

Also known as: Spiny pod borer

Scientific Name, Order and Family

Etiella zincknella (Treitschke)

(Lepidoptera; Pyralidae)

Taxonomical Position/Scientific Classification

Kingdom:	Animalia
Phylum:	Arthropoda
Class:	Insecta
Order:	Lepidoptera
Family:	Pyralidae
Genus:	Etiella
Species:	*E. zinckenella*

Biology and Life Cycle

- ☆ **Egg:** Eggs are laid singly (or) in groups preferably at the junction of the calyx and pod or on the pod surface. A female lays 47-178 eggs, which hatch in 5-6 days.
- ☆ **Larva:** The larva bores within the green pods and feeds on seeds. Larval period lasts for 10-13 days. Greenish initially, turns pink before pupation. It has 5 black spots on the prothorax.
- ☆ **Pupa:** When fully grown the larva drops to ground and forms a cocoon about 2.5 cm or so below ground or under dry leaves. Pupal duration lasts for 9-20 days depending on the climate.
- ☆ **Adult:** Brownish grey moth, Prothorax is orange in colour, forewings has a white stripe along the anterior margin. The moths pair 24-30 hour after emergence.

Marks of Identification

Moth – Stoutly built and is yellowish brown in colour, there is a black speak that forms 'V' shape mark and a dark area near the outer margin of each forewing. On the outer side of forewings, there is a black kidney shaped mark. Hindwings are whitish with a blackish band along the outer margin.

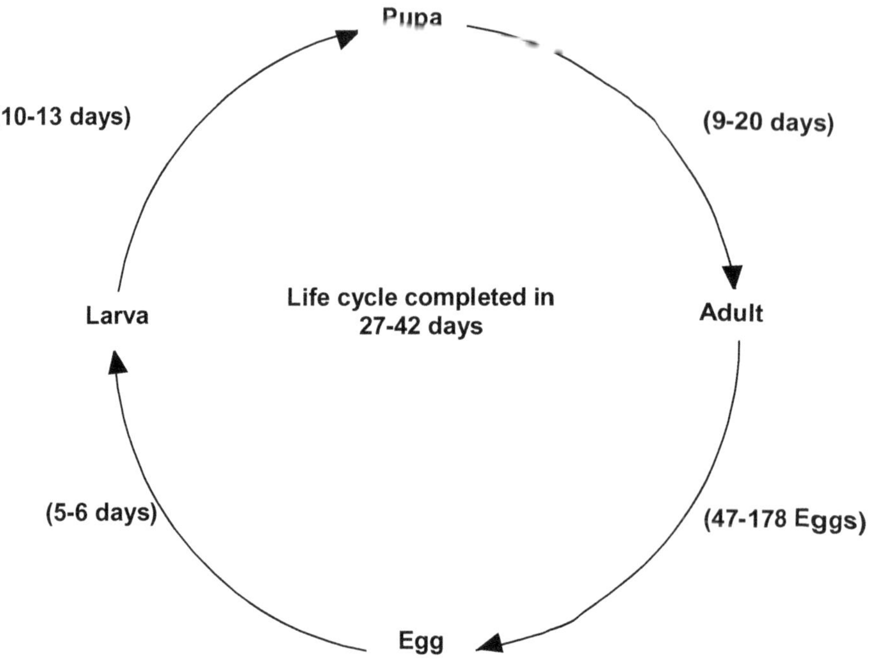

Figure 16: Life Cycle of *Etiella zincknella*.

Host Crops

Gram, tur, pea, mung, urd, lentil, soybean, cowpea, cotton, jwar, okra, maize, tomato, sunflower *etc.*

Nature and Symptoms of Damage

- ☆ The young larvae feed on the foliage by scarping and later stages, bore into the buds, flowers and pods and feed on developing seeds with their body hanging outside the pod.
- ☆ This results in dropping of flowers and young pods.
- ☆ Older pods marked with a brown spot where a larva has entered.
- ☆ A single larva can destroy 30-40 pods during its life cycle.

Economic Threshold Level (ETL)

One larva per five plants in the pod initiation stage or 10 per cent affected parts.

Management

(A) Preventive Measures

1. Deep ploughing after harvest of crop.

(B) Cultural Measures

1. Growing of marigold *i.e.* trap crop.
2. Intercropping with barley or maize, should be preferred over sole crop.

(C) Mechanical Measures

1. Install light traps/pheromone traps @ 5/ha.

(D) Chemical Measures

1. Apply chloropyriphos 20 EC @ 2 lt/ha. Spray should be initiated at the start of pod formation and repeat after 2 weeks if necessary.
2. After flower initiation, spraying of Quinolphos 25 EC @ 1 litre/ha or Carboryl 50 WP @ 2 kg/ha (Repeat the application after 15 days, if necessary).
3. Apply Spinosad 45 SP @ 250 ml/ha or Indoxacarb 14.5 SC @ 250 ml/ha.

(E) Biological Measures

1. Apply NPV @ 250 LE/ha (Spraying should be done in evening hrs).
2. Conserve natural enemies like *Tetrastichus* sp., *Bracon hebetor, Phanerotoma* sp. and *P. hendecasisella.*

2. Pod Fly

Scientific Name, Order and Family

Melanagromyza obtusa (Malloch)

(Diptera; Agromyzidae)

Taxonomical Position/Scientific Classification

Kingdom:	Animalia
Phylum:	Arthropoda
Class:	Insecta
Order:	diptera
Family:	Agromyzidae
Genus:	Melanagromyza
Species:	***M. obtusa***

Biology and Life Cycle

1. **Egg**: Eggs are deposited singly on outer surface of pod or on flower buds. A female fly lays 4 eggs per pod and 80 eggs in its life time. Hatching period is 2 to 4 days.
2. **Larva**: Maggot development takes place in 5 to 10 days. Maggot is creamy white in colour. Under abundant moisture condition, two broods can be seen in a year. Full grown maggot pupates inside larval grooves in pods.

3. **Pupa**: Pupation takes place inside the damaged pods. The pupal period lasts for 8 to 13 days.
4. **Adult**: Adult is a black fly with strong legs and ovate abdomen. Its eye are distinct, wings are clear veined, brownish yellow at their bases. Small black fly thrusts its minute eggs into the tissues of the tender pod and flower buds. Fly pierces pericarp with ovipositior and lay eggs which are seen like needles projecting inwards from the pods. One generation or total life cycle is completed in 15 to 32 days.

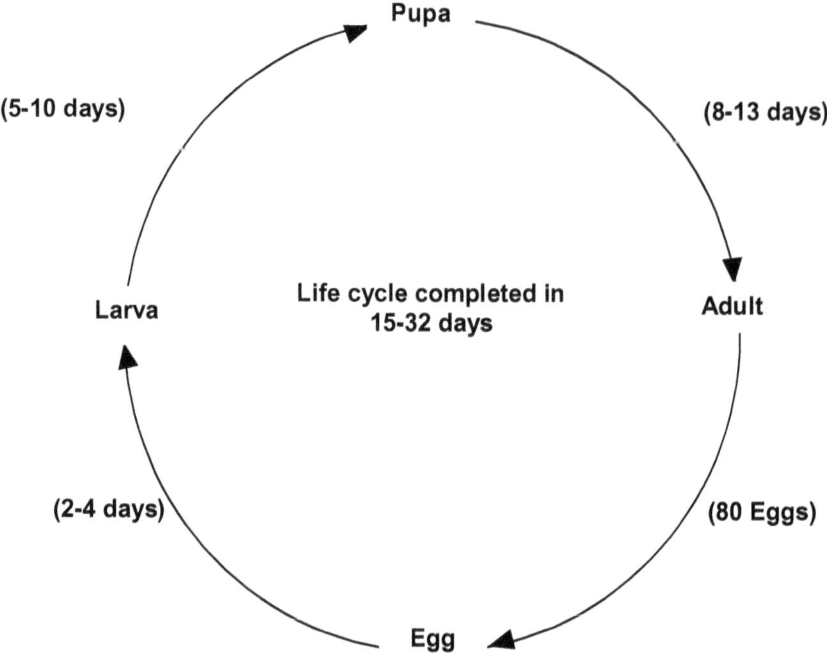

Figure 17: Life Cycle of *Etiella zincknella.*

Marks of Identification

Adult is a small sized metallic black coloured fly measuring about 2.75 mm in length. Adults have pubescent eyes, strong legs and transparent wings. Newly hatched maggots are 0.61 mm in length, 0.12 mm in breadth and whitish in colour. Fully-grown maggot is creamy white in colour, 1.25 to 1.5 mm in breadth and 3.75 to 4 mm in length. Pupae measure about 2.5 mm in length and 1.5 mm in breadth.

Host Crops

Red gram, cowpea, safflower, soybean, okra, mothbean, black gram and horse gram.

Nature and Symptoms of Damage

☆ Newly hatched maggot bores into the young pods and feeds on seed.

- ☆ The maggots first feed on the epidermis of seed by making galleries superficially just below the epidermis. Later on, they feed deeper into the seeds.
- ☆ The partially damaged seeds are further infected by bacteria and fungi causing them to decay.
- ☆ The heavily damaged grains are made unfit for consumption and for seed purpose.
- ☆ It is a major pest especially of red gram causing upto 60 per cent damage to pods.
- ☆ Maximum infestation is in October.

Diagnostic Symptoms

- ☆ Discolouration of the infested pods visible in green podded varieties.
- ☆ At the later stage of infestation, the holes about 1mm in diameter covered with a thin membrane readily seen on the infested pod.
- ☆ Exit holes visible after the adult emergence.

Economic Threshold Level (ETL)

2 maggots/100 pods or Presence of 5 per cent oviposition on pod basis.

Management

(A) Cultural Measures

1. Early sowing in endemic areas.
2. Removal of affected pods of first brood during winter.
3. Collection and destruction of infested plants parts in the initial stage of attack.

(B) Mechanical Measures

1. Collection and destruction of infested pods during the early stage of attack.
2. Collection of caterpillars by shaking shoots and their destruction in initial stages.
3. Bird perches 50/ha.

(C) Chemical Measures

1. Spray crop with Dimethoate 30 EC @ 0.03 per cent (500 ml/500 litres water/ha) during pod formation.
2. Foliar sprays with dimethoate 2 ml/l are effective against larva.
3. Spray Carboryl 50 WP @ 2kg/ha or Quinolphos 20 EC @ 1 lt/ha (Initiation at pod formation and repeat after 15 days).

(D) Biological Measures

- Pre-pupal stage is parasitized by *Euderus agromyzae* and pupa is parasitized by *Euderus lividus.*

3. Plume Moth

Also known as: Tur Pod caterpillar

Scientific Name, Order and Family

Exelastis atomosa Walsingham

(Lepidoptera; Pterophoridae)

Taxonomical Position/Scientific Classification

Kingdom:	Animalia
Phylum:	Arthropoda
Class:	Insecta
Order:	Lepidoptera
Family:	Pterophoridae
Genus:	Exelastis
Species:	***E. atomosa***

Biology and Life Cycle

1. **Egg**: Minute, single eggs, total about 17 to 19 are deposited by female on tender parts of the plant or terminal shoots having buds and flowers. Egg period is 4-5 days.
2. **Larva**:Larval development takes place in 10 to 25 days. Larva pupates outside the pod on its surface or in the entrance hole.
3. **Pupa**: Pupation is on pod surface or burrows of infested pods. Pupa is also fringed with short hairs. Pupal period is 4-8 days. Pupa looks like larva except for the colour which is brown.
4. **Adult**: Moth is slender, less than 12 mm long and are grey with long narrow wings. The forewings are divided into two parts and hindwings into three parts and provided with a fringe like border. One life cycle is completed in 17 to 42 days.

Marks of Identification

Adult is 12 mm long, delicate moth with plumose light brown wings. Forewings are divided into 2 parts and hindwings cut into 3 parts and are provided with a fringe like border. Caterpillar is 13 mm in length when full-grown and greenish brown in colour. Larva has short hairs and spines on the body. Pupa is covered with thin hairs. Eggs are oval in shape and greenish in colour.

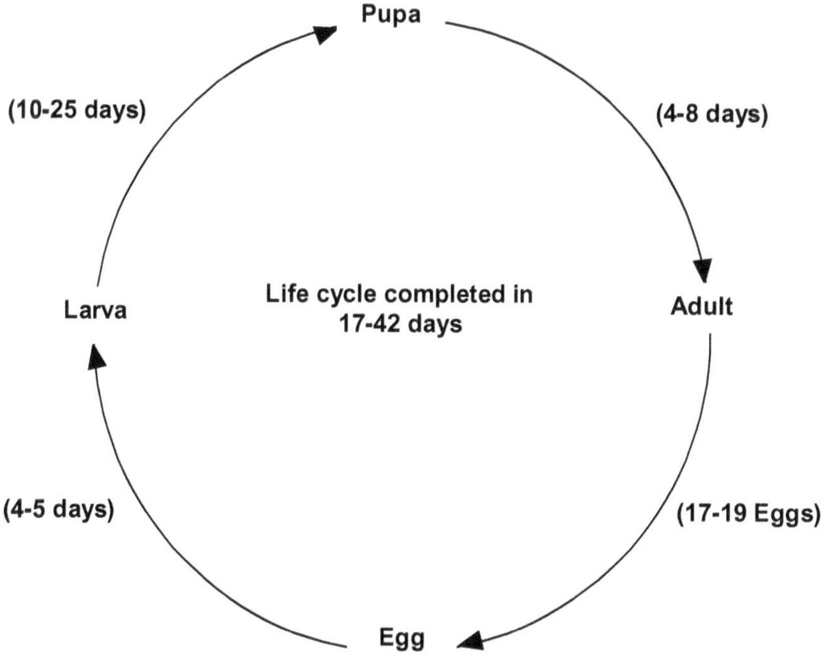

Figure 18: Life Cycle of *Exelastis atomosa.*

Host Crops

It is specific pest of red gram but also infests lablab and horse gram.

Nature and Symptoms of Damage

- Tiny caterpillar scrapes the pod surface and cuts a hole and thrusts the head into it and feeds on seed by remaining outside. The caterpillars bore into green pods and feed on the developing seeds which are more or less completely devoured or eaten away.
- Also feeds on flower buds.
- This pest is usually found at flowering and known to cause heavy damage to redgram. Attack by this pest can cause severe bud, flower and pod drop. The larva never enters inside the pod and feeds remaining outside the pod.
- Larvae never enter the pod completely.
- The damage results in the following symptoms:
  - Small hole on seeds.
  - Dropping of flower buds and flowers in severe cases.
  - Completely eaten and devoured seeds.

Economic Threshold Level (ETL)

5 larvae per plant.

Management

(A) Cultural Measures

1. Deep ploughing after harvest of crop.
2. Avoid growing leguminous crops consecutively in same field. Follow proper crop rotation.
3. Growing of marigold *i.e.* trap crop.
4. Intercropping with barley or maize, should be preferred over sole crop.

(B) Mechanical Measures

1. Collection and destruction of infested pods during the early stage of attack.
2. Collection of caterpillars by shaking shoots and their destruction in initial stages.
3. Bird perches 50/ha.

(C) Chemical Measures

1. Field applications (two) of the following insecticides at two weeks interval starting from pod formation. First at pod formation and second two weeks after first application:

   Phenthoate 0.07 per cent or Quinalphos 0.05 per cent or Quinalphos 1.5 D @ 20 kg/ha or Carbaryl 10 D @ 20 kg/ha.
2. Dimethoate 30 per cent EC 1237 ml/ha or Emamectin benzoate 5 per cent SG 220 g/ha or Indoxacarb 15.8 per cent SC 333 ml/ha or Chlorantraniliprole 18.5 SC 150ml/ha or Spinosad 45 per cent SC 125-162 ml/ha or NSKE 5 per cent twice followed by triazophos 0.05 per cent or Neem oil 2 per cent or Phosalone 0.07 per cent (Spray fluid 625 ml/ha).

(D) Biological Measures

1. Ha NPV 3 x1012 POB/ha in 0.1 per cent teepol.
2. Conserve larval parasitoids; *Apanteles paludicolae, Diadegma sp.*

4. Pod Bug

Also known as: Pod Sucking Bug

Scientific Name, Order and Family

Cavigralla gibbosa Spinola

(Hemiptera; Coreidae)

Taxonomical Position/Scientific Classification

Kingdom:	Animalia
Phylum:	Arthropoda
Class:	Insecta
Order:	Hemiptera
Family:	Coreidae
Genus:	Cavigralla
Species:	***C. gibbosa***

Biology and Life Cycle

1. **Egg**: Female lays on pods, less frequently on leaves or flower buds in bunches of 5-25. Eggs are dark in colour and 60-65 eggs in its life time. The incubation period is of 7-8 days.
2. **Nymph**: The nymph period is of 22-24 days. The Nymph having five instars.
3. **Adult**: The female adult completed its life cycle in 38 days, whereas male is in 45 days. The total life cycle of the insect varies from 70-75 days.

Marks of Identification

Adult is greenish brown in colour and having spines on the pronotum and femur. The young nymph is reddish in colour at later stage, its looking greenish

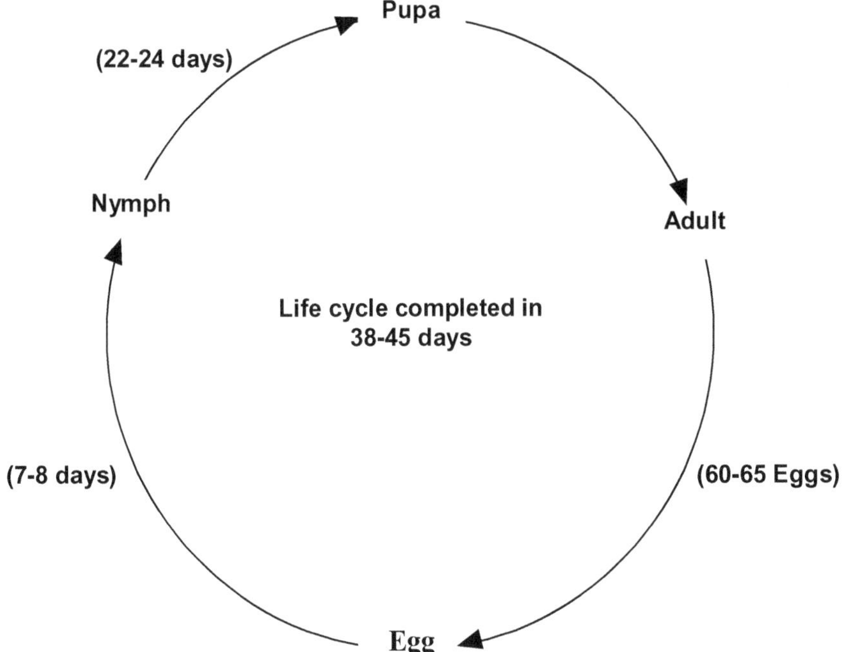

Figure 19: Life Cycle of *Cavigralla gibbosa*.

brown. The eggs are smooth and shiny sculptured lays in clusters. The nymph and adult is also brownish grey with more robust bug. Freshly laid eggs brown in colour and changed to blackish dark brown before hatching. They have five distinct nymphal instars and adults are yellow to solemon in colour and changed to brown within few hours.

Host Crops

Tur, cowpea, chickpea, cluster bean.

Nature and Symptoms of Damage

Hundreds of nymphs and adults suck sap from the shoots, leaves, flower buds and pods. Major damage is done to the green pods before the maturity of the crop. The attacked pods show pale yellow patches. The grains in the pods become shrivelled and small in size resulting in considerable yield losses. Shoots fade, pods shrivel and seeds with dark patch loose germination capacity due to the feeding of bugs.

Economic Threshold Level (ETL)

1 egg mass/plant

Management

(A) Cultural Measures

1. Physical shaking of the infested plants over the vessels of oil and water or oily cloth help reduce the population.

(B) Mechanical Measures

1. Collection of egg mosses, nymphs and bugs and their destruction by dipping into kerosinized water.

(C) Chemical Measures

1. Dusting with Carbaryl 10 D @ 10 kg/acre or foliar spray with Dimethoate 2 ml/l or Acephate 75 SP @ 1 kg/ha.
2. Spraying with one of the insecticide after flower initiation: Quinolphos 25 EC @ 1 lt/ha or Chloropyriphos 25 EC @ 1.5 lt/ha or Polytrene (Profenophos 40 per cent + Cypermethrin 4 per cent) 44 EC @ 1 lt/ha.

(D) Biological Measures

1. Use of Predators: *Antilochus coqueberti.*
2. Parasitic mite: *Bachartia* sp.
3. Egg parasite: *Gryon antenstidae.*

(D) PESTS OF GRAM/CHICKPEA

1. Pod Borer

Scientific Name, Order and Family

Helicoverpa armigera Hubner

(Lepidoptera; Noctuidae)

Taxonomical Position/Scientific Classification

Kingdom:	Animalia
Phylum:	Arthropoda
Class:	Insecta
Order:	Lepidoptera
Family:	Noctuidae
Genus:	Helicoverpa
Species:	***H. armigera***

Biology and Life Cycle

1. **Egg:** Yellowish shiny, spherical in shape, sculptured eggs are laid singly on tender parts of plants. Each female lays 300-400 eggs. Egg period is 2-4 days.
2. **Larva**: shows colour variation from greenish to brown. Green with dark brown grey lines laterally on the body with lateral white lines and also has dark and pale bands. Larval period is 18 to 25 days.
3. **Pupa**: Pupa is dark brown and Pupal period is 6 to 12 days. Pest pupates in soil, leaf, pod and crop debris. Hibernation takes place in the pupal stage in soil.
4. **Adult**: Moth is stout with dark yellow olive grey or brown wings crossed by a dark band near outer margin and a dark spot near costal margin of forewings and hindwings pale with a dark apical border. Number of generations completed in a year is eight.

Marks of Identification

Full grown caterpillar is cylindrical 40 – 48 mm in length with variable colour, dark green or reddish brown or brownish and marked with a white broken lines and a prominent white line along lower part of sides. Forewing grey to pale brown with V shaped speck. Hindwings are pale smoky white with a broad blackish outer margin.

Host Crops

Gram, tur, pea, cotton, tobacco, tomato, safflower, sunflower and sorghum.

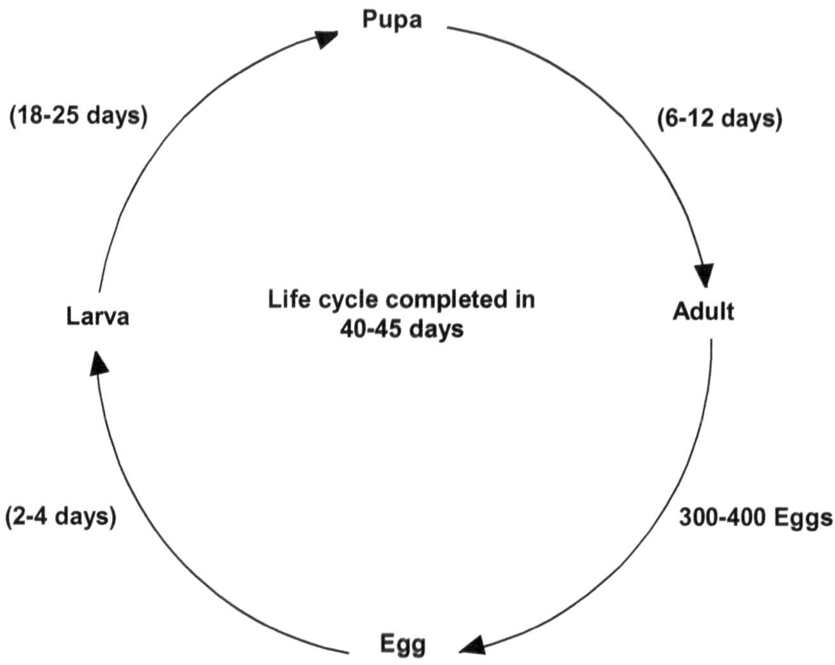

Figure 20: Life Cycle of *Helicoverpa armigera*.

Nature and Symptoms of Damage

- The young larvae feed on foliage for couple of days and subsequently bore into pods of leguminous crops by thrusting their head inside and keeping rest of the body outside. They feed on grain.
- Larva is migratory in habit and hence it damages so many plants and pods.
- In Bengal gram, the pest causes serious damage during March when the crop is in the pod formation stage.
- Red gram varieties which mature early during November are generally more severely damaged by this pest.
- The young caterpillars feed on the tender foliage and as they grow they bore into the pods and destroy the seeds, while feeding it thrusts its head inside the pod leaving the rest of its body outside.
- The basic symptoms are:
  - Defoliation in early stages
  - Larva's head alone thrust inside the pods and the rest of the body hanging out.
  - Pods with round holes

Economic Threshold Level (ETL)

10 per cent of affected pods.

Management

(A) Cultural Measures

1. Use small or medium sized seed varieties like JG-74, JG-315 *etc.*
2. Mixed cropping with linseed, mustard and wheat (In N. India) and with coriander crop (in S. India) should be preferred over solo crop.
3. Collect and destroy infested plant parts *i.e.*, leaves, pods *etc.*
4. Deep ploughing after harvesting of the crop to expose hibernated pupae and larvae to sun and predators.
5. Trap crops like Lubia (chawli), marigold along the border of main crop may be grown.

(B) Mechanical Measures

1. Hand collection of larvae and their mechanical destruction.
2. Use light traps and pheromone traps. Erect pheromone traps @ 10/ha to monitor the pest. Light traps during August-September and November – December.

(C) Chemical Measures

1. Spraying of Polytrene 44 EC (Profenophos 40 per cent + Cypermethrin 4 per cent) @ 1 lt/ha or Fenvelrate 20 EC @ 500 ml/ha or Quinalphos 25 EC 0.05 per cent (1000 ml/500 litre/ha) or Quinalphos 1.5 D @ 20 kg/ha or Carbaryl 10 D @ 20 kg/ha.
2. Spraying of Chlorpyriphos 2.5 ml/lt at initiation of flowers or Quinalphos 2 ml/lt or acephate 1.5 g/lt at flowering and fruiting using 750-1000 lt. of spray fluid with High Volume sprayer.
3. In severe incidence, indoxacarb 1 ml/lt or spinosad 0.3 ml/lt.
4. Avoid application of insecticide before flowering of chickpea.

(D) Biological Measures

1. Apply NPV @ 250 L.E./ha.
2. Release of *Trichogramma chilonis* @ 50,000 egg parasite per ha at the appearance of the pest. This parasite is available in the market in the form of Tricho-cards.
3. Other Important biocontrol agents are :

   Solitary larval parasite: *Campoletis chlorideae* (Parasitoid)

   Larval parasite: *Bracon kitcheneri*

   Larval predator: *Andrallus spinidens*

2. Cut Worm

Also known as: Dark sword-grass moth, Black cutworm, Greasy cutworm.

Scientific Name, Order and Family

Agrotis ypsilon Hufnagel

(Lepidoptera; Noctuidae)

Taxonomical Position/Scientific Classification

Kingdom:	Animalia
Phylum:	Arthropoda
Class:	Insecta
Order:	Lepidoptera
Family:	Noctuidae
Genus:	Agrotis
Species:	***A. ypsilon***

Biology and Life Cycle

1. **Egg:** The egg is white in color initially, but turns brown with age. It measures 0.43 to 0.50 mm high and 0.51 to 0.58 mm wide and is nearly spherical in shape, with a slightly flattened base. The egg bears 35 to 40 ribs that radiate from the apex; the ribs are alternately long and short. The eggs normally are deposited in clusters of 10 to 15 on foliage. Females may deposit 1200 to 1900 eggs. Duration of the egg stage is 3-6 days.
2. **Larva:** There are five to nine instars, with a total of six to seven instars most common. Head capsule widths are about 0.26-0.35, 0.45-0.53, 0.61-0.72, 0.90-1.60, 2.1-2.8, 3.2-3.5, 3.6-4.3, and 3.7-4.1 mm for instars one through eight, respectively. Head capsule widths are very similar for instars one through four, but thereafter those individuals that display eight or nine instars show only small increments in width at each molt and eventually attain head capsule sizes no larger than those displaying only six or seven instars. The larval period is lasts for 30-34 days.
3. **Pupa:** Pupation occurs below ground at a depth of 3 to 12 cm. The pupa is 17 to 22 mm long and 5 to 6 mm wide and dark brown. Duration of the pupal stage is normally 12 to 20 days.
4. **Adult:** The adult is fairly large in size, with a wingspan of 40 to 55 mm. The forewing, especially the proximal two-thirds, is uniformly dark brown. The distal area is marked with a lighter irregular band, and a small but distinct black dash extends distally from the bean-shaped wing spot. The hind wings are whitish to gray, and the veins marked with darker scales. The adult preoviposition period is about seven to 10 days. Moths select low-growing broadleaf plants preferentially for oviposition, but lacking these will deposit eggs on dead plant material.

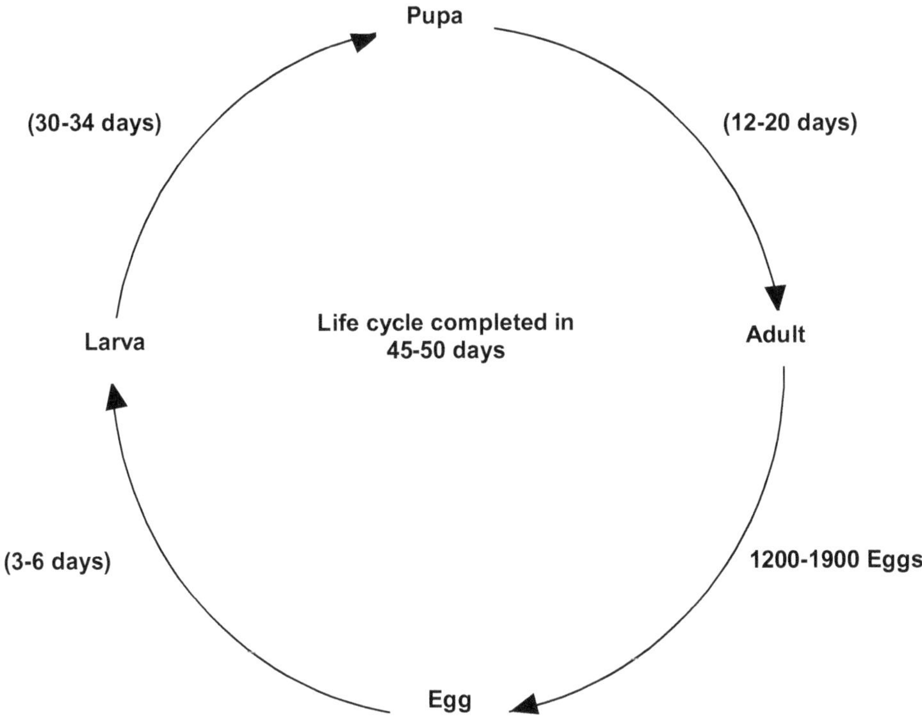

Figure 21: Life Cycle of *Agrotis ypsilon.*

Marks of Identification

1. **Adult**: 45 mm wingspan, brown fore wings with a clearer area on the apical quarter. Each bears a clear uniform spot which is extended by a slender black triangle. The hind wings are extremely pale beige. Orbicular is tear-shaped; reniform has a distinct black wedge- or dagger-shaped black marking on its outer margin; claviform is small, dark, oblong, and filled with dark scales. There is a zigzag line of pale scales on a dark background in the subterminal area. The male antennae are plumose (feathered), and the female antennae are filiform.
2. **Larva**: 45 mm, grey with, on each segment, 4 black spots disposed in a trapezium. The head is yellowish-brown.

Host Crops

Chickpea, potato, cereals, tobacco, maize, mustard, linseed, sugarcane, cucurbits, okra, brinjal, cabbage, beet, asparagus, lettuce, onion, chicory, tomato, *etc.*

Nature and Symptoms of Damage

- ☆ Young larvae feed on the epidermis of young leaf as they grow their habit changes.

☆ During day time, they live in cracks and holes in the ground and come out at night and feed the plants by cutting their tender stem, either below the surface or above the ground.

☆ The cut brances are sometime seen to have been drag into the holes where the larvae are eaten at leisure (free time).

☆ Early instar of larvae feeding on tender leaves of seedling plants. 3rd to 7th instars become negatively phototaxic and feed mostly at night. Damage from these instars is usually observed as a cutting of young seedlings, often causing death of the cut seedlings. Sometimes wilting is observed because of partial cutting.

☆ A larva will often cut one plant, quickly move on to other plants and continue cutting. Relatively small populations of cutworms are capable of destroying entire stands of some crops, such as cotton or maize.

☆ When seedlings are too large to be cut, foliar feeding may reduce plant vigour and subsequent yield.

☆ The yield reduction may vary to 30-40 per cent.

Management

(A) Preventive Measures

1. If possible, avoid planting crops in fields with a known history of cutworm problems.
2. Avoid planting crops (especially maize) following long standing pastures, meadows, lucerne or red clover.
3. Plough in the autumn and use shallow tillage to keep down late autumn and early spring vegetation (where conservation practices allow).
4. Monitor larvae with larval cutworm bait traps.
5. Low mow grass to remove eggs, disposing of cuttings at a distance.
6. Topdressing with sand does not kill larvae but deters them from travelling.

(B) Cultural Measures

1. Weed hosts on outlying areas are often preferred sites of oviposition and serve as food for the younger larvae. Thus destruction of these weeds may help reduce black cutworm populations.
2. Deep ploughing of fields between crops turn up larvae and pupae to the soil surface making them susceptible to predators and sun.
3. Depending on the crop, flooding of the infested field may be a feasible control method in some cases.
4. In small gardens, manual collection of the larvae is sufficient for control.

(C) Mechanical Measures

1. Light traps are most effective in the summer and autumn, but the late season generations generally pose little threat to crops.

2. Pheromone traps are more effective during the spring flight, when larvae present the greatest threat to young plants. Trap color affects moth capture rate, with white and yellow traps capturing more than green traps.

(D) Chemical Measures

1. **Poison baits**: Wheat bran/rice bran @ 10 kg + Chloropyriphos dust (1.5 per cent) @ ½ kg + Jaggery or gur @ 1 kg.

   Method of preparation – following method should be followed:

   I. In the morning, soak the wheat bran in the water. After that at evening, mix the insecticide and gur with little quantity of mustard oil and lemon juice with soaked wheat bran.

   II. This bait should be spread in the field during evening time.

   III. This is very effective in controlling cut worm.

2. The pest can be control by the application of chloropyriphos 1.5 per cent dust @ 25 kg/ha or carboryl 5 per cent dust @ 35 kg/ha.

(E) Biological Measures

1. **Parasites**: *Archytas cirphis, Chaetogaedia monticola, Eucelatoria armigera, Hyposoter exiquae, Meteorus laphygamae, Pseudamblyteles koebelei,* and *Pterocormus rufiventris.*
2. **Predator**: *Calosoma blaptoides tehuacanum.*
3. **Wasps**: *Apanteles marginiventris, Campoletis argentifrons, Campoletis flavicincta, Hyposoter annulipes* and *Ophion flavidus.*
4. **Larvae parasitized** by *Meteorus leviventris.*
5. **Other parasitoids:** *Archytas cirphis, Carcelia Formosa, Eucelatoria armigera, Euphorocera claripenni, Lespesia archippivora.*

(E) PESTS OF PEA

1. Aphid

Common name/Also known as: Green aphid or Green dolphin, Pea louse, and Clover louse.

Scientific Name, Order and Family

Acyrthosiphon pisum (Mordvilko)

(Hemiptera; Aphididae)

Taxonomical Position/Scientific Classification

Kingdom:	Animalia
Phylum:	Arthropoda
Class:	Insecta

Order:	Homoptera
Family:	Aphididae
Genus:	Acyrthosiphon
Species:	***A. pisum***

Biology and Life Cycle

1. **Egg**: Approximately 0.85 mm long, the light green egg turns a shiny black before hatching. Eggs are very rare in North Carolina.
2. **Nymph**: The immature aphid is smaller than, but similar to, the larger wingless adult. It requires four molts to reach the adult stage.
3. **Adult**: The long-legged pea aphid adult is light to deep green with reddish eyes. It has a body length of 2.0 to 4.0 mm. antennae about as long as the body. The cornicles (a pair of tailpipe-like structures projecting from the abdomen) of this aphid are characteristically long and slender. Both winged and wingless forms are seen. The total life cycle is completed in about 1 week.

Pea aphids have multiple generations per year and overwinter as eggs in alfalfa, clover or vetch. In the spring, nymphs hatch from eggs and appear similar to the wingless adult but smaller. Nymphs molt four times and mature into adults in 10 to 14 days. Pea aphids can reproduce rapidly when temperatures are around 65°F and relative humidity near 80 per cent. Infestations can originate from local alfalfa fields or can migrate in from the southern states.

Marks of Identification

Adults are 4 mm long and may be yellow, green or pink. There are dark bands around the antennae segments. They have long siphunculi (tube-like projections on either side at the rear of the body). Adults may have wings. Nymphs are similar but smaller in size.

Host Crops

Field peas, sweet peas, beans, sweet clover, alfalfa.

Nature and Symptoms of Damage

- Pea aphids extract sap from the terminal leaves and stem of the host plant.
- Their feeding can result in deformation, wilting, or death of the host depending upon the infestation level.
- Plants that survive heavy infestations are short and bunchy with more lightly colored tops than those of healthy plants.
- Wilted plants appear as brownish spots in the field. Moreover, plants are often coated with shiny honeydew secreted by the aphids, and cast skins may give the leaves and ground a whitish appearance.

☆ Pulse crops are especially susceptible in the flowering and early pod stage can result in lower yields due to less seed formation and smaller seed size.

☆ These aphids also transmit the pea enation mosaic and the yellow bean mosaic viruses.

Economic Threshold Level (ETL)

3-4 aphids/stem tip.

A series of economic threshold based on the height of the plant and the number of aphids per stem are helpful to determining when chemical control is necessary. They are as follows: 40 to 50 aphids per 25 cm (10 inch) or shorter stem; 70 to 80 aphids per 25 to 38 cm (10 to 15 inch) stem; 100 aphids per 50 cm (20 inch) stem.

Management

(A) Cultural Measures

1. Early sowing in a short period and cultivation of early maturing varieties enable to reduce the damage due to aphids.
2. Use of resistance varieties.
3. Destroy infested plants and crop residues after harvest.
4. Remove alternate hosts such as mustard from the fields.
5. Avoiding high dose of nitrogen fertilizers which favorable to aphid reproduction.

(B) Mechanical Measures

1. Trap the aphids with the help of yellow sticky traps.

(C) Chemical Measures

1. One or twice spraying of Dimethoate 30 EC @ 600 ml/ha or Oxydemeton methyl 25 EC @ 600 ml/ha or Acephate 75 WP @ 300 g/ha.

(D) Biological Measures

1. Lady beetles are perhaps the best known aphid predator, but big eyed bugs, damsel bugs, minute pirate bugs and syrphid flies use them as food during different stages of development.
2. Parasitic wasps lay eggs inside aphid bodies, causing the aphid's death. Aphids attacked by parasitic wasps will be black or bronze and have a perfectly round exit hole in their bodies where the young wasp emerged.
3. Coccinellids were the most abundant predators, followed by syrphid larvae.
4. Fungal parasites (*Entomophthora* sp.) are also important natural enemies of the pea aphid.

2. Pod Borer

Common name: Limabean Pod Borer, Pea Pod borer, Legume Pod Moth.

Scientific Name, Order and Family

Etiella zincknella (Treitschke)

(Lepidoptera, Pyralidae)

Taxonomical Position/Scientific Classification

Kingdom:	Animalia
Phylum:	Arthropoda
Class:	Insecta
Order:	Lepidoptera
Family:	Pyralidae
Genus:	Etiella
Species:	***E. zinckenella***

Biology and Life Cycle

1. **Egg:** The female lays oval, shiny white, 0.3x0.3 mm, elliptical eggs singly or in small groups (2-6) on the young or developing pods. Each female lays 50-200 eggs. Egg period or incubation period is 4-5 days.
2. **Larva**: Tiny larvae are green in colour while full grown larvae are rosy with have a purplish tinge. Larval period is of 10-27 days.
3. **Pupa**: Pupation takes place in soil in earthern cocoons or in soil debris. Pupal period is 9-22 days.
4. **Adult**: Moth is gray with a wing exponds of 25 mm. the forwings have dark marginal lines with having pale brownish with yellow colour scales. The hindwings are pale in colour.

Marks of Identification

Body length 8-11 mm, wingspan 19-27 mm. Wings longer than abdomen, folding as roof. Forewing is yellow- or brown-grayish with characteristic light stripe along fore edge, with orange spot on basal third, and with dark fringe. Hindwings are light gray, with dark venation and dark double line near fringe; the fringe is long and light in color. Top of abdomen with a tuft of golden-yellow hairs. Coloration of larvae is variable, from dirty greenish-gray to reddish; body length to 15-22 mm. Pupa is brilliant, brown, fine punctured, to 7-10 mm in length; cocoon is thick, white, and usually covered with soil particles. Number of generations per year reaches three, though the third generation can be facultative. Overwinters as larva. Development of egg lasts 4-21 days, of larva 19-40 days, of pupa 12-18 days, depending on temperature. Life span of adult is 20 days, average fecundity is about 100-300, maximum 600, eggs.

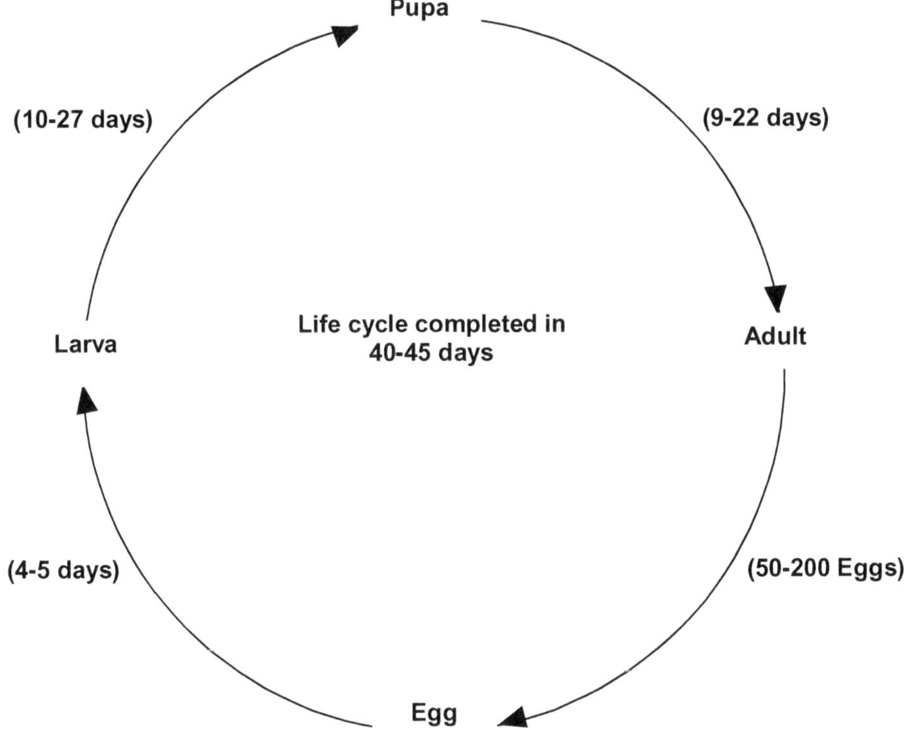

Figure 22: Life Cycle of *Etiella zincknella.*

Host Crops

Green peas, Field peas, Lentil, Cowpea, Redgram, Sunhemp, Mung and Urd.

Nature and Symptoms of Damage

- ☆ The young larvae feed on the foliage by scarping and later stages, bore into the buds, flowers and pods and feed on developing seeds with their body hanging outside the pod.
- ☆ This results in dropping of flowers and young pods.
- ☆ Older pods marked with a brown spot where a larva has entered.
- ☆ One or two larvae may be seen in a pod. After feeding, the larvae moves into the another pod.
- ☆ A single larva can destroy 30-40 pods during its life cycle.
- ☆ The reduction in the yield may be upto 5 per cent.

Economic Threshold Level (ETL)

10 per cent affected parts.

Management

(A) Preventive Measures

1. Deep ploughing after harvest of crop.

(B) Cultural Measures

1. Growing of marigold *i.e.* trap crop.
2. Intercropping with barley or maize, should be preferred over sole crop.

(C) Mechanical Measures

1. Install light traps/pheromone traps @ 5/ha.

(D) Chemical Measures

1. After flower initiation, spraying of Quinolphos 25 EC @ 1 litre/ha or Carboryl 50 WP @ 2 kg/ha (Repeat the application after 15 days, if necessary).
2. Apply Chloropyriphos 20 EC @ 2 lt/ha. Spray should be initiated at the start of pod formation and repeat after 2 weeks if necessary.

(E) Biological Measures

1. Apply NPV @ 250 LE/ha (Spraying should be done in evening hrs).
2. Conserve natural enemies like *Tetrastichus* sp., *Bracon hebetor, Phanerotoma* sp. and *P. hendecasisella.*

(F) PESTS OF LATHYRUS

1. Thrips

Scientific Name, Order and Family

Caliothrips indicus

(Thysanoptera, Thripidae)

Taxonomical Position/Scientific Classification

Kingdom:	Animalia
Phylum:	Arthropoda
Class:	Insecta
Order:	Thysanoptera
Family:	Thripidae
Genus:	Caliothrips
Species:	*C. indicus*

Biology and Life Cycle

1. **Egg**: A female can lay 20-40 eggs singly in leaves, flower tissue and fruits. The incubation period lasts from 6-15 days.
2. **Nymph**: Nymphs are minute and brown in colour. The nymph stage is typically found on leaves, in buds and flowers and at the base of some vegetable fruits. Nymphal period lasts from 4-10 days having 2 instars. The pre-pupal period is of about 1-5 days and pupal period varies from 1-6 days. The pupation takes place in soil or in hiding places on host plants such as the bases of leaves; adults on leaves, in buds and flowers.
3. **Adult**: Adult female is black. Male is smaller and light black or brown in colour. Wings are fringed with brown bands. The overall life cycle lasts from 44 days at 15°C to 15 days at 30°C.

Marks of Identification

1. **Adult** is tiny, slender insect with narrow fringed wings. Males are 1.2–1.3 mm long, pale yellow, with a narrow abdomen rounded at the end. Females are 1.6-1.7 mm long, yellow to brown, with a more rounded abdomen ending in a point.
2. **Nymphs** (larvae) are yellowish-white.

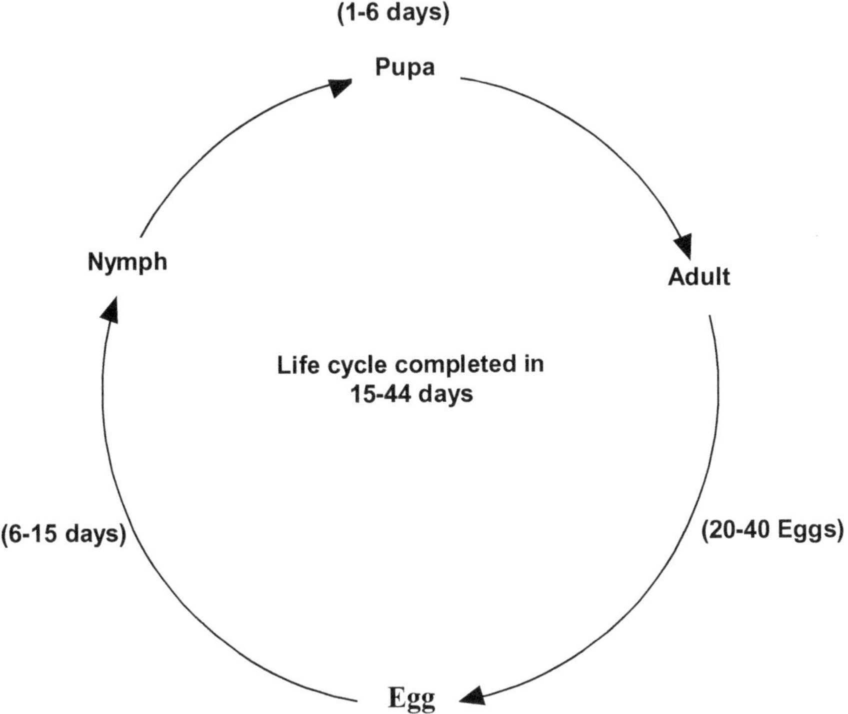

Figure 23: Life Cycle of *Caliothrips indicus*.

Host Crops

Lathyrus, Pea, Tomato, Cauliflower, Knol-khol, Chilly, Brinjal, Cotton, Linseed *etc.*

Nature and Symptoms of Damage

- ☆ Damage is caused by both nymph and adult by sucking sap from leaves, flowers, pods *etc.*
- ☆ Symptoms of damage is different as tender leaves show yellowish green patches on the upper surface; brown necrotic areas and silvery sheen on the lower surface.
- ☆ The leaves of attacked plant become curles, wrinkled and gradually dry out.
- ☆ Lower leaves showing white spots or streaks intermingled with black excreta on the surface.
- ☆ They also transmit peanut bud necrosis.
- ☆ In severe infestation, the plant bears very few pods.

Economic Threshold Level (ETL)

When 25 per cent of the leaves are damaged and live thrips are found in the field or 5 thrips per terminal shoot.

Management

(A) Cultural Measures

1. Sanitation of the fields such as weeds and debris, certain weeds those belongs to compositae and solanaceae family with yellow flowers not only attract the adults but serve as reservoir for viruses trasmitted by adults.

(B) Mechanical Measures

1. Colored sticky traps or water traps are useful for monitoring thrips. The color spectrum of sticky traps influenced their efficacy. Blue, yellow or white colors are used especially and it seems bright colors attract more thrips than darker ones.

(C) Chemical Measures

1. Spraying of Quinolphos 25 EC @ 500 ml/ha or Dimethoate 30 EC @ 300 ml/ha or Oxydemeton methyl 25 EC @ 300 ml/ha is very effective.

(D) Biological Measures

1. Larvae of *Aeolothrips colloris* feed on the larvae of thrips.
2. Larvae of *Chrysoperla predacious* feed on the larvae of thrips.

Pest Management of Oilseeds

(A) PESTS OF GROUNDNUT

1. Aphid

Also known as: Bean aphid or Cowpea aphid or Black legume aphid.

Scientific Name, Order and Family

Aphis craccivora (Koch)

(Hemiptera; Aphididae)

Taxonomical Position/Scientific Classification

Kingdom:	Animalia
Phylum:	Arthropoda
Class:	Insecta
Order:	Hemiptera
Family:	Aphididae
Genus:	Aphis
Species:	*A. craccivora*

Biology and Life Cycle

Aphis craccivora is a small species of aphid. The female has a glossy black or dark brown body with a prominent cauda (tail-like protrusion), and legs in some shade of brown or yellow. The antennae have six segments and these segments, cauda and cornicles are pale proximally (close to the body) and dark distally (further from the body). The adults do not have wax on their dorsal surface but the nymphs are lightly dusted with wax. Winged females are up to 2.2 mm (0.1 in) long and have cross-barring on the abdomen. Wingless females are a little smaller.

The eggs hatch in early spring and the first larvae are known as fundatrix (stem mothers) and feed at first. These aphids are all female and reproduce by parthenogenesis, producing nymphs which moult four times over the course of eight to twelve days. By the end of April, winged females have migrated to other host plants, often Acacia, and later to cotton, on which crop this pest does much damage. It may move back to alfalfa later in the year. A female aphid lives for 9 to 25 days and can produce from 25 to 125 young during its life. There may be up to twenty generations in the year. By November, winged forms have developed and eggs are laid before winter sets in.

Marks of Identification

The apterae of *Aphis craccivora* is dark brown with (usually) a very solid black shiny carapace from the metanotum to abdominal tergites 6. Many North American and a few southern European Aphis craccivora populations have a reduced sclerotic shield. Their siphunculi very rarely have any trace of constriction before the flange, and are 1.17-2.18 times the length of the rather tapering cauda. The body length is 1.16-2.3 mm.

Nymphs and Adult – dark coloured with cornicles in the abdomen.

Host Crops

It is a polyphagous pest. Redgram and other pulses, citrus etc are other host plants.

Nature and Symptoms of Damage

- ☆ Direct damage caused by *A. craccivora* occurs on the seedling, vegetative and flowering stages of plants. Aphids prefer to feed on young leaves, shoots, flowers and immature seed pods.
- ☆ Both nymphs and adults suck sap from tender or under surface of leaves and shoots resulting in twisting of leaves, poor pod development, devitalization of plants and sooty mould.
- ☆ The removal of sap weakens the plant, causing poor and stunted growth, leaf curling and distorted leaf growth, wilting and reduced resistance to drought conditions, all resulting in yield losses.
- ☆ Damage due to irritants and toxins, produced by aphid feeding on the leaves and growing points, manifests itself in necrosis and other adverse reactions.
- ☆ When the attack occurs at the pod formation stage, the yield is reduced considerally.
- ☆ The infestation on the groundnut crop usually occurs 4-6 weeks after sowing.

☆ The groundnut aphid also transmits groundnut rosette virus and sometimes groundnut stunt virus diseases.

Economic Threshold Level (ETL)

1 per Pod or 22 aphids per plant. However, the economic damage of aphids varies with the stage of plant, with most damage caused if aphids infest the growing points of groundnut early in the plant's development

Management

(A) Cultural Measures

1. Use of early and dense sowings. Early sowings allow plants to start flowering before aphids appear, while dense sowings provide a barrier to aphids penetrating in from field edges.
2. Virus-infected plant material should be removed after harvest and any volunteer plants or weeds that harbour viruses should also be destroyed.
3. Intercropped with barley or linseed minimizing the incidence of *A. craccivora* in India.

(B) Mechanical Measures

1. Handpicking and destruction of various insect stages and the affected plant parts.
2. Monitoring should be done during seedling, flowering and pegging stages of the crop. Monitor undersides of the leaves and yellow traps can use for monitoring aphids.

(C) Chemical Measures

1. As soon as, the pest appears at growing point, Spray any one of the following insectcides: Dimethoate 30 EC @ 425 ml/ha or Malathion 50 EC @ 625 ml/ha or Oxydemeton methyl 25 EC @ 200 ml/ha.
2. Dusting with Malathion 5 per cent D or Quinolphos 1.5 per cent D or Phosalone 4 per cent D @ 20 kg/ha.

(D) Biological Measures

1. Release of aphid parasites *i.e. Lysiphlebus* sp. and *Diaraetiella* sp.
2. Release of aphid predators including lady beetles, lacewings, bigeyed bugs, damsel bugs, and syrphid flies. Important predators include coccinellid beetles, *e.g. Cheilomenes sexmaculata* and *Coccinella septempuncta*, syrphid larvae, *e.g. Ischiodon scutellaris*, neuropteran larvae, *e.g. Micromus timidus*, and a predatory diptera, *e.g. Aphidoletes aphidimyza*.
3. Spiders may also be important in some areas.

4. Treatment with fungal pathogens include *Fusarium pallidoroseum* on cowpeas against *A. craccivora* on cowpea showed that an increase in spore concentration resulted in a corresponding decrease in aphid population.

2. Leaf Minor

Scientific Name, Order and Family

Stomoperyx nertaria (Meyrick)

(Lepidoptera; Gelechidae)

Taxonomical Position/Scientific Classification

Kingdom:	Animalia
Phylum:	Arthropoda
Class:	Insecta
Order:	Lepidoptera
Family:	Gelechidae
Genus:	Stomoperyx
Species:	***S. nertaria***

Biology and Life Cycle

1. **Egg**: Shiny and sculptured eggs are laid singly on tender leaves and tender shoots. Eggs are laid upto 100-400 mostly in the period from August to 15 September. Incubation period is 3 to 4 days.
2. **Larvae:** Full grown caterpillar is greenish with a small dark head. Larval period is 9 to 17 days.
3. **Pupa**: Pupation is inside the blotch mine and emerges as an adult in about 4 days.
4. **Adult**: Moth is very small with dark brown wings and small distinct white spot on forewings. One life cycle is completed in 15 to 28 days, with many generations in a year.

Marks of Identification

Adult is small, 8 to 10 mm long moth with wing expanse of 20 mm. Wings are grayish with a pale white dot on each of the forewings. Caterpillars are small, 6 to 8 mm long, 1 mm broad, cylindrical tapering posteriorly, brown or light green in colour with dark head and prothorax. Pupa is shining blackish brown in colour and measure about 4 to 5 mm in length and 1 to 1.5 mm in breadth.

Host Crops

Groundnut, Pigeonpea and Soybean.

Nature and Symptoms of Damage

The newly hatched **larva/caterpillar** mines into tender leaflets during early stage

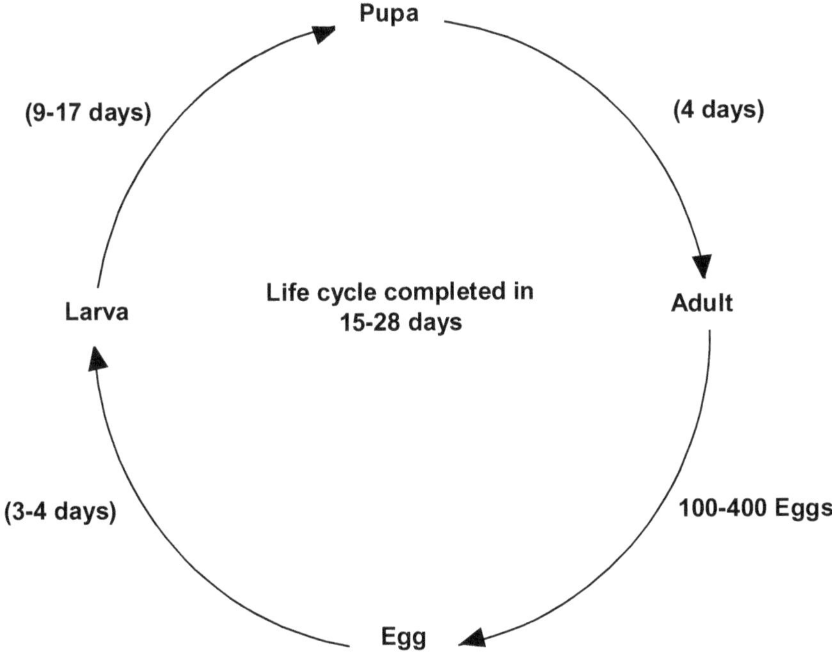

Figure 24: Life Cycle of *Stomoperyx nertaria*.

of their growth or it webs together adjacent leaflets and feeds on the tissue. The mined leaves within a few days show brown streaks.The leaflets get distorted and as a result, leaves dry and the plant wither adversely affecting the yield. Damage results in:

- ☆ Mining of larvae in the upper epidermis of leaves which causes in characteristic blotches.
- ☆ Folded leaves.
- ☆ Drying of affected leaves and withering of plants.
- ☆ Severly infested field looks as if burnt from a distance.

Economic Threshold Level (ETL)

2 live larvae/plant or 10 per cent mined leaves in the central whorl of plant or 1 larva/m linear row.

Management

(A) Cultural Measures

1. Regular monitoring and surveillance
2. Collection and destruction of the larvae and infested plant parts.
3. Crop rotation with a non leguminous crop to avoid out breaks of the pest.
4. Raising soybean as trap crop.

(B) Mechanical Measures

1. Setting of light traps/pheromone traps.

(C) Chemical Measures

1. Spraying the crop with 0.01 per cent cypermethrin or 0.01 per cent fenvalerate or 0.0025 per cent decamethrin or 0.2 per cent carbaryl or 0.05 per cent quinalphos or dusting the crop with quinalphos 1.5 D or phosalone 4 D or malathion 5 D @ 20 kg/ha.
2. Foliar sprays with acephate 1 g/l or chlorpyriphos 2.5 ml/l or Dusting or spraying with quinalphos 1.5 D @ 10 – 12 kg/ac or 2 ml/l

(D) Biological Measures

1. Use of Egg parasitoids *i.e. Trichogramme* sp.
2. Use Larval parasitoids *i.e. Bracon hebetor, Cotesia* sp.

3. White Grub

Also known as: Root grub

Scientific Name, Order and Family

Holotrichia consanguinea Blanchard

(Coleoptera; Scarabaeidae)

Taxonomical Position/Scientific Classification

Kingdom:	Animalia
Phylum:	Arthropoda
Class:	Insecta
Order:	Coleoptera
Family:	Scarabaeidae
Genus:	Holotrichia
Species:	***H. consanguinea***

Biology and Life Cycle

1. **Egg**: A female lays 50 to 70 eggs in soil at the depth of 10 cm soon after the onset of monsoon. The freshly laid eggs are milky white. Incubation period is 10 to 12 days.
2. **Larva**: Larval period is of about 8-10 weeks.
3. **Pupa**: The pupa is semicircular and creamy white. Pupal stage lasts for about 20 to 24 days. Pupation takes place in soil.
4. **Adult**: Grubs moult twice and become full-grown in 5 to 8 months. Adults formed during November-December remain in soil and emerge

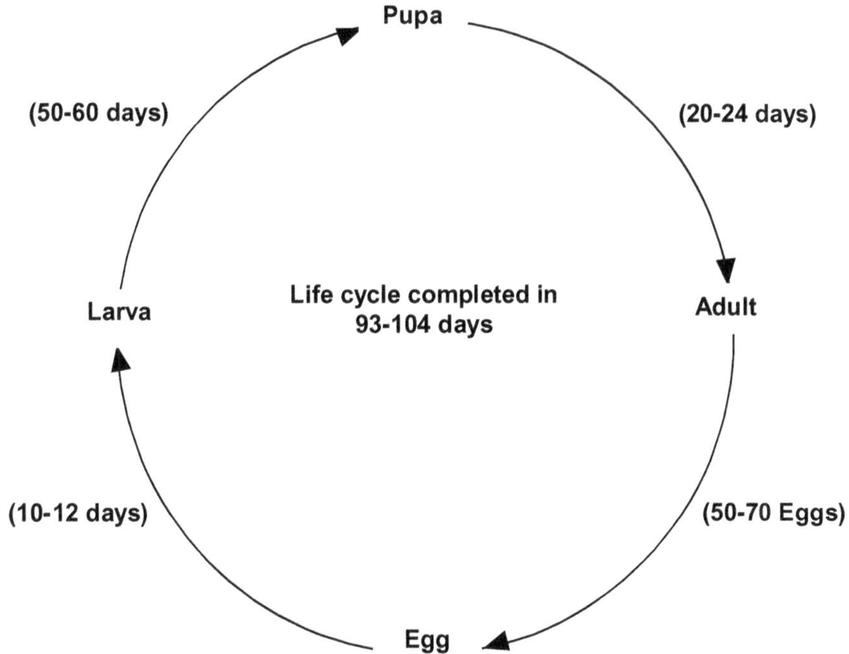

Figure 25: Life Cycle of *Holotrichia consanguinea*.

during first pre-monsoon shower in May or June. Since there is maximum emergence of adults during May/June, the pest is sometimes known as May/June beetle. Pest has only one generation in a year. Longevity of adult is 93 to 104 days.

Marks of Identification

Adult is dull brown coloured beetle, having light brown coloured abdomen and dark brown legs. Body measures about 18 to 22 mm in length and 7 mm in breadth. The newly hatched grubs are dirty white in colour and measure about 9 to 10 mm in length and 2 to 2.5 mm in breadth. The full-grown grubs are creamy white in colour having dark brown coloured head and prominent three pairs of legs. The grubs measure about 46 mm in length and 6 to 7 mm in breadth. These grubs are scarabaei form having 'c' shaped body. The posterior end of the grub is shining white dorsally. The eggs are milky white, while the pupae are semicircular and creamy white in colour.

Host Crops

Groundnut, cotton, tobacco, sorghum, maize, bajara, chilli, tomato, potato, sugarcane *etc.*

Nature and Symptoms of Damage

The damage is caused by both the grubs and adults. Both of them are nocturnal in habit. The grubs on hatching feed on the nodules of fine rootlets of leguminous crop. They may even girdle the tap root system. The crops like groundnut and tobacco which have tap root system are more susceptible to grub damage. On the other hand, plants like sorghum, maize, bajara, sugarcane *etc.*, which have an adventitious root system, can withstand the larger grub damage. The infested plants turn pale, show wilted appearance and finally dry up. The attacked plants can be easily pulled out from the soil. Adult beetles feed on the tender foliage of certain plants like mango, cluster bean, citrus, fig, ber, java, plum, pear, peach, tamarind, gulmohar, banyan *etc.*, during night and defoliate them.

- ☆ The grubs feed on the nodules and the fine rootlets of plants, and may also girdle the main root.
- ☆ Due to this type of damage, groundnut plants which have a tap root system are highly susceptible to the attack.
- ☆ An attacked plant becomes pale, gives a wilted appearance, and finally dried up.
- ☆ Adult beetles feed on foliage of trees like neem, banyan, java, mango, plum, cluster fig at night and defoliate them.

Economic Threshold Level (ETL)

20 or more beetles per tree or 0.14 larvae/m^2.

Management

(A) Cultural Measures

1. Deep ploughing the field twice during May and June. It would help to expose the resting pest stages in soil.
2. Flooding the fields.
3. Manual destruction of infested plant stems.
4. Treat the seed before sowing with 12.5 ml of Chloropyriphos 20 EC per kg of kernel.
5. Growing varieties of crops having vigorous or dense root system. *e.g.* Co 935 - variety of sugarcane.

(B) Mechanical Measures

1. Install light traps in the field; with the onset of rains the beetles come out of the soil and attract to the light so count the number of beetles per day. Also, dig 10 pits per ha in 100 × 100 × 20 cm and count the number of beetles per pit every day.
2. Collect the beetles by light traps during night hours and destroy them in kerosinised water.
3. The grubs and pupae present in soil can be collected and destroyed while undertaking weeding and interculturing operations.

(C) Chemical Measures

1. Spraying the host trees immediately after the first shower of monsoon or when there are 20 or more beetles per tree with carbaryl 50 WP at concentration of 0.1 per cent. Repeat the spray three weeks after first application.
2. Apply quinalphos 5 G or phorate 10 G or carbofuran 3 G @ 25 kg/ha in the soil or mix them thoroughly by harrowing or drill them into the soil.

(D) Biological Measures

1. Grub is parasitized by *Scolia* aureipennis, *Metarhizium anisopliae* Beauvaria (fungus).
2. Green muscardine fungi - *Metarrhizium anisopliae*
3. Bacteria - *Scolia* spp.
4. Parasitoid: Parasitic wasp (*Tiphia* spp.) Entomopathogenic nematode.
5. Predators: Ground beetle, ants, wasps.

4. Red Hairy Caterpillar

Scientific Name, Order and Family

Amsacta moorei

(Lepidoptera; Arctiidae)

Biology and Life Cycle

1. **Egg**: The creamy or light yellowish eggs are laid in groups mostly on the under surface of leaves, on clods, stones, dry twigs *etc.* Single female lays 300-1000 eggs. Incubation period is 3-4 days.
2. **Larvae**: Full grown caterpillars of both these species are reddish brown with black bands on either end and have long reddish brown hairs all over the body arising on warts. The head and prothorax are red. Larval period is 40- 50 days.
3. **Pupa**: The grown up larva burrows into the moist soil and pupates in earthen cell at a depth of 10-20 cm. mostly along field bunds and in moist and shady areas under trees in the field. The insect undergoes pupal diapause in the soil till next year.
4. **Adult**: Adults are medium sized moths. In *A. albistriga* forewings are white with brownish streaks all over and yellowish streaks along the anterior margin and hind wings white with black markings. A yellow band is found on the head. In *A. moorei* all markings are red in white wings. On receipt of heavy rains, about a month after sowing in *kharif* season, white moths with black markings on the hind wings emerge out from the soil in the evening hours.

Marks of Identification

1. **Adult**: The anterior marginal streak of forewings and the band on the head are red in colour.

Host Crops

They are polyphagous, also feeds on sorghum, cowpea, cotton, fingermillet, castor, cotton *etc.*

Nature and Symptoms of Damage

- ☆ The larvae feed on the leaves gregariously by scraping the under surface of tender leaflets leaving the upper epidermal layer intact in early stages.
- ☆ Later they feed voraciously on the leaves and main stem of plants. They march from field to field gregariously. Often it results in total loss of crop.
- ☆ Severely affected field looks as though they are grazed by cattle. Sometimes it results in the total loss of pods.
- ☆ They also feed on sorghum, cotton, finger millet, castor, pulses and cowpea, *etc.*

Economic Threshold Level (ETL)

8 egg masses/100 meter.

Management

(A) Cultural Measures

1. Deep summer ploughing after harvest to expose diapausing pupae.
2. Placing shoots of Jatropha or Ipomoea on bunds to attract migrating larvae and spraying on shoots.
3. Growing cowpea and castor as trap crops.
4. Grow cowpea or red gram as an intercrop to attract adult moths to lay more eggs.

(B) Mechanical Measures

1. Organize campaign to collect and destroy the pupae after summer ploughing on receipt of showers.
2. Set up 3-4 light traps and bonfires immediately at the onset of rains at 4 weeks after sowing in the rainfed season. Setting bonfires or light traps to attract the moths up to 11.00 P.M.
3. Collect and destroy egg masses in the groundnut, cowpea and redgram.
4. Collect and destroy gregarious early instar larvae on lace like leaves of inter crops *viz.*, red gram and cowpea.
5. Organize campaign by involving school children (or) general public to collect and destroy the migrating grown up caterpillars from the field.

6. Dig out a trench around the field to avoid the migration of caterpillars, trap larvae and kill them.

(C) Chemical Measures

1. Trenching around the field and dusting with carbaryl dust @ 250 gm/one meter length.
2. For grown up caterpillars - spray Dichlorvos 625 ml/ha or Chlorpyriphos 1250 ml/ha in 375 litres of water.
3. Spraying with Dimethoate 2ml/l.
4. Poison baiting for late instars with rice bran 10 kg + jiggery 1 kg + Quinalphos 1 litre.

(D) Biological Measures

1. Use nuclear polyhedrosis virus @ 250 LE/ha.
2. Natural enemies include a predatory pentatomid bug attacking larvae.
3. Release of larval parasites *i.e. Apanteles flavipes, A. creatonoti, Exorista civiloides, Sturnia inconspicuella.*

(B) PESTS OF SESAMUM

1. Gall Fly

Also known as: Gingelly gall fly

Scientific Name, Order and Family

Asphondylia sesami

(Diptera; Cecidomyiidae)

Taxonomical Position/Scientific Classification

Kingdom:	Animalia
Phylum:	Arthropoda
Class:	Insecta
Order:	Diptera
Family:	Cecidomyiidae
Genus:	Asphondylia
Species:	***A. sesami***

Biology and Life Cycle

1. **Egg:** Female fly lays eggs singly in the flower buds, flowers and developing capsules. The fecundity is 50-60 eggs. The egg period is 2-4 days.

2. **Larva:** Maggots are tiny whitish, legless and with body tapering exteriorly. The maggot is found inside the flowers. The larval period is about 14-21 days.
3. **Pupa:** It pupates inside the malformed capsule (gall). The fly emerges from galls in 6-10 days.
4. **Adult**: Adult is a small mosquito like fly. The total life cycle is completed in 23-37 days. there are several generations in a year.

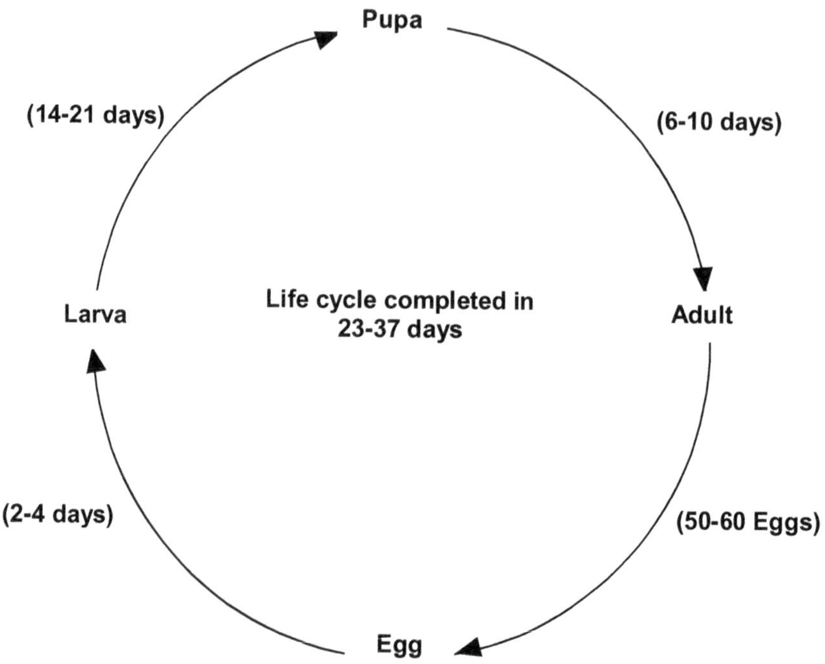

Figure 26: Life Cycle of *Asphondylia sesami.*

Marks of Identification

The adult insect is a minute mosquito like fly, measuring about 2.5 mm in length, with a wing span of 5 mm. There is a single pair of wings, which are thin and transparent. The maggots are small and dirty white in colour.

Host Crops

Sesame and Cluster bean only.

Nature and Symptoms of Damage

Maggots feed on the ovary (flower buds) and results in the malformation of pods without proper setting of seeds. At a later stage, the larvae infest the sesame fruit capsule making an entrance hole on the lateral side and feeding on the seeds inside the capsule; they leave excreta on the seeds. Flowers and young capsules with gall like swelling is the typical symptom of attack. The pest is quite serious and the loss has been reported upto the extent of 20 per cent. In severe infestation, the crop may be a complete failture.

- ☆ Maggots feed inside the floral bud.
- ☆ Leading to formation of gall like structure which do not develop in to flower/capsules.
- ☆ The affected buds wither and drop.

Economic Threshold Level (ETL)

2-5 galls/plant or 10 per cent damage.

Management

(A) Cultural Measures

1. Controlling weeds beneath or around the plants and removing fallen fruit from fields (bury at least 15 cm deep or burn them) may reduce the infestation of gall midge.
2. Grow resistant varieties like N-32, RT-46, Swetha til, RT103, RT-108, RT-125 *etc.*
3. Field should be ploughed to expose pupae and larvae of pests for predation. Sunlight radiation also kills soil pathogens.
4. Fields should have a good drainage of excess rain water.
5. Soon after harvesting, straw and stubbles should be burnt to avoid the carry over of the pests.
6. In *kharif*, early sowing in the first week of July and use of early maturing varieties like Uma, RT-46 and JT-7 reduces the crop losses due to pests and diseases.
7. Intercropping of sesame with mungbean, or moth bean, pearl millet and groundnut reduces the damage due to gall fly and leaf roller/capsule borer.
8. Early sowing (after onset of monsoon) manages Cercospora leaf spot and Alternaria. However, delayed sowings up to 3 weeks after onset of monsoon is good for phyllody management.
9. If a field is of no further economic use, destroy completely or remove all reproductive structures at least twice weekly.

(B) Mechanical Measures

1. Clipping off the gall, pick and burn the shedded buds and not allowing any stray plant to grow in off season.
2. Collect the galls and count number of larvae.
3. Color traps present a potential tactic for monitoring soybean pod gall midge or mating disruption (RADA, 2010).

(C) Chemical Measures

1. Spray with Dimethoate 30 EC @ 0.02 per cent or Carbaryl 50 WP @ 2kg/ha or Imidacloprid 17.8 SL @ 100ml/ha or Phosalone 1.0 L, Quinalphos 1.0 lt or Dichlorvos 500 ml/ha in 700 lt water per hectare.

(D) Biological Measures

1. Larval parasite: *Eurytoma dentipectus.*
2. Parasitoids: *Pteromalus fasciatus etc.*
3. Predators: Spider, ladybird beetle, lacewing *etc.*
4. Use larval parasites of gall fly like *Eurytoma dentipectus* Ghan, *Bracon hebetor.*

2. Howk Moth

Also known as: Dead head moth/Sphinx moth/Leaf eating caterpillar

Scientific Name, Order and Family

Acherontia styx (Westwood)

(Lepidoptera; Sphingidae)

Taxonomical Position/Scientific Classification

Kingdom:	Animalia
Phylum:	Arthropoda
Class:	Insecta
Order:	Lepidoptera
Family:	Sphingidae
Genus:	Acherontia
Species:	*A.styx*

Biology and Life Cycle

1. **Egg:** A female moth lays globular eggs singly on the under surface of leaves. The egg/incubation period is 2-5 days. Fecundity is about 160-175 eggs.
2. **Larva**: The larva is stout, green with yellowish oblique stripes and curved anal horn. The larval period lasts for 60-70 days.
3. **Pupa**: It pupates in earthern cocoon in soil. The pupal period lasts 14-21 days and 7 months in summer and winter respectively.
4. **Adult**: The adult moth is giant hawk moth, brownish with a characteristic skull marking on the thorax and violet yellow bands on the abdomen. Hind wings yellow with black markings. This insect completes three generations per year. The total life cycle is completed in 76-96 days.

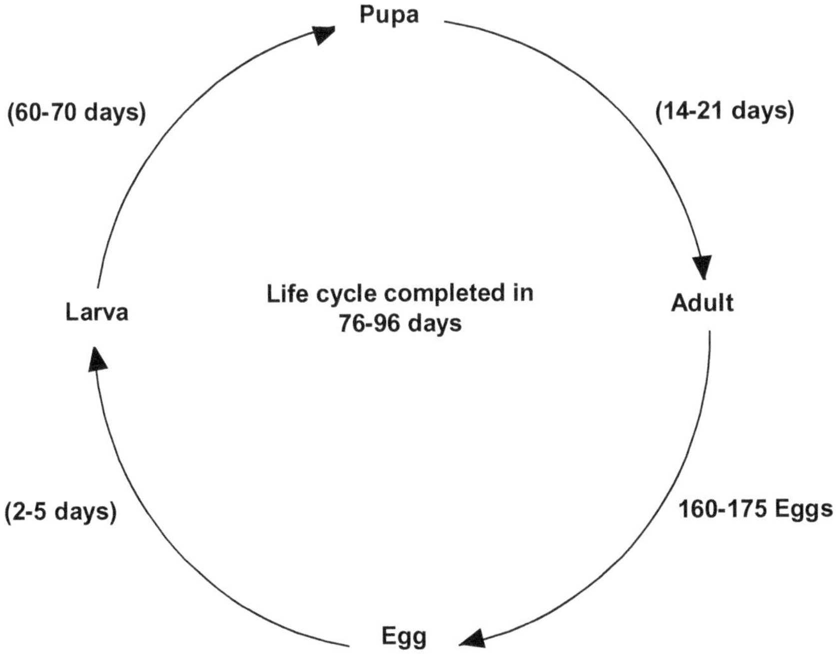

Figure 27: Life Cycle of *Acherontia styx*.

Marks of Identification

- **Adult**: Brownish giant hawk moth, thorax with a characteristic skull marking, abdomen has violet and yellow band, forewings are dark brown and Hind wings are yellowish with 2 black lines.

Host Crops

Sesame, potato, brinjal and jasmine.

Nature and Symptoms of Damage

The damage is caused by the larvae which feed voraciously on leaves and defoliate the plants. The moth is also harmful as it sucks honey from the honey combs in apiaries.

- The young larvae roll together a few top leaves and feed them.
- In the early stage of infestation, the plant dies without producing any branch or shoot.
- In later stage of attack, infested shoots stop growing.
- At flowering, larvae feed inside the flowers and on capsule formation, larvae bore into capsule and feed on developing seeds.

Economic Threshold Level (ETL)

1-2 larvae/plant

Management

(A) Cultural Measures

1. Field should be ploughed to expose pupae and larvae of pests for predation. Sunlight radiation also kills soil pathogens.
2. Fields should have a good drainage of excess rain water.
3. Soon after harvesting, straw and stubbles should be burnt to avoid the carry over of the pests.
4. In kharit early sowing in the first week of July and use of early maturing varieties like Uma, RT-46 and JT-7 reduces the crop losses due to pests and diseases.
5. Deep ploughing in the winter exposes the pupae for predation to insectivorous birds.
6. Intercropping of sesame with mungbean, or moth bean, pearl millet and groundnut reduces the damage due to gall fly and leaf roller/capsule borer.

(B) Mechanical Measures

1. Hand picking the larvae in the initial stage of attack and destroy by keeping in kerosene oil.
2. Bird perching: Perching provide shelter to birds, which readily predate on the caterpillars of insects and thus checks the insect population.

(C) Chemical Measures

1. Spray Carboryl 50 WP @ 2kg/ha.
2. Two rounds of dusting with phosalone 4 per cent or Malathion 5 per cent, first at 30 DAS and second at 45 DAS.

(D) Biological Measures

1. Egg parasite – *Agrmmatus* sp.
2. Larval parasite – *Cotesia* sp.
3. Use egg parasite Anastatus acherontiae.
4. Predators - Lacewing, ladybird beetle, reduviid bug, spider, red ant, black drongo (King crow), common mynah, big-eyed bug (*Geocoris* sp.), pentatomid bug (*Eocanthecona furcellata*), preying mantis *etc.*
5. Use larval parasite like *Sarcophaga* sp., *Zygobothria ciliate walp, Apanteles acherontiae.*

3. Leaf Webber

Common name: Sesamum leaf roller/Til leaf caterpillar

Scientific Name, Order and Family

Antigastra catalaunalis (Dup.)

(Lepidoptera; Pyralidae)

Taxonomical Position/Scientific Classification

Kingdom:	Animalia
Phylum:	Arthropoda
SubPhylum:	Hexapoda
Class:	Insecta
Order:	Lepidoptera
Family:	Pyralidae
Genus:	Antigastra
Species:	***A. catalaunalis***

Biology and Life Cycle

1. **Egg**: A female lay about 15-300 greenish minute eggs on the under surface of the apex of the tender leaves or on flowers at night. The egg period is 4-5 days.
2. **Larva**: Fully grown pale green larva with black head and dots all over the body measures 20 mm in length. Larvae become fully grown in 10 to 11 days under going 5 instars. The larval period is 11-16 days.

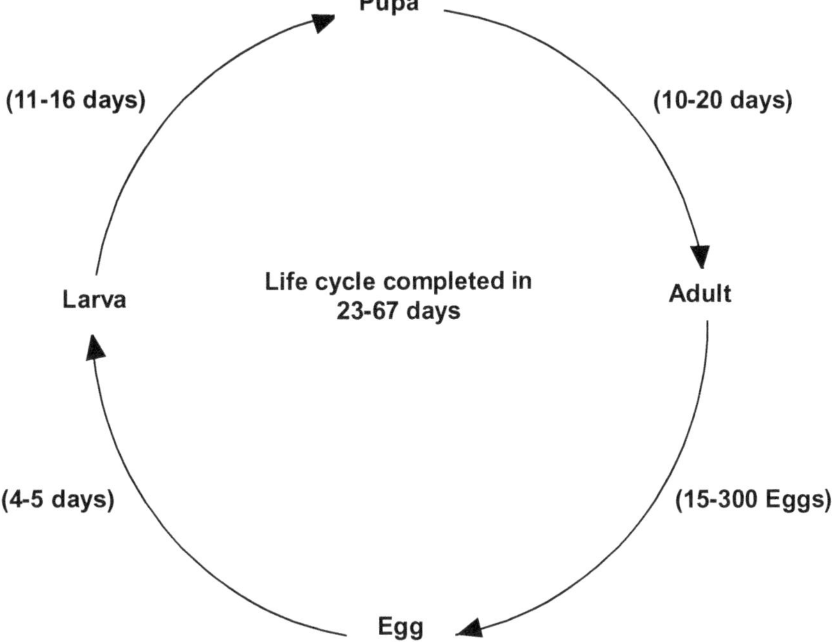

Figure 28: Life Cycle of *Antigastra catalaunalis*.

3. **Pupa**:. It pupates within the webbing, under fallen leaves or in soil cervices in a thin transparent white silken cocoon. The pupation period is about 10-20 days.
4. **Adult**: Moth is brown with yellowish brown wings. Adult moths are small in size with orange brown forewings and pale yellow transparent hind wings. Adult longevity was 6 to 8 days. The total cycle was completed in 67 days during winter and 23 days in summer.

Marks of Identification

1. **Larvae**: Young larva is pale yellow gradually become greenish in colour with black head having short white hairs.
2. **Adult**: Medium sized moth with reddish yellow forewings and dark brown markings on wing tips.

Host Crops

Sesame, Antirrhinum and Duranta.

Nature and Symptoms of Damage

- Larva webs the top leaves together and bore the tender shoots in the vegetative phase. Flowers and young capsules are bored at reproductive stage.
- Young larvae are less frequent on pods than on other plant parts. They feed externally by making a loose web, which sticks several leaves together. The larvae feed on leaves and young shoots. Excreta (frass) remain between the leaves and the loose web. At a later stage, the larvae infest the sesame fruit capsule making an entrance hole on the lateral side and feeding on the seeds inside the capsule; they leave excreta on the seeds.
- Pest attacks both the summer and Kharif crops right from the beginning of the growth and if the infestation occurs at very early stage, the plants dies without producing any branch or shoot and a single caterpillar could destroy two to three plants in a week. If infestation occurred at a later stage, the infested shoot remains without further growth.
- The nature and damage of insect is in short:
  - The young larvae roll together a few top leaves and feed them.
  - In the early stage of infestation, the plant dies without producing any branch or shoot.
  - In later stage of attack, infested shoots stop growing.
  - At flowering, larvae feed inside the flowers and on capsule formation, larvae bore into capsule and feed on developing seeds.

Economic Threshold Level (ETL)

2 webbed leaves/M^2 or 10 per cent damage.

Management

(A) Cultural Measures

1. Culture of sesame like EH7, 57, 84, 105, 106 and 156 should be encouraged as these are observed to be completely resistant against *A. catalaunalis.*
2. Sowing the crop during June and July will escape from leaf webber damage and delayed sowing resulted in a significantly higher level of damage to leaves, flowers and pods and poor yield.
3. Pest load found to be less in Kharif crop as compared to late sown.
4. Intercropping with pigeon pea and black gram, green gram, cluster bean, sorghum and pearl millet found to be significantly reduced the leaf webber damage.

(B) Mechanical Measures

1. Collection and destruction of infested parts of plants further minimize the damage of caterpillar.
2. If possible, manual collection and destruction of larvae will reduce the population build up.

(C) Chemical Measures

1. Spray of carbaryl 50 WP@ 2 kg/ha or Fenvelrate 20 EC @ 250 ml/ha. Two sprays of Quinalphos 0.05 per cent at 30 and 45 days after sowing will control the pest effectively.
2. Spraying with Deltamethrin 2.8 EC @ 400 ml/ha or Dimethoate 30 EC 500 ml or Methyl parathion 50 EC 500 ml or Cypermethrin 10 EC @ 500 ml/ha.
   - Spray should be done twice at pest appearance on flowering stage and pod formation stage.

(D) Biological Measures

1. The most important parasite is *Trathala flavoorbitalis*; up to 20 per cent of larvae may be infected.
2. Conservation of existing natural enemies (spiders, coccinellid beetles, predatory stink bugs, preying mantids, black ant) and parasitoids (Braconids and Ichneumonids) through ETL based (2 webbed leaves/sq. m or 10 per cent damage) application of botanical insecticides and safer chemicals.
3. Augmented release of parasitoids *viz., Trathala flavoorbitalis* (Behera, 2011) and *Apanteles* sp. and the predators like *Chrysoperla carnea* also would reduce the population build up.

(C) PESTS OF LINSEED

1. Bud Fly

Also known as: Gall midge of linseed.

Scientific Name, Order and Family

Dasineura lini (Barnes)

(Diptera; Cecidomyidae)

Taxonomical Position/Scientific Classification

Kingdom:	Animalia
Phylum:	Arthropoda
Class:	Insecta
Order:	Diptera
Family:	Cecidomyidae
Genus:	Dasineura
Species:	***D. lini***

Biology and Life Cycle

1. **Egg**: The female lays 29-103 smooth, transparent eggs in the folds of 8-17 flowers or in tender green buds, either singly or in clusters of 3-5. The eggs hatch in 2-5 days.
2. **Larva**: Just after emergence, the larvae are transparent, with a yellow patch on the abdomen. They pass through four instars in 4-10 days and when full-grown become deep pink and measure about 2 mm in length.
3. **Pupa**:. The full-grown maggots drop to the ground, prepare a cocoon and pupate in the soil. The pupal period lasts 4-9 days.
4. **Adult**: The adult of this gall-midge is a small orange fly. A generation is completed in 10-24 days. There are four overlapping generations during the season.

Marks of Identification

1. **Egg:** The eggs are smooth and transparent.
2. **Larva:** The full grown larvae have also transparent dark pinkish in colour with yellow patch on their abdomen.
3. **Adult:** The adults are mosquito like small fly tendency to hover near the tops of plants and relatively slow flight speed distinguish them from other flies.

Host Crops

Linseed (Monophagous)

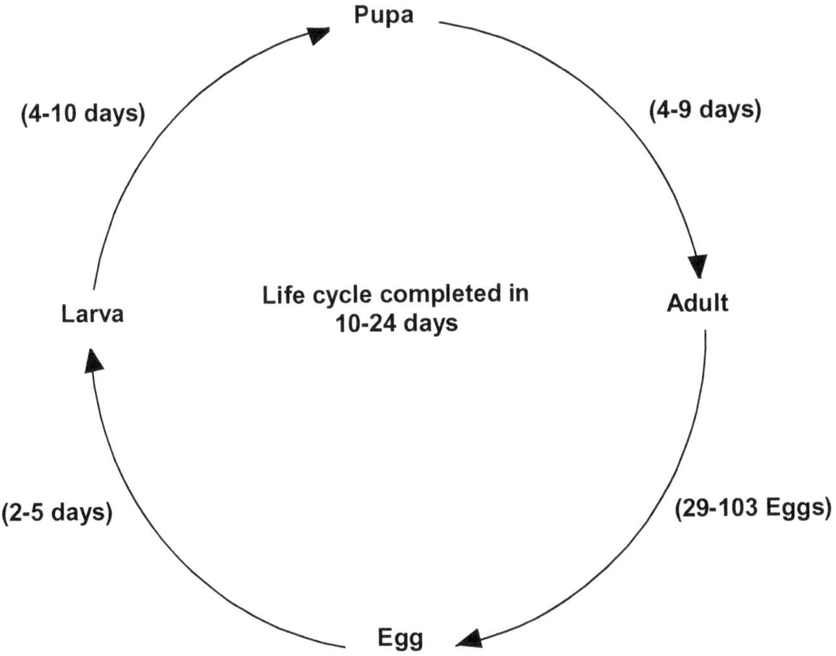

Figure 29: Life Cycle of *Dasineura lini*.

Nature and Symptoms of Damage

- Bud fly is a major limiting factor in the production of flax for seed in India.
- The damage is caused by maggots, which feed on the flower buds and prevent their proper opening.
- Damage is caused by the larvae (maggot) feed on flower buds and prevent from proper opening. Consequently the seed does not set properly.
- Due to their feeding, galls are produced and there is no pod formation. The incidence of their pest goes up to 20 per cent.

Economic Threshold Level (ETL)

5 per cent bud infestation or 2 insects per 10 flax.

Management

(A) Cultural Measures

1. Early sowing of the crop, since the attack is more severe on late sown crop.
2. The infested buds should be removed and destroyed to reduce further incidence of the pest.
3. Deep summer ploughing of the fields exposes the bud fly maggots.

4. Sowing the pest resistant and high yielding varieties like Garima, Shubhra, Sweta, Laxmi-27, Kiran, Parvati, Padmini, Shekhar and Sheela in first fortnight of October to get higher seed yield with moderate bud fly infestation.
5. Intercropping of linseed with chickpea (3:1) or mustard (5-6:1) reduces the bud fly infestation.
6. Use recommended dose of fertilizer (60-80 kg N, 40 kg P_2O_5/ha) with two irrigations at 35 and 65 days after sowing under irrigated conditions.
7. Put bamboo pegs @ 40-50/ha for increasing the activity of natural enemies especially predatory birds.

(B) Mechanical Measures

1. The adult flies can be killed by using light traps.
2. The flies are also attracted in day-time to molasses or gur added to water.

(C) Chemical Measures

1. The crop may be sprayed with Carbaryl 50 WP @ 1.125 kg/ha in 600-750 litres of water/ha.
2. Spraying of Imidacloprid 17.8 SL @ 100 ml/ha at the time of bud initiation. As per need, repeat the application after 15 days.
3. Dusting of Carbaryl 5 per cent @ 15-20 kg/ha

(D) Biological Measures

1. Use of *Systasis dasyneura* parasite on maggots. Conserve larval parasitoids *viz.*, *Systasis dasyneurae* Mani (Miscogasteridae), *Elasmus* sp. (Elasmidae), *Eurytoma* sp. (Eurytomidae), *Torymus* sp. (Torymidae) and *Tetrastichus* sp. (Eulophidae).
2. Use larval parasite *Pteromalus fasciatus* Thomas

2. Linseed Caterpillar

Also known as: Beet armyworm

Scientific Name, Order and Family

Spodoptera exigua

(Lepidoptera; Noctuidae)

Taxonomical Position/Scientific Classification

Kingdom:	Animalia
Phylum:	Arthropoda
Class:	Insecta
Order:	Lepidoptera

Family:	Noctuidae
Genus:	Spodoptera
Species:	***S. exigua***

Biology and Life Cycle

1. **Egg**: Eggs are laid in clusters of 50 to 150 eggs per mass. Normal egg production is about 300 to 600 per female. Eggs are usually deposited on the lower surface of the leaf, and often near blossoms and the tip of the branch. Eggs hatch in 2-3 days during warm weather.
2. **Larva:** There are usually 5 instars (takes 8- 10 days to reach 3rd instar). 1st and 2nd instars are pale green to yellow; 3rd instar has pale stripes; 4th instar is darker dorsally with dark lateral stripes; 5th instar larvae are variable in color being green dorsally or pink or yellow ventrally with a white stripe laterally. Dark spots and dashes usually present. The spiracles are white with a narrow black border. The body is practically devoid of hairs and spines. The larval period lasts for 15-20 days.
3. **Pupa**: The pupa is light brown in color and measures about 15 to 20 mm in length. Pupation occurs in the soil. The chamber is constructed from sand and soil particles held together with an oral secretion that hardens when it dries. Duration of the pupal stage is 5-7 days during warm weather.
4. **Adult:** The moths are moderately sized, the wing span measuring 25 to 30 mm. The forewings are mottled gray and brown, and normally with an irregular banding pattern and a light colored bean-shaped spot. The hind wings are a more uniform gray or white color, and trimmed with a dark line at the margin. Mating occurs soon after emergence of the moths, and oviposition begins within two to three days. Oviposition extends over a three to seven day period, and the moths usually perish within 9- 10 days of emergence.

Marks of Identification

1. Egg is cylindrical, greenish to white in color, laid in groups on leaves and covered with cottony whitish scales.
2. **Larva (worms):** Younger larvae are pale green or yellow in color. Older worms are darker when viewed from above and have a dark stripe along the side
3. **Pupa**: The pupa are light to dark brown in colour and found in the soil.
4. **Adult/Moth:** The outerwings are mottled gray and brown, normally with an irregular banding pattern and a light colored beanshaped spot.

Host Crops

Asparagus, bean, beet, broccoli, cabbage, cauliflower, celery, chickpea, corn, cowpea, eggplant, lettuce, onion, pea, pepper, potato, radish, spinach, sweet potato, tomato, and turnip, alfalfa, corn, cotton, peanut, safflower, sorghum, soybean, sugarbeet, and tobacco.

Nature and Symptoms of Damage

- ☆ Larvae feed on both foliage and fruit and it is regarded as a serious defoliator of flower crops and cotton.
- ☆ Young larvae feed gregariously and skeletonise foliage. As they mature, larvae become solitary and eat large irregular holes in foliage. They also burrow into the crown or centre of the head on lettuce, or on the buds of brassica crops.
- ☆ Beet armyworms eat irregular holes in foliage, eventually skeletonizing the leaves. They can eat tender young transplants to the ground and defoliate older plants. They burrow into heading vegetables, such as lettuce and cabbage. Beet armyworms also leave gouges in tender fruit, especially tomatoes.

Economic Threshold Level (ETL)

5-8 larvae (worms)/25 plants

Management

(A) Cultural Measures

1. Practice crop rotation and field sanitation.
2. Use grass barriers around fields.
3. Use varieties which are more tolerant to the pest.
4. Controlling broad leaf weeds and rapid disposal of crop residues after harvesting may reduce population buildup.

(B) Mechanical Measures

1. Hand-pick and destroy eggs masses and worms.
2. Use pheromone traps for pest monitoring and mass trapping.

(C) Chemical Measures

1. Spray the crop with Quinolphos 25 EC @ 625 ml/ha or Carbaryl 50 WP @ 1.125 kg/ha or Malathion 50 EC @ 1.0 litre in 600·750 litres of water per ha.

(D) Biological Measures

1. Among the most common parasitoids are *Chelonus insularis*, *Cotesia marginiventris*, and *Meteorus autographae*.
2. Predators frequently attack the eggs and small larvae; among the most important are minute pirate bugs (*Orius* spp.), Bigeye bugs (*Geocoris*

spp.), Damsel bugs (*Nabis* spp.) and a predatory shield bug (*Podisus maculiventris*).

3. Pupae are subject to attack, especially by the red imported fire ant, *Solenopsis invicta* Buren. Fungal diseases, *Erynia* sp. and *Nomurea rileyi*, and a nuclear polyhedrosis virus also inflict some mortality.

3. Thrips

Scientific Name, Order and Family

Thrips lini

(Thysanoptera, Thripidae)

Taxonomical Position/Scientific Classification

Kingdom:	Animalia
Phylum:	Arthropoda
Class:	Insecta
Order:	Thysanoptera
Family:	Thripidae
Genus:	Thrips
Species:	*T. lini*

Marks of Identification

Thrips are very small 0.5-2.0 mm long with narrow bodies and promonent legs and antennae. They have four narrow wings that are fringed with long hairs and are strong fliers. Immature stage similar in appearance to adults.

Host Crops

Cabbage, wheat, barley, oat, onion, garlic, tobacco, pea, beans, lucerne *etc.*

Nature and Symptoms of Damage

- ✰ Thrips can be problems in linseed where they can stop normal capsule development.
- ✰ Feeding by nymphs and adults on the growing points and young leaves of older plants either kill or cause abnormal cell division with distorts.
- ✰ Feeding by thrips can cause flax foliage to assume a red spotted appearance.
- ✰ Heavily infested plants are often short which can reduce the length of their fiber but usually does not affect its quality.
- ✰ The damaged flower often drops prematurely or reduce the number of seeds. Capsule on infested plants burst open before ripening causing reduction in their weight and in number of seeds.

Economic Threshold Level (ETL)

8-10 Adults or Nymphs per Leaf

Management

(A) Cultural Measures

1. Crop rotation is effective form of control. Non susceptible crops like cereals, oilseeds rape, oat or alternative crops like potato, beans can be sown with linseed.
2. Early sowing in growing season can reduce thrips injury particularly where *T. linarius* is common.
3. Deep ploughing during simmer, thrips are very susceptible to drowning by heavy rains when they are burrowing into the soil to overwinter.

(B) Mechanical Measures

1. Use of yellow sticky traps to attract the insects.

(C) Chemical Measures

1. On the main crop, spraying with Dimethoate 0.03 per cent or Methyl demeton 0.02 per cent at the time of infestation.
2. We can also use recommended dose of Lamda cyhalothrin or Acephate or Alphamethrin or Beta cyfluthrin or Deltamethrin or Acetamiprid.

(D) Biological Measures

1. Biological control of the thrips by Anthocoridae, Lygaidae and predator mites will be effective.

4. Jassids

Also known as: Linseed leafhopper

Scientific Name, Order and Family

Empoasea kerri var. *motti*

(Homoptera, Cicadellidae/Jassidae)

Taxonomical Position/Scientific Classification

Kingdom:	Animalia
Phylum:	Arthropoda
Class:	Insecta
Order:	Hemiptera
Family:	Cicadellidae
Genus:	Empoasca
Species:	***E. kerri***

Biology and Life Cycle

1. **Egg**: The female lays eggs singly inside the veins of leaves. The fecundity is 25-60 eggs. The egg period completed in 4-11 days.
2. **Nymph**: Nymphal period varies from 24-25 days with 5 instars.
3. **Adult**: Male adult lives for 18-30 days and female lives for 20-34 days.

Marks of Identification

Adults are active, elongate, yellowish green in colour, wedge shaped insects and walked diagonally in a characteristic manner.

Host Crops

Linseed, Potato and Brinjal.

Nature and Symptoms of Damage

Both adults and nymphs suck sap from young leaves, mostly from the lower surface. The first symptom of attack is a whitening of the veins. Chlorotic (yellow) patches then appear, especially at the tips of leaflets, probably caused by a reaction between the jassids salivary secretion and plant sap. Under severe infestation, the leaf tips become necrotic in a typical V shape, giving the crop a scorched appearance known as 'hopper burn'

Economic Threshold Level (ETL)

25 per cent leaves are damaged or presence of 5-10 per cent nymphs/adults per plants.

Management

(A) Cultural Measures

1. Crop rotation with non host crop.
2. Intercropping with pearl millet and avoidance of groundnut, castor intercrops.
3. Use balance dose of nitrogen fertilizers.
4. Planting of tolerant variety.

(B) Mechanical Measures

1. Use of yellow sticky traps to attract and kill the insects.

(C) Chemical Measures

1. Spraying of Imidacloprid 17.8 SL @ 100 ml/ha or Oxydemeton methyl 25 EC 600 ml/ha or Dimethoate 30 EC @ 625 ml/ha.

(D) Biological Measures

1. Predators: lady bird beetls, preying mantids, freen lace wings, long horned grasshoppers, spiders.
2. Entomophagous pathogen: *Metarhizium anisopliae*

(D) PESTS OF SAFFLOWER

1. Aphids

Scientific Name, Order and Family

Uroleucon carthami

Dactynotus carthami

(Homoptera; Aphididae)

Taxonomical Position/Scientific Classification

Kingdom:	Animalia
Phylum:	Arthropoda
Class:	Insecta
Order:	Homoptera
Family:	Aphididae
Genus:	Uroleucon
Species:	***U. carthami***

Biology and Life Cycle

Females reproduce viviparously and parthenogenetically and on an average a single female produce about 30 young ones.

1. **Nymph**: Nymphs are reddish brown in colour. Nymphs moult four times to become adult in 7-9 days. The nymphs are smaller than the adult and wingless.
2. **Adult**: Large sized, black, soft body bright with pear-shaped bodyand conspicuous cornicle. Females reproduce viviparously and parthenogenetically and on an average, single female produces about 30-56 nymphs in a life cycle of 2-17 days.

Marks of Identification

The adults are black coloured, but nymphs are reddish brown and soft bodied insects measuring 1.5 to 2 mm in length. Alate and apterous forms are commonly seen an aphid colony. The alates are found in the beginning of the season (November-December) and again towards the maturity of the crop, while apterous forms are abundantly found between the two extremes mentioned above.

Both winged and wingless forms are common, the former being noticed in the beginning of season or towards the maturity of crop. Adult is soft bodied, shining black coloured insect measuring about 1.5 to 2 mm in length. Wings when present are thin and transparent and are held like a roof over the body.

Host Crops

Safflower, niger, citrus, cornflower, calopsis, dahlia.

Nature and Symptoms of Damage

- ☆ During pre-flowering stage, both nymphs and adults suck the sap from lower surface of leaves and tender shoots and impair the vitality of plant.
- ☆ In case of severe infestation, entire plant may be covered with soft-bodied blackish insects. It gives blackish appearance to the infested plant.
- ☆ Moreover, honey dew secreted by the aphids facilitates the occurrence of black sooty mold on foliage, which hampers the photosynthetic activity of the plant.
- ☆ Aphid infested plants remain stunted and may die before maturity sometimes without any seed set. Finally the plants dry up.
- ☆ Yield losses cause by aphids is about 40-50 per cent and infestation may occur on 30-45 days old crop.
- ☆ The pest is active from November to December.

Economic Threshold Level (ETL)

50-60 aphids/5cm twig/plant.

Management

(A) Cultural Measures

1. **Use of resistant varieties**: Genotype having dwarf plant type with thin and wiry peduncles, tiny capitula and small sized seeds were also found to have a high degree of resistance to aphids *i.e.* Bhima and A-1, GMU-4608, GMU-46O9, GMU-4610, GMU-4625, GMU-4627, JLSF-295, JSLF-406, JSLF-409, SSF-139, SSF-141, SSF-282, SSF-428, GMU-4625677-1.
2. **Effect of date of sowing**: Date of sowing is crucial for aphid management. Sowing during mid or second fortnight of September, under drought prone conditions and September to mid October in irrigated conditions, was found effective to keep aphid population below economic threshold in India.
3. **Avoid late sowing**: Crop should be sown early *i.e.* in the last week of September to first week of October.
4. If the attack is observed in the border rows take control measures:Maintain 2 or 3 rows of Maize and Sorghum around the fields.
5. Application of balanced fertilizer, intercropping and mixed cropping reduces the pest population.
6. Intercultural operations like harrowing; hoeing can reduces the weeds which are host plants for safflower aphids.

(B) Mechanical Measures

1. Install yellow traps for monitoring aphids.

(C) Chemical Measures

1. Apply following insecticides as soon as pest appears - Carbaryl 50 WP 0.1 per cent or Thiometon 25 EC 0.03 per cent or Dimethoate 30 EC 0.03 per cent or Quinalphos 25 EC 0.05 per cent.
2. Dusting of Quinalphos 1.5 D @ 20 kg/ha. Dusts are to be used only when spraying is not possible.

(D) Biological Measures

1. The safflower aphid has been reported to be parasitised by an endoparasite, *Pseudendaphis* sp. (Diptera: Cecidomyiidae) to the extent of 66.6 per cent in the first week of March.
2. Some of the Coccinellids *i.e. Brumoides suturalis* Fabr., *Coccinella septempunctata* Linn., *Menochilus sexmaculata* (Fab.) and *Scymnus nubilus* have also been reported predating on aphids.
3. Other natural enemies *viz., Sphaerophoria indiana* (Big.), *Coccinella arcuata* (Fab.), *Micromus cinearis* (Hagen), *Ischiodon scutellaris* (Fab.), *Harmonia octomaculata* (Fab.) and *Coccinella repanda* (Thomb), *Aphelinus* sp. and *Aphidencyrtus* sp. Mayr have also been found predating on aphids (Parlekar, 1987).
4. Release of Chrysoperla eggs/grubs @ 1-2/plant.
5. Conserve *Aphidencyrtus aphidivorus, Micromus cinearis, Ischiodon scutellaris, Harmonia octomaculat* and *Aphelinus* sp.

2. Bud Fly

Also known as: Sesamum Capsule fly.

Scientific Name, Order and Family

Acanthiophilus helianthi

(Diptera; Tephritidae)

Taxonomical Position/Scientific Classification

Kingdom:	Animalia
Phylum:	Arthropoda
Class:	Insecta
Order:	Diptera
Family:	Tephritidae
Genus:	Acanthiophilus
Species:	***A. helianthi***

Biology and Life Cycle

1. **Egg**: The female lay eggs in cluster of 6-44 withine the flower. The incubation period is of about 1 day in April.
2. **Larva**: Maggot is dirty white in colour. Larval period varies from 6-7 days. Young maggots start feeding on the florets and the thalamus. Within one week they grow to the full and attain a size of 5 x 1.5 mm.
3. **Pupa**: Pupation takes place inside the buds. Pupal period varies from 7-8 days. The adults emerge out of the bud through the holes made by the maggots before they pupate.
4. **Adult:** Flies are ash coloured with light brown legs. The adults are active from March to May. Three generations are completed during a crop season.

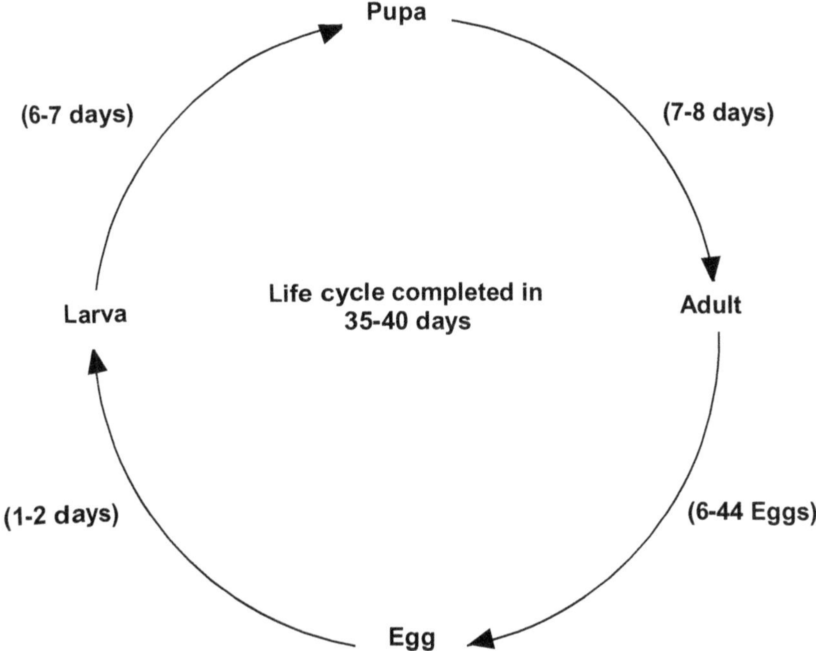

Figure 30: Life Cycle ot *Acanthiophilus helianthi.*

Marks of Identification

Capsule fly or safflower bud fly adults are ash colored; black obscured by a dense gray microtrichia and wing with characteristic diffused pattern. They have light brown legs. Maggot is dirty white in colour.

Host Crops

Sesame and Safflower.

Nature and Symptoms of Damage

☆ The damage is caused by maggot/larvae.

- ☆ Newly hatched larvae feed on the floral parts including thalamus and soft parts of the capsules.
- ☆ Affected buds show small bore holes.
- ☆ The infested buds rotten with foul smelling ooze coming out of the apices.
- ☆ The pest causes reduction in the yield of safflower seed.

Economic Threshold Level (ETL)

No action threshold has been set for safflower bud fly.

Management

(A) Cultural Measures

1. Early removal of destruction of infested buds is helpful in cheking the spread of pest.
2. Timely Sowing of the crop; clean cultivation and using resistant varieties are cultural techniques for controlling safflower bud flies.

(B) Mechanical Measures

1. Modified Steiner traps baited with methyl eugenol, water traps and sticky traps can be used to monitor bud flies and their immigration to the crop.

(C) Chemical Measures

1. At the time capsule formation – Spraying of Imidacloprid 17.8 SL @ 100 ml/ha or Odydemeton methyl 25 EC @ 600 ml/ha or Dimethoate 30 EC @ 600-650 ml/ha or Malathion 50 EC @ 1.00 litre/ha or Phosphomidon 100 EC@ 150-200 ml/ha. About 600 to 650 litre of water is needed for spraying in about one hectare area.

(D) Biological Measures

1. Conserve larval parasitoids *viz., Ormyrus* sp. (Ormyridae), *Eurytoma* sp. (Eurytomidae) and *Pachyneuron muscarum* (Braconidae) and predator *Chrysopa virgestes* (Chrysopidae).
2. Some bio-control agents such as *chrysopa* sp., *Orymurus* sp., *Eurytoma* sp., *Stenomalus muscarum* (L.), *Syntomopus* sp., *Bracon* sp., *Pronotalia* sp. and *Antistrophoplex conthurnatus* (Masi) can reduce the population build up.

(E) PESTS OF SUNFLOWER and NIGER

1. Bihar Hairy Caterpillar

Also known as: Jute Hairy Caterpillar

Scientific Name, Order and Family

Spilosoma obliqua

(Lepidoptera; Arctiidae)

Taxonomical Position/Scientific Classification

Kingdom:	Animalia
Phylum:	Arthropoda
Class:	Insecta
Order:	Lepidoptera
Family:	Arctiidae
Genus:	Spilosoma
Species:	***S. obliqua***

Biology and Life Cycle

1. **Egg**: Pest breeds from March to April and again from July to November. Adult female lays 400-1000 light green, spherical eggs in clusters on the underside of the leaves. Egg period 4-7 days.
2. **Larva**: Larva is orange coloured with broad transverse bands with tuft of yellow hair that are dark at both end. Larval instars are 7 and period is 25-45 days. Early instars are gregarious and later instars disperse in search of food.
3. **Pupa**: Pupation takes place in plant debris or soil and pupal period 7-15 days. The pest goes in pupation during May-June in summer and during December-February in winter.
4. **Adult**: The moth measures about 50 mm across the wing spread. Adults have crimson coloured body with black dots. Wings pinkish with numerous black spots. Adult lives for 7 days.

Marks of Identification

Moth medium sized and of pinkish-yellowish colour with black dots on both pairs of wings. The larvae is redish-black coloured on both sides with middle portion yellowish. The larva has brownish coloured hairs all over the body.

Host Crops

Sesamum, lobia, urd, mung, linseed, mustard, sunflower, jwar, brinjal, tomato, cotton, maize, cauliflower *etc.* therefore, it is Polyphagous pest.

Nature and Symptoms of Damage

After hatching, the larvae feed on chlorophyll from both side of leaves. They are voracious feeder and attack the crops in army fashion. In case of severe infestation, larvae completely defoliate the crop and affects badly on plant stand. In 1975, this pest made hawack in Bihar, hence the name Bihar Hairy caterpillar. The pest is active during June to November and March to April.

Economic Threshold Level (ETL)

Economic threshold is when 20-25 per cent defoliation is observed.

Management

(A) Cultural Measures

1. Cleaning of bunds during mansoon season.
2. Uproot the damaged plants along with the young larvae at the gregarious phase and burry under the soil.
3. Pre-monsoon deep ploughing (two to three times) will expose the hibernating pupae to sunlight and predatory birds.
4. Timely sowing and clean cultivation.
5. Removal and destruction of alternate weed hosts which harbor the hairy caterpillars.
6. Use of well rotten manures.
7. Intercropping with pigeon pea at a row ratio of 2:1 is effective in reducing the insect attack.

(B) Mechanical Measures

1. Mass collection and destruction of leaves with eggs.
2. Mass collection and destruction of early instar larvae.
3. Collection of infested leaves which show characteristic drying symptoms will reduce the population to a great extent because of the gregarious nature of young larvae.
4. Use of light traps for attracting moths.
5. Monitor the flight intensity of the male moth by using pheromone traps 5/ha and determine the percentage of crop defoliation.

(C) Chemical Measures

1. Spray of Quinalphos 25 EC @ 2.0 ml/l or Dichlorvos 100 EC @ 1.0 ml/l or Fenvalerate 20 EC @ 1.0 ml/lt or dusting with Fenvalerate 0.4 per cent @ 15 kg/ha should be done in the evening.
2. Spot application of Chlorpyriphos 20 EC 1.0 ml/litre of water are highly effective for the control of gregarious phase larvae.
3. Digging trench around the field and dusting them with carboryl 10 per cent dust prevents the migration of caterpillars from one field to another.
4. The young caterpillars can be killed easily by dusting the infested crop with Malathion 5 per cent @ 25 kg/ha.

(D) Biological Measures

1. Conserve the natural bio control population of spiders, long horned grasshoppers, praying mantis, robber fly, ants, green lace wing, damsel flies/dragon flies, flower bugs, shield bugs, lady bird beetles, ground beetles, predatory crickets, earwigs *etc.*

2. Use of NPV (nuclear polyhedrosis virus) on cloudy days @ 500 LE/ha will be effective.
3. Spraying of *Bacillus thuringenesis* is also advocated @ 400 g/ha or 1 g/l.
4. Conserve the braconids parasites.

(F) PESTS OF MUSTARD

1. Aphid

Scientific Name, Order and Family

Lipaphis erysimi (Kalt.)

(Hemiptera; Aphididae)

Taxonomical Position/Scientific Classification

Kingdom:	Animalia
Phylum:	Arthropoda
Class:	Insecta
Order:	Hemiptera
Family:	Aphididae
Genus:	Lipaphis
Species:	***L. erysimi***

Biology and Life Cycle

The aphid population generally makes its appearance sometimes during winter and it continues to breed parthenogenetically till the end of spring when winged individuals are produced and large-scale dispersal takes place. The population, however, dwindles mostly due to climatic reasons and practically disappears for the whole of the summer and also most of the autumn.

- ☆ **Aphids:** are small, soft-bodied, pearl-shaped insects that have a pair of cornicles (wax-secreting tubes) projecting out from the fifth or sixth abdominal segment.
- ☆ The nymphal period is completed in 7-10 days.

Marks of Identification

Both nymph and adult are louse like, pale-greenish insects abundant from December to March. During summer, it is believed to migrate to the hills. The pest breeds parthenogenetically and the females give birth to 26-133 nymphs. They grow very fast and are full-fed in 7-10 days. About 45 generations are completed in a year. Cloudy and cold weather (20°C or below) is very favourable for the multiplication of this pest. The winged forms are produced in autumn and spring, and they spread from field to field and, from locality to locality.

Host Crops

Cruciferous oilseeds like toria, sarson, raya, taramira and Brassica vegetables like Cabbage, Cauliflower, Knol-khol *etc.*

Nature and Symptoms of Damage

- ✰ Both the nymphs and adults suck cell-sap from leaves, stems, inflorescence or the developing pods.
- ✰ Vitality of plants is greatly reduced. The leaves acquire a curly appearance, the flowers fail to form pods and the developing pods do not produce healthy seeds.
- ✰ The aphid also secretes honeydew which leads to growth of black fungus or mould and hampus the photosynthetic activity of the plant.
- ✰ The infected filed looks sickly and blighted in appearance
- ✰ The yield of an infested crop is reduced to one-fourth or one-fifth.

Economic Threshold Level (ETL)

50-60 aphids per 10 cm terminal portion of the central shoot or when an average of 0.5-1.0 cm terminal portion of central shoot is covered by aphids or when plants infested by aphids reach 40-50 per cent.

Management

(A) Cultural Measures

1. Sow the crop early wherever possible, preferably up to third week of October.
2. Apply recommended dose of fertilizers.
3. Use tolerant varieties like JM-1 and RK-9501.
4. The crop sown before 20th October escapes the damage.

(B) Mechanical Measures

1. Destroy the affected parts along with aphid population in the initial stage.
2. Set up yellow stick trap to monitor aphid population.

(C) Chemical Measures

1. Foliar spray of Oxydemeton methyl 25 EC or Dimethoate 30 EC @ 600 ml or
2. Spray application of metasystox 0.05 per cent or imidacloprid 0.01 per cent or acetamiprid @ 0.01 per cent.
3. Quinalphos 25 EC or Chlorpyriphos 20 EC @ 950-1500 ml in 600-1000 L of water per ha depending on the stage of the crop.
4. Granular insecticides - Phorate 10 G @ 10 kg or Carbofuran 30 G @ 33 kg per ha followed by a light irrigation.

(D) Biological Measures

1. Conserve parasitoids *Ischiodon scutellaris, Diaeretiella rapae* and *Lipolexis gracilis.*
2. Predators *viz., Syrphus serarius, Brinckochrysa scelestes, Coccinella septempunctata* and *Menochilus sexmaculatus.*
3. Entomopathogens *viz., Entomophthora coronata* and *Cephalosporium aphidicola.*
4. Conserve the following natural enemies: Ladybird beetles *viz., Coccinella septempunctata, Menochilus sexmaculata, Hippodamia variegata* and *Cheilomones vicina* that are most efficient predators of the mustard aphid. Adult beetles may feed on an average of 10-15 adults/day.
5. A number of entomogenous fungi, *Cephalosporium* spp., *Entomophthora* and *Verticillium lecanii* infect aphids.
6. Several species of syrphid fly *i.e., Sphaerophoria* spp., *Eristallis* spp., *Metasyrphis* spp., *Xanthogramma* spp. and *Syrphus* spp. are predating on aphids
7. The braconid parasitoid *Diaretiella rapae,* a very active bio-control agent cause the mummification of aphids
8. The lacewing, *Chrysoperla carnea* predates on the mustard aphid colony
9. Predatory bird *Motacilla cospica* is actively feeding over aphids in February-March.

2. Saw Fly

Scientific Name, Order and Family

Athalia proxima lugens (Klug)

(Hymenoptera; Tenthridinidae)

Taxonomical Position/Scientific Classification

Kingdom:	Animalia
Phylum:	Arthropoda
Class:	Insecta
Order:	Hymenoptera
Family:	Tenthridinidae
Genus:	Athalia
Species:	***A. p. lugens***

Biology and Life Cycle

1. **Egg:** A female lays 20 to 100 eggs with an average of 35, singly and inserted into the under surface of leaf tissue (near the leaf margin with its saw like ovipositor). Egg period is 4-8 days.

2. **Larvae**: Dark green larvae have 8 pairs of abdominal prolegs. There are five black stripes on the back, and the body has a wrinkled appearance. A full-grown larva measures 16-18 mm in length. The larvae feed exposed in groups of 3-6 on the leaves during morning and evening. They remain hidden during the day time and, when disturbed, fall to the ground and feign death. There are 7 instars with a larval period of 16-35 days.
3. **Pupa**: Pupation is in water proof oval cocoons in soil and the pupal period is 11-31 days. The over-wintering takes place in pupal stage which lasts for about 14 weeks.
4. **Adult**: The adults are small orange yellow insects with black markings on the body and have smoky wings with black veins. The mustard sawfly breeds from October to March and undergoes pupal diapause during summer. The adults emerge from these cocoons early in October. They live for 2-8 days. Life cycle is completed in 31-34 days. It completes 2-3 generations from October to March.

Marks of Identification

1. **Adult**: Adult is 8 to 10 mm long, small insect having black head and thorax, orange coloured abdomen and two pairs of translucent smokey wings with black veins. The adult is not able to fly long distance. They hop from one leaf to another leaf or from one plant to another plant. Female adult has a saw like ovipositor and hence the common name "Mustard sawfly".

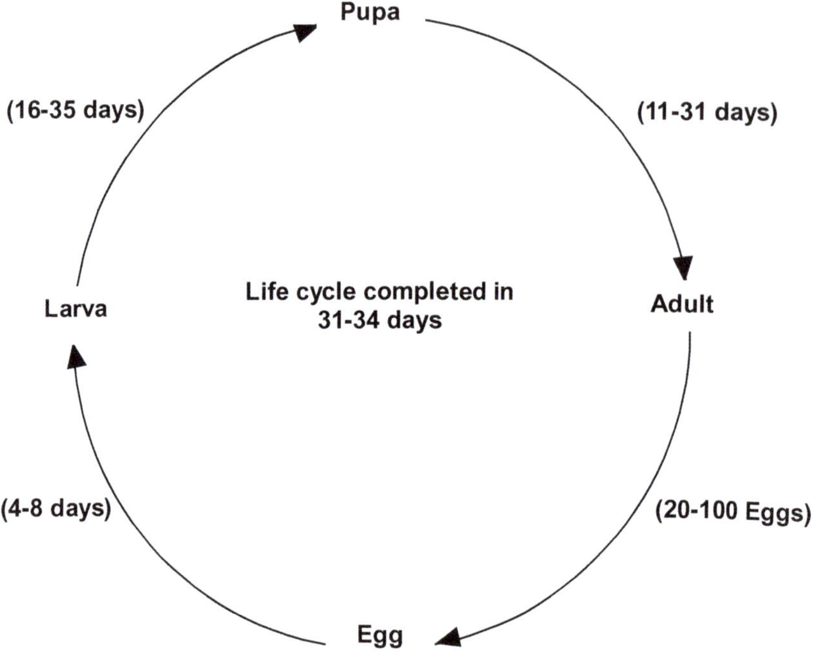

Figure 31: Life Cycle of *Athalia proxima lugens.*

2. **Larvae**: The newly hatched larva is smooth, cylindrical, greenish grey in colour and 2 to 3 mm long. Full grown larva is smooth, cylindrical, greenish black in colour and 16 to 20 mm in length. Larva possesses 8 pairs of prolegs in addition to 3 pairs of thoracic legs. Larvae have tendency to curl and drop down on the ground on being touched. They look and behave like caterpillars.

Host Crops

Mustard, Toria, Rapeseed, Cabbage, Cauliflower, Knolkhol, Turnip, Radish *etc.*

Nature and Symptoms of Damage

- ☆ The larvae/grub alone are destructive.
- ☆ The early stage larvae bite holes into leaves preferring the young growth and skeletonize the leaves completely. Sometimes, even the epidermis of the shoot is eaten up. Although the seedlings succumb; the older plants, when attacked, do not bear seed.
- ☆ The larvae on hatching nibble the margins of tender leaves and later on bite holes in the leaves. The larvae are diurnal in habit. They feed generally during early morning and evening hours and remain practically motionless at night.

Economic Threshold Level (ETL)

No action threshold has been set for sawflies in mustard.

Management

(A) Cultural Measures

1. Give first irrigation 3-4 weeks after sowing as it reduces the bug population significantly.
2. Light ploughing in between the rows. Operation helps in exposing the larvae and pupae for destruction.
3. Summer ploughing to destroy the pupa.
4. Early sowing should be done.
5. Maintain clean cultivation.
6. Apply irrigation in seedling stage is very crucial for sawfly management because most of the larvae die due to drowning effect. Severe cold reduces pest load.

(B) Mechanical Measures

1. The larvae can be hand picked from infested plants during dawn and dusk and destroyed mechanically if area under crop is limited.

(C) Chemical Measures

1. Spraying of Malathion 50 EC @ 1.0 litre or Quinalphos 25 EC @ 625 ml in 500-600 L of water per ha once in October and again in March-April.
2. Spray the crop with Malathion 50 EC @ 0.05 per cent or Dichlorvos 76 EC @ 0.05 per cent or Formothion @ 0.05 per cent.

(D) Biological Measures

1. Conserve *Perilissus cingulator* (parasitoids of the grubs), and the bacterium *Serratia arcescens* which infect the larvae of sawfly.

3. Painted Bug

Scientific Name, Order and Family

Bagrada cruciferarum (Krik.), *B. hilaris* (Burm.)

(Hemiptera; Pentatomidae)

Taxonomical Position/Scientific Classification

Kingdom:	Animalia
Phylum:	Arthropoda
Class:	Insecta
Order:	Hemiptera
Family:	Pentatomidae
Genus:	Bagrada
Species:	***B. cruciferarum***

Biology and Life Cycle

1. **Egg**: These bugs lay oval, pale-yellow eggs singly or in groups of 3-8 on leaves, stalks, pods and sometimes on the soil. Eggs may be laid during day or night. A female bug may lay 37-102 eggs in its lifespan of 3-4 weeks. Egg period is 3-5 days during summer and 20 days during December.
2. **Nymph**: The full-grown black nymphs are about 4 mm long and 2.66 mm broad. There are five nymphal instars with duration of 22-34 days.
3. **Adult**: Sub-ovate, black adult bugs are 3.71 mm long and 3.33 mm broad with a number of orange or brownish spots. It is active from March to December and during this period all the stages can be seen. It passes the winter months of January and February in the adult stage under heaps of dried oilseed plants lying in the fields. Longevity of adult is 16 to 18 days. About 6 to 8 generations are completed in a year.

Marks of Identification

1. The nymphs are beautifully patterned with a mixture of black, white and orange colour. Nymphs measure about 1.5 to 4.5 mm in length.

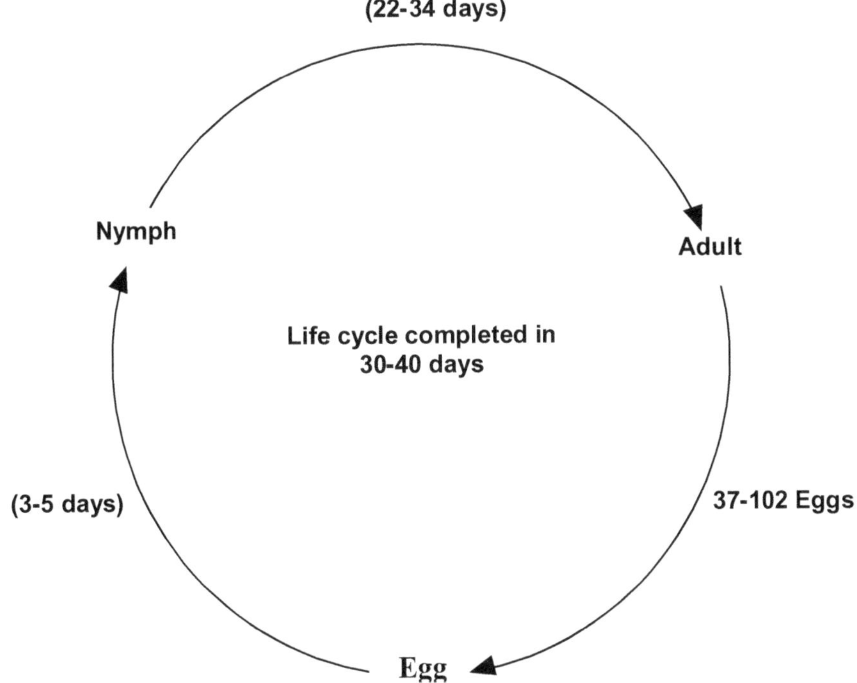

Figure 32: Life Cycle of *Bagrada cruciferarum*.

2. The body of adult bugs is also painted with combination of black and orange colour like nymphs, that's why pest has earned the common name of "Painted bug". Males measure about 6 to 7 mm in length while females measure about 7 to 8 mm in length.

Host Crops

Cabbage, Cauliflower, Knolkhol, Turnip, Radish, Rice, Sugarcane and Coffee.

Nature and Symptoms of Damage

- ☆ Both nymphs and adults suck the cell sap from the tender leaves causing yellowing of leaves which gradually dry up and ultimately fall down exposing the plant parts to secondary invasion of bacteria and fungi.
- ☆ The infested plants wither, wilt and finally dry up.
- ☆ The nymphs and adult bugs also excrete a sort of resinous material which spoils the pods.
- ☆ Due to severe infestation of pest, there is qualitative and quantitative reduction in the yield.

Economic Threshold Level (ETL)

One bug/10 ft.^{2} on plants in the early growth stage increasing to 3 bug/10 ft.^{2} on mature plants.

Management

(A) Cultural Measures

1. Give first irrigation 3-4 weeks after sowing as it reduces the bug population significantly.
2. Clean cultivation *i.e.* removal and destruction of alternate hosts.
3. Deep ploughing so that the eggs of painted bugs are destroyed.
4. Early sowing is needed to avoid pest attack.
5. Irrigate the crop during four week after sowing to reduce pest attack.
6. Quick threshing of the harvested crop should be done.
7. Burn the remains of mustard crop so that the stages of insect do not reach the next year crop.

(B) Mechanical Measures

1. Hand collection and destruction of bugs.
2. Monitor egg masses; if parasitized egg masses (black ones) are found, treatment for newly hatched nymphs might not be necessary.
3. Also, monitor by beating or by shaking the vines on the cloth and count number of nymphs and adults.

(C) Chemical Measures

1. The bugs usually congregate on the leaves and stem, which can be jerked to dislodge them and killed in kerosinised water.
2. Spray the crop with dimethoate 30 EC @625 ml in 600-700 liter water.
3. Spray Malathion 50 EC @ 1.0 litre or Quinalphos 25 EC @ 625 ml in 500-600 litres of water per ha once in October and again in March-April.
4. Spray 0.05 per cent dichlorvos/DDVP or 0.1 per cent Malathion or dust the crop with carbaryl 10 D @ 20 to 25 kg/ha.

(D) Biological Measures

1. Conserve egg parasitoid *Gryon* sp. and the adult parasitoid *Alophora* sp. (Tachinidae).

(G) PESTS OF CASTOR

1. Semilooper

Scientific Name, Order and Family

Achaea janata Linn.

(Lepidoptera; Noctuidae)

Taxonomical Position/Scientific Classification

Kingdom:	Animalia
Phylum:	Arthropoda
Class:	Insecta
Order:	Lepidoptera
Family:	Noctuidae
Genus:	Achaea
Species:	***A. janata***

Biology and Life Cycle

1. **Egg**: Female moth lays about 90 to 300 eggs (maximum 622 eggs) blue green rounded and ridged eggs singly @ 1 to 6 eggs per leaf, scattered over the lower surface of leaves. Eggs are greenish and 0.9 mm in diameter. Hatching period is 2 to 5 days.
2. **Larvae**: Larval development is completed within 11 to 15 days. Full-grown larva falls down on ground and pupates under fallen leaves on the soil.
3. **Pupa**: Pupation takes place in the soil or among fallen dried leaves. Pupa is ashy gray in colour, 19 to 25 mm in length and 6 to 8 mm in breadth. Pupal period is 10 to 15 days.
4. **Adult**: The moths on emergence feed on soft fruits of citrus, mango *etc.* Pest completes as many as 5 to 6 generations in a year.

Marks of Identification

1. **Adult** is a pale reddish or grayish brown moth with wavy lines on the forewings and black and white blotches on the hind wings and three large white spots on the outer margins. The wing expanse of adult is 60 to 70 mm. Full-grown caterpillar is 68 mm long, smooth and grayish brown in colour with pale white stripes.
2. **Caterpillar** is a semilooper, long, smooth, greyish brown in colour. The first pair of prolegs is reduced and as such a semilooper. Caterpillar posess red or whitish side stripes. Full grown larva has black head, a red spot on the black loop and red anal tubercles and measures 60-70 mm in length.

Host Crops

This is a polyphagous pest infesting Castor, Pomegranate, Rose, Citrus, *Euphorbia* spp., *Cassia fistula*, *Cassia auriculata*, *Ziziphus* spp., *Ficus benghalensis*.

Nature and Symptoms of Damage

The damage is caused by both the caterpillar and adult moth. The caterpillars feed voraciously on castor leaves. Feeding from the edges inwards, leave behind only the mid rib and the stalk. The damage is maximum in August, September and October. The adult of this species are fruit sucking moths and cause serious damage to citrus crop.

The caterpillar feed voraciously on castor leaves from margin inwards leaving behind only the midribs and the stalks. Maximum damage is caused by the second and third instar caterpillars. Due to excessive loss of foliage the seed yield is reduced very considerably. The peak period of pest activity is from August to September. Adults are also harmful but they never feed on castor plant. Adults suck the juice from ripened fruits of citrus, mango, pomegranate *etc.*

The caterpillar feeds sparingly at first and feeds voraciously during later stages leaving only mid rib and veins.

- ☆ Defoliated leaves,
- ☆ in severe cases only mid rib and veins of the leaves

Management

(A) Cultural Measures

1. Intercropping with cluster bean, cowpea, black gram, groundnut (1:2 proportions) reduce semilooper infestation and buildup natural enemies like *Microlitis* sp., *coccinellids* and spiders.

(B) Mechanical Measures

1. Mechanical collection and destruction of caterpillars helps in minimizing the incidence of pest.
2. Install the light trap to trap the adults.

(C) Chemical Measures

1. Dusting the infested crop with Parathion 2 per cent dust @ 20-25 kg/ha.
2. Spray carbaryl 50 per cent WP 2 kg in 1000-1200 L water/ha.
3. Foliar spray with thiodicarb 1g/l or spinosad @ 0.33ml/l.

(D) Biological Measures

1. Use of *Telenomus* and *Tetrastichus* sp. parasitize the eggs.
2. Braconid parasite: *Micropletis ophiusae* acts as larval parasite whose cocoons may be seen attached to the ventral aspect of the posterior end of the host caterpillar.
3. Erection of bird perches @ 10/ha.
4. Application of B.t 1 g/lt.

2. Capsule Borer

Also known as: Capsule and shoot borer

Scientific Name, Order and Family

Conogethis punctiferalis (G.)

(Lepidoptera; Pyralidae)

Taxonomical Position/Scientific Classification

Kingdom:	Animalia
Phylum:	Arthropoda
Class:	Insecta
Order:	Lepidoptera
Family:	Pyraustidae
Genus:	Conogethes
Species:	***C. punctiferalis***

Biology and Life Cycle

1. **Egg**: A female lays eggs in small group of 2 to 3 on inflorescence and seed capsules. Hatching period is of about 6-7 days.
2. **Larvae**: Larva measures 24 mm when fully grown. Larva is pale green with pinkish tinge and fine hairs with dark head and prothoracic shield. Larva lives under a cover of silk, frass and excreta. Larval period is 12-16 days.

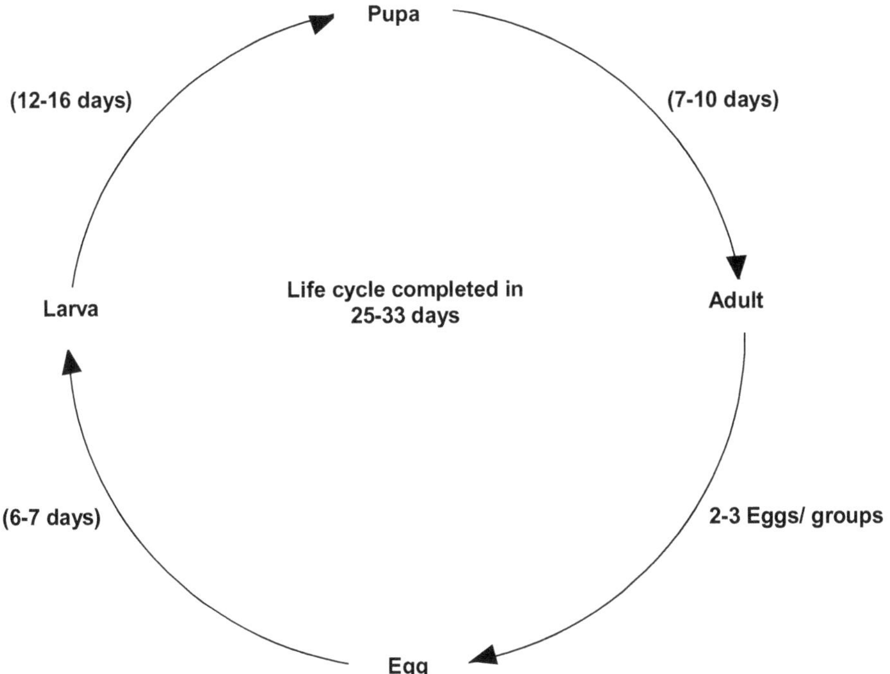

Figure 33: Life Cycle of *Conogethis punctiferalis*.

3. **Pupa**: Pupation ocurs inside the damaged stem or capsule, in a thin silken cocoon. Pupal period is 7-10 days.
4. **Adult**: Adult is medium sized with small black dots on pale yellow wings. Total life history takes 25-33 days with three generations per year.

Marks of Identification

Moth is medium sized having bright orange yellow coloured wings with numerous black dots or spots. The body length is about 10 mm while the wingspan is 22 mm. Female moth lays pinkish oval, flat eggs singly or in groups on tender parts of plant and developing capsules. Incubation period is 6 to 7 days.

Full grown caterpillars measure 15 to 25 mm in length, reddish brown in colour with black blotches all over the body and a pale stripe on the lateral side.

Host Crops

Castor, mango, sorghum ears, guava, peaches, cocoa, pear, avacado, cardamom, ginger, turmeric, mulberry, pomegranate, sunflower, cotton tamarind and hollyhock.

Nature and Symptoms of Damage

The damage is caused by the **caterpillar**, which bores into the main stem of young plant and ultimately into the capsules. The borer is distributed throughout India where castor is grown. Larvae bore into the shoots as well as capsules and destroy them. Occasionally the larva is found at the junction of the petiole with the lamina and rarely in thick mid rib.

The symptoms are:

- ✰ Frassy matter at the bored shoots
- ✰ Webbed seed capsules covered with dark excreta.

The pest attacks the crop during flowering and at the time of maturity of the capsules. When attack is on young plants, caterpillars bore into terminal shoots causing shoots to wither and dry away. The frass and pelleted excreta may be noticed around entry holes made by larvae. The inflorescence is also badly damaged. Attacked inflorescence withers and dries up. Pest also attacks the older plants by feeding on greenish coat between the warts of capsule and enters the capsule at the pediceller end. At the site of entry, a silken gallery is made in which excreta and frass get accumulated. The gallery forms a protective covering for the larva. The larvae then bore into capsule and feed on inner contents. As many as, 20 caterpillars could bore into capsule and tender shoot in a branch. Pest damages about 23 per cent of capsules and is active from August to March-April.

Management

(A) Cultural Measures

1. Pest incidence can be minimized by promptly collecting and destroying infested shoots and capsules.

2. Clearing previously damaged orchards and fields of debris scraping off the fruit tree bark in which *C. puntiferalis* larvae are methods which and burning crop stems after harvest are methods which may reduce the overwintering populations.

(B) Mechanical Measures

1. Use of light trap and sugar vinager traps (containing sugar, vinegar water and insecticide) in orchards and fields may assist in the control of *C. puntiferalis* adults.
2. Covering the fruits with paper bags to avoid from laying eggs and boring.
3. Use of pheromones to attract the adults.

(C) Chemical Measures

1. Spraying the infested crop with Carbaryl 50 WP 2 kg 1000-1200 litres of water per hectare proved effective in controlling the pest.
2. As soon as incidence is noticed apply Fenthion 50 EC @ 0.02 per cent or Thiometon 25 EC @ 0.02 per cent or Phosalone 35 EC @ 0.02 per cent or Quinalphos 25 EC @ 0.02 per cent.
3. Sprayings should be commenced from the time of formation of inflorescence and again after 20 days. Insecticides like Dimethoate 2 ml/l or Methyl demeton 2 ml/l are recommended.

(D) Biological Measures

1. Larval parasite *Cotesia* sp. and *Theronia* sp. are released in the field.
2. Parasitoids: *Bracon brevicornis, Brachymeria euploeae.*

3. Tussck Hairy Caterpillar

Scientific Name, Order and Family

Notolophus posticus

(Lepidoptera; Lymantriidae)

Taxonomical Position/Scientific Classification

Kingdom:	Animalia
Phylum:	Arthropoda
Class:	Insecta
Order:	Lepidoptera
Family:	Lymantriidae
Genus:	Notolophus
Species:	***N. posticus***

Biology and Life Cycle

1. **Egg**: Females lay 109-656 cream coloured subspherical eggs in mass on the leaves of castor. Egg period is of 7 days.
2. **Larvae**: Larval period is of about 16 to 19 days. Larva has brown head with a pair of long pencils of hair pointing forward from prothorax, tuft of yellowish hairs laterally on first two abdominal segment and dorsally on first four abdominal segments and long brown hairs dorsally from 8th abdominal segment. A fully grown larva is 27 mm in length.
3. **Pupa**: It pupates in transparent silken cocoon inside leaf roll. Pupation period of male is 7 days and female is of 4 days.
4. **Adult**: Male is winged and female being apterous, sluggish cling to the cocoon after emergence. Males are attracted to the females at dusk.

Marks of Identification

☆ The larva is brown with yellowish and white tufts of hairs from 8th segment onwords. The females have wingless and usually cling to pupal cases and lay white eggs in groups.

Host Crops

Castor (Monophagous).

Nature and Symptoms of Damage

The newly emerged larvae scraped leaves and feeding on the paranchymatus tissue of the leaves. Later on, the caterpillar feed voraciously on the castor leaves leaving behind only the midrib and stalk. The larvae are more active during morning and evening hours.

Management

(A) Cultural Measures

1. Sowing the trap crops like cucumber or cowpea before sowing the main crop all along the field boeders attract the migrating cateroillar and facilitate mechanical killing of the larvae by jerking into kerosinized water.
2. Digging trench around and killing the trapped larvae.

(B) Mechanical Measures

1. Set light trap @ 1/acre and kill traped moths.
2. Place twigs of jatropa, *Ipomea* or *Calotropis* on the field border to attract the migrating caterpillar and kill the feeding larvae mechanically.

(C) Chemical Measures

1. Spraying of Quinolphos 7 ml/liter or Dichlorvos 76 EC 700 ml/ha or Chorpyriphos 6 ml/ha is effective to control hairy caterpillars.

(D) Biological Measures

1. *Bacillus thuriengiensis* var kurstaki @ 400 gram in 200-300 liter of water per acre.

Pest Management of Fibre Crops

(A) PESTS OF COTTON

1. Pink Boll Worm

Scientific Name, Order and Family

Pectinophora gossypiella S.

(Lepidoptera; Gelechidae)

Taxonomical Position/Scientific Classification

Kingdom:	Animalia
Phylum:	Arthropoda
Class:	Insecta
Order:	Lepidoptera
Family:	Gelechiidae
Genus:	Pectinophora
Species:	***P. gossypiella***

Biology and Life Cycle

1. **Egg**: Female moth lays upto 456 eggs (on an average 125) singly on tender parts of the plants *i.e.* floral buds, bracts and young bolls. The eggs are elongate, flattened and white when freshly laid. Eggs turn brownish at hatching. Eggs hatch within 4 to 25 days.

2. **Larvae**: The larva is pink in colour with dark brown head. The full grown larva measures about 15 mm in length. Larval period varies from 8 to 41 days. There are two types of generations, the short cycle generation and the long cycle generation. In the first type the full grown larvae pupate soon after becoming full-fed and emerge as adults. In the second type, the full-fed larvae enter into the resting stage remaining quiescent for 8 to 10 months before they pupate and emerge out as moths.
3. **Pupa**: Full-grown larva pupates inside the soil or in boll, often in seed hollowed out by larva. Pupation occurs in flimsy cocoon. Pupal period lasts for about 6 to 20 days depending on the season.
4. **Adult**: Adult is small, 5 to 6 mm long dark brown moth with wing expanse of 12.5 mm. Adult has numerous small black spots on the wings. About 4 to 9 generations are completed in a year.

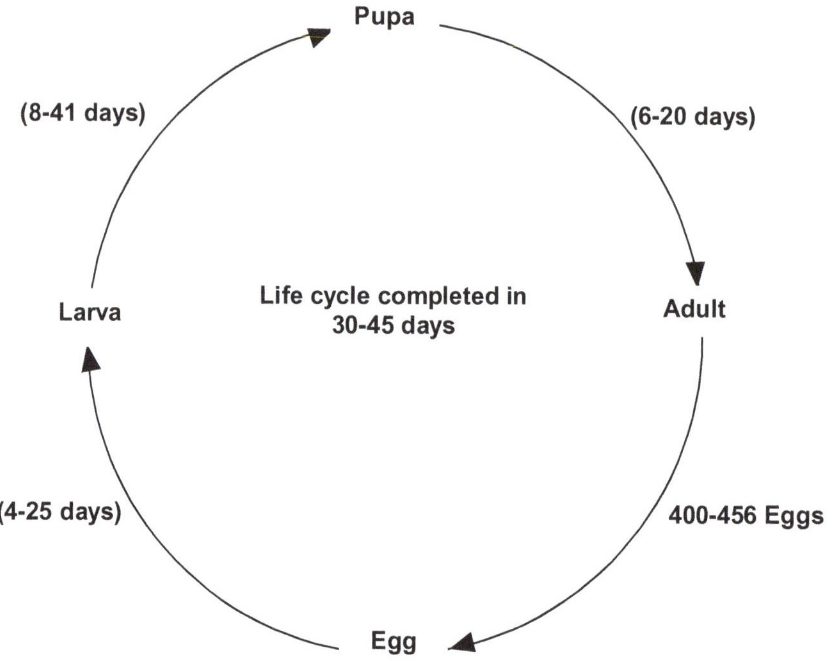

Figure 34: Life Cycle of *Pectinophora gossypiella*.

Marks of Identification

2. **Larvae**: The freshly hatched larvae are white and turn pink as they grow older. Larva is full grown in 25 – 30 days. The full grown, uniformly pinkish larvae measures about 8-16 mm with dark brown head and prothoracic shield. The larva undergoes generally only 3 moults.
2. **Adult**: Moth is small about 5-6 mm in length and has wing span of 12.5 mm. Body is dark brown in colour with numerous small black spots on

the wings. The first segment of the antenna bears 5-6 long stiff hairs and the palpi are pointed and curved upwards. The moths are active during night. Female lays flattish scale like whitish eggs singly on various parts of young shoots. However, half developed bolls are preferred when available.

Host Crops

Cotton, bhendi, ambadi, hollyhock and other malvaceous crops.

Nature and Symptoms of Damage

- The caterpillar by its feeding activities damages buds, opened flowers, young and developing bolls, grown up bolls and seeds of cotton.
- The larva may move from bud to bud destroying several buds. **"Rosette" flowers and bursting of immature bolls** is the characteristic sign of pest infestation.
- The larvae do most spectacular damage to practically mature cotton bolls which they enter mostly at such a tiny stage of just hatched larvae that their entry holes get healed and in which they remain, devouring both seed and fiber forming tissues. The infestation at times is
- so severe that up to 10 larvae are found in each boll and 75-100 per cent bolls are found infested.
- The attacked buds and young bolls drop. The larger bolls may remain on the plants but their lint is spoiled, the seeds are destroyed. Poor germination capacity of seeds in the attacked boll.
- Ginning percentage, oil contents and spinning qualities of the lint are impaired.
- The extent of damage caused by this pest may be anything between 75-100 per cent depending upon the presence and abundance of the insect. The shedding of bolls to the extent of 60 per cent and infestation of bolls to the extent of 50 per cent has been reported.

Economic Threshold Level (ETL)

5 Larvae/100 bolls or 10 per cent infested flowers or bolls with live larvae or 8 moths per trap (glossyplure pheromone traps) per day for 3 consecutive days. The number of traps should be 5 per ha.

2. Spotted Bollworm

Also known as: Spiny Bollworm

Scientific Name, Order and Family

Earias vittella Fb.,

Earias insulana Boisd.

(Lepidoptera; Noctuidae)

Taxonomical Position/Scientific Classification

Kingdom:	Animalia
Phylum:	Arthropoda
Class:	Insecta
Order:	Lepidoptera
Family:	Noctuidae
Genus:	Earias
Species:	***E. insulana***

Biology and Life Cycle

1. **Egg**: Female moth lays about 200 to 400 eggs singly or in group of 2 or 3 on tender parts of the plants. Eggs are bluish green in colour and spherical in shape. The eggs hatch in 2 to 9 days.
2. **Larvae**: The caterpillars of both the species have a number of black and brown spots on the body and hence the name spotted boll worm. Full grown larva measures 14 mm in length. Larval period is completed in 9 to 25 days.
3. **Pupa**: Pupa is reddish brown in colour. Pupation takes place on plant, fallen buds and bolls or in soil in a dirty white boat shaped cocoon of tough silk. Pupal stage lasts from 6 to 25 days depending upon the climate, the average period being one to three weeks.

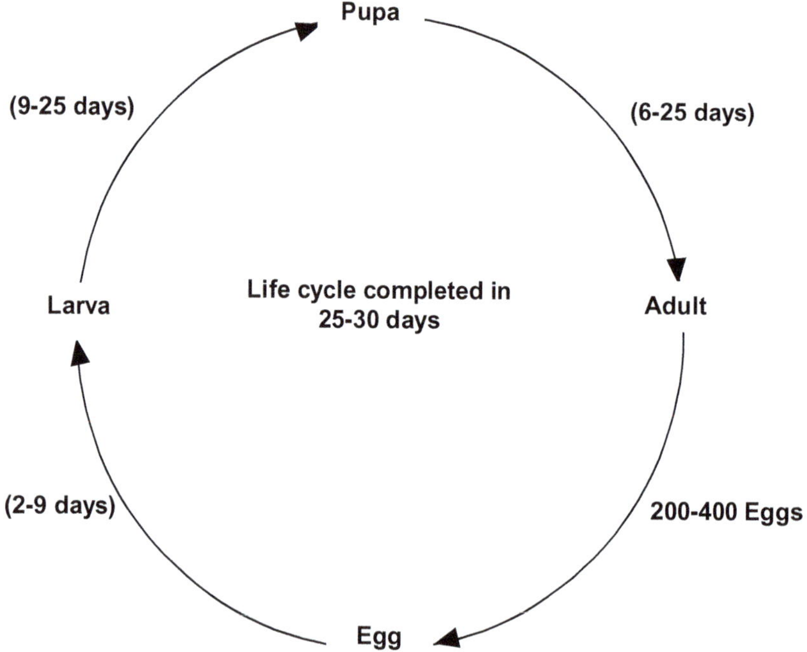

Figure 35: Life Cycle of *Earias vittella.*

4. **Adult**: Adult life varies from 8 to 25 days. 7 to 12 generations are completed by the pest in a year. *Earias* species do not hibernate in winter.

Marks of Identification

1. **Larvae**: The caterpillars of both the species are brownish white and have a dark head and prothoracic shield. They have number of black and brown spots on the body and hence the name **"Spotted bollworm".**
2. **Adult**: Adult of *Earias vittella* has pale white forewings with a broad greenish band in the middle while adult of *E. insulana* has completely greenish coloured forewings. Adults of both species measure about 10 mm in length and 25 mm across the wings.

Host Crops

Cotton, okra, ambadi, hollyhock and several other malvaceous plants (polyphagous).

Nature and Symptoms of Damage

- ☆ The spotted bollworm appears about 6 weeks after sowing and initially damages the tender shoot by boring into it resulting in "**drying of central shoots**" which withers and drops down.
- ☆ The larvae later bore into the flower buds, squares and bolls. The larva inserts its head inside the boll and feeds by filling the boll with excreta. A larva may move out and feed on another bud or boll. The feeding causes severe shedding of early formed flower buds and bolls.
- ☆ In the beginning of season, the caterpillars bore into growing tender shoots of the plant. The attacked shoots wither and droop down and ultimately die. When the flower bud appears, caterpillars bore into them. Later they also bore into bolls which show holes plugged with excreta.
- ☆ The damaged flower buds, squares and young bolls fall down. However, bigger bolls do not shed. They open prematurely and lint obtained from them is of inferior quality.
- ☆ In case of heavy infestation, internal contents of the bolls are totally destroyed by the boring caterpillar.
- ☆ About 60 per cent shedding of bolls and 50 per cent withering of shoots have been noticed. Loss in kapas has been estimated to be around 20 per cent.
- ☆ The damage results in:
    - ❒ Presence of wilting, withering and drooping or drying of tender shoots in early stage of crop growth.
    - ❒ Presence of bored flower buds (squares), bored bolls with larval frass at the entrance holes
    - ❒ Premature dropping of affected bolls.

- ❐ Premature opening of damaged bolls, which remain on plants.
- ❐ Presence of badly damaged tissues including lint and seed in damaged bolls.

Economic Threshold Level (ETL)

10 per cent damaged shoot (or) 5 per cent damaged bolls or 1 larva per plant from 20 randomly counted plants.

3. American Bollworm

Also known as: Old world bollworm/Heliothis Borer

Scientific Name, Order and Family

Helicoverpa armigera (Hb.)

(Lepidoptera; Noctuidae)

Taxonomical Position/Scientific Classification

Kingdom:	Animalia
Phylum:	Arthropoda
Class:	Insecta
Order:	Lepidoptera
Family:	Noctuidae
Genus:	Helicoverpa
Species:	***H. armigera***

Biology and Life Cycle

1. **Egg**: Female moth deposits 400 to 500 spherical, yellowish eggs singly on tender parts of plants, flower buds and green bolls. The egg period lasts for 2-4 days.
2. **Larvae**: Caterpillars are of varying colour, initially brown and later turn greenish with darker broken lines along the side of the body. Body covered with radiating hairs. When full grown, they measure 3.7 to 5 cm in length. The larval period lasts for 18-25 days.
3. **Pupa**: Full grown larva falls down on the ground and pupates in earthen cocoon in soil. Pupal period varies from 16-21 days. Pest hibernates as pupae in soil.
4. **Adult**: Moth is stout, medium sized with brownish/greyish forewings with a dark cross band near outer margin and dark spots near costal margins, with a wing expanse of 3.7 cm. Pest is active from November to March. There may be several generations in a year.

Marks of Identification

Adult is stout, medium sized yellowish brown moth with a wing expanse of

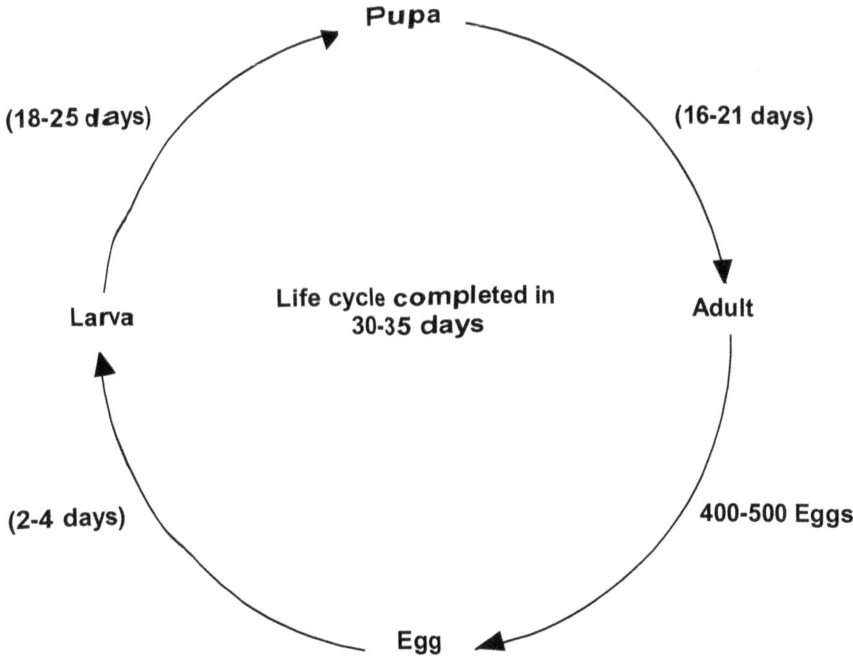

Figure 36: Life Cycle of *Helicoverpa armigera.*

37 mm. The forewings are light brown with dark black dots. Hindwings are smoky white with a blackish spot at the border. The full-grown caterpillar is 35 to 45 mm long, greenish brown in colour with yellowish or dark gray lateral lines on the body. Pupa is dark brown and 11 to 14 mm long. Eggs are yellowish white, dome-shaped and 0.4 to 0.5 mm in diameter. Larva is easily identified from its habit of feeding by inserting head and anterior half portion of its body inside the boll leaving rest of its body outside.

Host Crops

Cotton, bhendi, ambadi, hollyhock, citrus, groundnut, linseed, safflower, millets, tomto, tobacco, soybean, sunflower, gram, red gram, sorghum *etc.*

Nature and Symptoms of Damage

- ☆ Caterpillar feeds on tender foliage for couple of days and later bores into the squares, flower buds, young developing bolls by thrusting its head and anterior half part of the body within, leaving rest of the body exposed.
- ☆ Pest is migratory in habit. The caterpillar damages so many flower buds and flowers during its life cycle. It also feeds on the seed and causes substantial damage.
- ☆ The damage results in:
  - ☆ Fed leaves shoot and buds.

- ✰ **"Flared or open" squares**. Bolls are bored at the base of flower buds which are hollowed out, bracts of damaged flower buds spread out and curl down wards.
- ✰ Premature boll opening and shedding

Economic Threshold Level (ETL)

10 per cent of damaged buds (or) 5 per cent of damage bolls or one egg/plant or one larva/10 plants.

Management (for all bollworms)

(A) Preventive Measures

1. Immediately after harvest, collect infested flowers, bolls and other plant debris and destroy them mechanically.
2. Avoid growing summer bhendi and other malvaceous plants during off seasons as they serve as alternate host plants.
3. Fumigate the cotton seed with carbon disulphide 4.5 ml/ft^2 space or heat the cotton seed upto 120° to 140° F.
4. Soak the cottonseed in suspension of 0.1 per cent dimethoate for 1, 4 and 6 hours, respectively for killing the hibernating larvae of pink bollworm present within the seed.

(B) Cultural Measures

1. As infestation of spotted bollworm is observed, remove and destroy infested shoots along with larvae.
2. Hand collection of larvae, shed materials like bolls, squares, bracts *etc* and their destruction.
3. Destruction of plants, crop residues and alternative weed hosts which harbour pests in off season.
4. Collection and destruction of infested shoots, squares and bolls and the fallen material.
5. Adopting crop rotation.
6. Deep ploughing in summer..
7. Intercultivation with sorghum, greengram, cluster bean, jowar *etc*.

(C) Mechanical Measures

1. Setting of light traps and pheromone traps @ 12/ha.

(D) Chemical Measures

1. As soon as the threshold level of 5 per cent boll damage appears in the field, apply the following insecticides :

(a) Synthetic pyrethroids: Cypermethrin 10 EC/25 EC @ 0.0075 per cent or Fenvalerate 20 EC @ 0.0125 per cent or Decamethrin 2.8 EC @ 0.0025 per cent or Permethrin 50 EC @ 0.02 per cent or Fluvalinate 90 EC @ 0.0092 per cent.

(b) Conventional pesticides: Carbaryl 50 WP @ 0.2 per cent or Quinalphos 25 EC @ 0.05 per cent or Phosalone 35 EC @ 0.05 per cent or dusting with Carboryl 10 D @ 20 kg/ha.

(E) Biological Measures

1. Release *Trichogramma chilonis* or *T. acheae* (egg parasitoid) @ 1 to 1.5 lakh/ha at weekly interval. Release parasitoid 3 to 6 times in a season.
2. Release *Chrysopa* larvae @ 10,000 eggs or larvae/ha, 2 times starting from 75 to 90 days at an interval of 30 to 35 days.

   Chelonus blackburni – Exotic egg-larval parasitoid.

   Bracon krikpatrickii – Exotic larval parasitoid.

   Apanteles angaleti – Indigenous larval parasitoid.
3. Application of HaNPV @ 250 LE/ha or NSKE 5 per cent (Neem Seed Kernel Extract) have also been found effective against American bollworm.

IPM for Bollworms

1. Deep ploughing exposes and eliminates hibernating insects and expose pupae to sun and predating birds.
2. Balanced organic fertilization keeps crop healthy and tolerant to pest attack.
3. Border crop with jowar, maize in 2 or 3 rows not only serves as a barrier for migration of insect pests but also pollen of maize helps in attraction of beneficial *Chrysoperla* to the field.
4. Growing trap crops @ 100 plants/acre. Castor as an ovipositional trap crop against *S. litura*: egg masses, gregarious larvae of *S. litura* on castor should be removed once in a week and destroyed.
5. Marigold as ovipositional trap crop against *Helicoverpa,* Okra (bhendi) against spotted bollworm moths.
6. Spray marigold/okra plants with endosulfan or phosalone to minimize larval population.
7. Keep 10 – 15 pheromone traps/ha to attract male moths. Egg scouting from square initiation stage is desirable.
8. ETLS – PBW: 8 moths, ABW: 10 moths, *S. litura:* 20 moths, SBW: 15 moths per trap per night.
9. Topping (removal of leaf terminals) for 80 – 100 days old crop during October – November since tender leaves and tips are preferred for egg laying.

10. Spray 5 per cent neem seed kernel extract (NSKE) (soak 10 kg neem seed powder in 200 litres of water for 24 h and filter through muslin cloth) to repel moths from egg laying and to kill eggs and early instar larvae.
11. Install 'T' shaped or long dried twigs as bird perches to attract predatory birds @ 20/ac.
12. Spray NPV @ 200 LE/ac in combination with jaggery 1 kg, sandovit 100 ml or Robin Blue 50 g thrice at 10-15 days interval on observing the eggs or first instar larvae in the evening hours. The diseased larvae die after 4 – 5 days showing tree top symptoms.
13. Spray commercially available B.t formulations (DIPEL, DELFIN, BIOBIT, HALT) @ 400 g or 400 ml/ac against *Helicoverpa*.
14. On the basis of ETLs spray the following insecticides, chlorpyriphos 2 ml/l or quinalphos 2 ml/l carbaryl 3 g/l or triazophos 2 ml/l or thiodicarb 1.5 g/l or profenophos 2 ml/l.
15. Mixing mustard oil with chlorpyriphos 1:4 improves toxicity.
16. In the entire schedule of spray, do not spray synthetic pyrethroids for more than two rounds.

4. Cotton Jassids

Also known as: Cotton leaf hopper

Scientific Name, Order and Family

Amrasca biguttula (Ishida)

(Hemiptera; Cicadellidae)

Taxonomical Position/Scientific Classification

Kingdom:	Animalia
Phylum:	Arthropoda
Class:	Insecta
Order:	Hemiptera
Family:	Cicadellidae
Genus:	Amrasca
Species:	***A. biguttula***

Biology and Life Cycle

1. **Egg**: On an average leaf hopper lays about 15-29 eggs. The female inserts its eggs inside leaf veins in the parenchymatous layer between the vascular bundles and the epidermis. Incubation period is 4 to 11 days.
2. **Nymph**: Nymphs are also pale greenish in colour like the adults but are wingless and are found in large numbers on lower surface of leaves.

Nymphal period varies from 7 to 21 days. Nymphs moult 5 times. One life cycle is completed in 15 to 46 days.

3. **Adult**: Adult lives for 5 to 7 weeks. About 7 to 11 generations are completed in a year. It is a small insect, varying from less than 1 mm to about 3 mm. Its adult stage is subjected to seasonal changes in colour. It is reddish in winter and greenish yellow in summer. The adult is a wedge shaped insect about 3.5 mm in length. There is a black spot on each forewing and two small black spots on the vertex. The whole life cycle is completed in about 14-45 days. There are 8-10 overlapping generations.

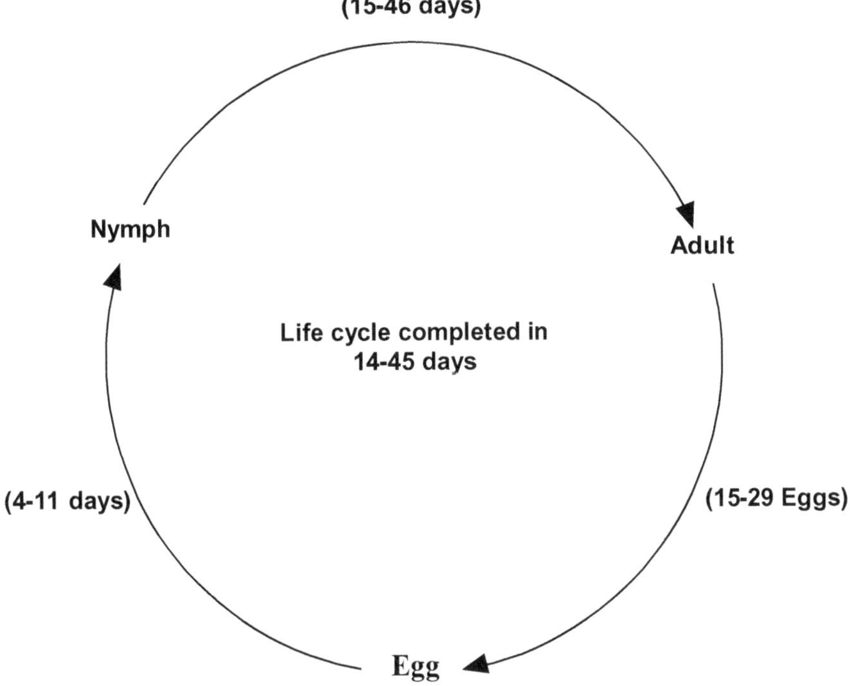

Figure 37: Life Cycle of *Amrasca biguttula*.

Marks of Identification

Adult is wedge shaped, 2 to 3.5 mm long greenish yellow coloured winged insect. There is a black spot on each of the forewings and two spots on the vertex. The nymphs are small, wingless and pale green in colour. They are found in large numbers usually on the lower surface of the leaves. The insect is characterized by its habit of walking diagonally in relation to the body.

Host Crops

Cotton, potato, brinjal, okra, sunflower, castor, cucurbits, ambadi and some wild plants like hollyhock.

Nature and Symptoms of Damage

- ☆ Injury to the leaves is caused by the nymphs and adults feeding on the sap and injecting toxic saliva into the leaf tissues which causes toxaemia. The edges of the infested leaf turn pale green, then yellow and finally brick red or brown in colour.
- ☆ The whole of the leaf gradually dries up after curling and drops down. The plant becomes stunted, the number of flowers, bolls and weight of seed produced are reduced.
- ☆ Due to continuous draining of sap from foliage characteristic symptom of **"Hopper burn"** appears. Quality of lint is also affected.
- ☆ The pest is active throughout the year. The pest causes serious damage during kharif season *i.e.* July to September. The irrigated crop suffers most. Excessive rainfall is favourable for multiplication of pest. Pest incidence is more pronounced under irrigated conditions.
- ☆ The major symptoms are:
  - ❒ Hopper burn *i.e.,* the leaf margins turning yellowish initially and subsequently turning reddish and curling up.
  - ❒ Stunted growth of the plant.
  - ❒ Brown necrotic patches on the leaves.

Economic Threshold Level (ETL)

2-3 nymphs or adults/leaf

Management

(A) Cultural Measures

1. Sow the crop early *i.e.* in the third week of May along with normal management of fertilizers and other agronomic practices.
2. Use cottonseed treated with carbofuran 50 SP before sowing.

(B) Mechanical Measures

1. Monitor the presence of leafhoppers by brushing the foliage; watching for adults and nymphs to jump and fly from plant to plant. Leaf hoppers are most sampled with a sweep net. Empty captured jassids into a container with 70 per cent alcohol (or methylated spirits), and express counts as leafhoppers per sweep (one sweep per row).

(C) Chemical Measures

1. Prepare solution of gum arabic by adding 25 gms of gum arabic in 85 ml of water. Pour 60 ml of this solution in 1 kg of linted or de-linted seed. Shake it thoroughly. Add 100 gms of carbofuran 50 SP to this seed of cotton. Again shake thoroughly. Dry this seed in shade for 12 hours before sowing.

2. If seed treatment is not given, apply phorate 10 G @ 1 to 1.5 g/spot at the time of sowing.
3. If seed treatment and phorate granules are not applied, give two sprays of following insecticide after appearance of the pest: Dimethoate 30 EC 0.03 per cent (500 ml/500 litres of water) or Methyl demeton 25 EC 0.02 per cent (320 ml/500 litres of water).
4. Apply following granular insecticides in soil at the time of planting: Carbofuron 3G @ 33 kg/ha or Phorate 10 G @ 10 kg/ha.
5. In initial stages of the crop before flowering, spraying of acephate 75 SP @ 1g/litre of water or Imidacloprid 17.8 SL @ 0.3 ml/litre of water.
6. Apply following insecticides in the field four times at 10 to 15 days interval starting from 25 days after planting, if soil application is not done with granules: Dimethoate 30 EC 0.05 per cent or Methyl demeton 25 EC 0.025 per cent or Thiometon 25 EC 0.025 per cent.

(D) Biological Measures

1. Natural enemies are not considered to have a significant effect on population of jassids although the parasitoid *Anagrus* sp. has been recorded, but it does not play any significant role in reducing the population.

5. Red Cotton Bug

Scientific Name, Order and Family

Dysdercus cingulatus Fb.

(Hemiptera; Pyrrhocoridae)

Taxonomical Position/Scientific Classification

Kingdom:	Animalia
Phylum:	Arthropoda
Class:	Insecta
Order:	Hemiptera
Family:	Pyrrhocoridae
Genus:	Dysdercus
Species:	***D. cingulatus***

Biology and Life Cycle

1. **Egg**: The female bug lays 100 to 300 eggs in cluster or in loose irregular masses in the moist soil or in the crevices of the grand. Eggs are round and light yellowish. Incubation period lasts 7 to 8 days.
2. **Nymph**: Nymphs moult 6 times in a total period of 30 to 35 days before becoming adults.

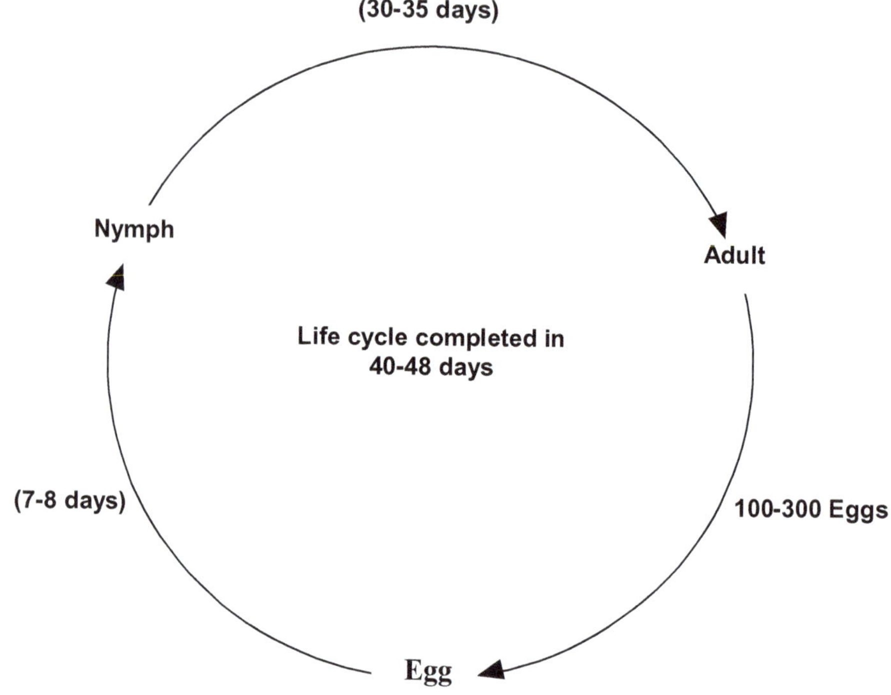

Figure 38: Life Cycle of *Dysdercus cingulatus*.

3. **Adult**: There are 5 generations in a year. Over-wintering takes place in adult stage. The life cycle is completed in 6 to 8 weeks.

Marks of Identification

Adult is medium sized, 12 to 13 mm long elongated slender bug. It is reddish black in colour with whitish stripes ventrally on the abdomen. The membranous portion of their forewings, antennae and scutellum are black. The freshly laid eggs are oval and pale yellow in colour but become orange at hatching. Nymphs are red, smaller than adults and wingless.

Host Crops

Cotton, okra, ambadi, hollyhock and some other malvaceous plants, maize, sorghum, wheat, pearl millet.

Nature and Symptoms of Damage

- Both nymphs and adults feed gregariously on foliage and green bolls by sucking the sap.
- The nymphs and adults suck sap from tender leaves, petioles and shoots in early stages and then infest flower buds and immature bolls and bolls that have just opened. Resulting plants loose their vigour and bolls open prematurely with stained lint.

- ✰ As a result, vigour of plant is reduced. The infested bolls open badly, quality of lint is reduced, oil content of seed is decreased and seeds become unfit for sowing or for oil extraction.
- ✰ Due to deposition of excreta and crushing of nymphs during ginning, the lint gets stained. Bugs may also introduce the bacterium, *Nematospora gossypii,* into the bolls which also causes staining of the lint. On account of this, red cotton bug is commonly called as **"cotton stainer".**

Management

(A) Cultural Measures

1. Deep ploughing at summer to expose soil inhabiting or resting stages of insects.
2. The infested leaves or bolls can be shaken in water and drowned.
3. Removal of weeds acting as alternative hosts of the bugs.

(B) Mechanical Measures

1. Collection and destruction of nymphs and adults of red cotton bug.

(C) Chemical Measures – Same as Cotton aphid

1. As soon as the pest appears, dusting of methyl parathion 2 D @ 20 to 25 kg/ha or or Carbaryl 10 D @ 25 kg/ha.
2. Spray phosphamidon 40 SL 600 ml/ha.

(D) Biological Measures

1. Predators: *Harpacter costalis* (Reduviid bug), *Ectomocoris tibialis, Rhinocoris fascipes, R. longifrons.* The predacious bug *Harpactor costalis* feed on nymphs and adults of red cotton bugs.

6. White Fly

Scientific Name, Order and Family

Bemisia tabaci

(Hemiptera; Aleyrodidae)

Taxonomical Position/Scientific Classification

Kingdom:	Animalia
Phylum:	Arthropoda
Class:	Insecta
Order:	Hemiptera
Family:	Aleyrodidae
Genus:	Bemisia
Species:	***B. tabaci***

Biology and Life Cycle

1. **Egg**: A single female lays 70-125 semi-elliptical, brown and stalked eggs singly on the undersurface of leaves, mostly on the top and middle leaves of plant. The insect can often breed parthenogenetically. The eggs are light yellow in the beginning but turn brown later on. Egg period or incubation period is 3-33 days (ranges from 3-5 days in summer and 5-33 days in winter).
2. **Nymph**: Nymphs are oval shaped, scale like, greenish yellow with marginal bristle like fringes. The nymphs remain stationary once they settle down. Nymphs moult thrice. Nymphal period lasts for 9-18 days (ranges from 9-14 days in summer and 17-73 days in winter).
3. **Adult**: Adult is minute insect measuring about 0.5-1.0 mm in length having white or grayish wings, a yellowish body and red medially constricted eyes. There are about a dozen overlapping generations in a year.

Marks of Identification

Adults are about 1 mm long with two pairs of white wings and light yellow bodies. Their bodies are covered with waxy powdery materials. Eggs are tiny and oval-shaped. Eggs at first are white then turn to brown. Nymphs first instar is 0.3 mm length and the second instar has legs pulled up under its body and other

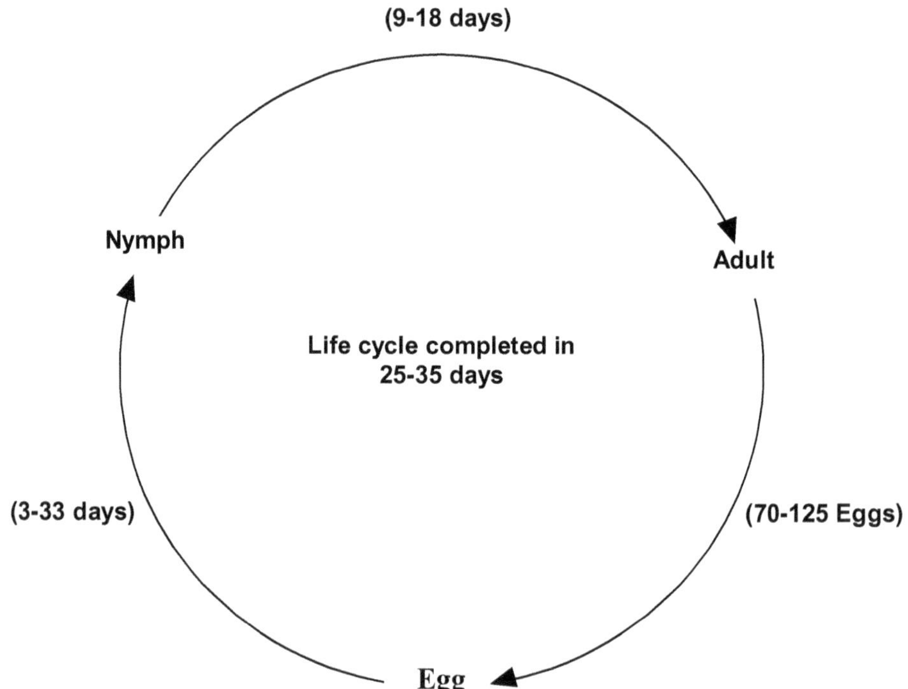

Figure 39: Life Cycle of *Bemisia tabaci*.

immature stages are sessile. The last nymphal instar develops red eye spots and often is called the pupal stage.

- **Adults** - Small, yellow bodied insects with two pairs of white wings which are densely covered with a waxy powder.
- **Nymphs and pupae** - Black and round or oval. Pupae have marginal bristles.

Host Crops

It is distributed in all cotton growing regions of the world. It also infests okra, coriander, papaya, radish, water melon, cucumber, chillies, brinjal, tomato, potato, tobacco etc

Nature and Symptoms of Damage

Both nymphs and adults suck sap from lower side of leaves resulting in:

- Chlorotic spots which later coalesce forming irregular yellowing of leaves which extends from veins to outer edges.
- Leave mottled and yellowish in colour.
- The vegetative growth retarded and boll formation seriously hampered.
- Shedding of the bolls accentuated and proper opening of the bolls interfered.
- Low quality lint and low oil content.
- Sooty mould development due to honey dew excretion on infested parts. It is vector of leaf curl virus and yellow mosaic virus.
- The maximum infestation on cotton occurs during July.

Economic Threshold Level (ETL)

- Nymphs or Adults per Leaf

Management

(A) Cultural Measures

1. Avoid the alternate, cultivated host crops of the white fly in the vicinity of cotton crop.
2. Growing cotton only once a year either in winter or summer season in any cotton tract.
3. Adopting crop rotation with non-preferred hosts such as sorghum, ragi, maize *etc.*, for the white fly to check the build up of the pest.
4. Removal and destruction of alternate weed hosts like *Abutilon indicum, Chrozophore rottlari, Solanum nigrum* and *Hibiscus ficulensus* from in the fields and neighbouring areas and maintaining field sanitation.

5. Timely sowing with recommended spacing, preferably wider spacing and judicious application of recommended dose of fertilizers, particularly nitrogenous and irrigation management is essential to arrest the excessive vegetative growth and pest build up.
6. Late sowing may be avoided and the crop growth should not be extended(beyond its normal duration.
7. Field sanitation may be given proper attention.
8. Cultivation of most preferred alternate host crops like brinjal, bhendi, tomato, tobacco and sunflower may be avoided.

(B) Mechanical Measures

1. Monitoring the activities of the adult white flies by setting up yellow pan traps and sticky traps at 1 feet height above the plant canopy and also in situ counts.
2. Collection and removal of whitefly infested leaves from the plants and those which were shed due to the attack of the pest and destroying them.

(C) Chemical Measures

Chemical control same as under cotton leaf hopper.

1. Spray of any one of the following insecticides: Quinolphos 25 EC @ 0.06 per cent or Chloropyriphos 20 EC @ 0.06 per cent or Methyl demeton 25 EC 500 ml/ha or Dimethoate 30 EC 500 ml/ha or Malathion 50 EC @ 0.06 per cent or Fenvelrate 20 EC @ 0.016 per cent.
2. Spraying of Imidacloprid 17.8 SL @ 0.3 ml/litre of water.
3. Acetamiprid(20 per cent SP 100 g/ha or Azadirachtin 0.15 per cent 500-1000 ml/ha or Buprofezin 25 per cent SC 1000 ml/ha or Clothianidin 50 per cent WDG 40-50 g/ha or Diafenthiuron 50 per cent WP 600 g/ha or Fipronil 5 per cent SC 250-340 ml/ha or Profenofos 50 per cent EC 1000 ml/ha or Thiacloprid 21.7 per cent SC 500-600 ml/ha or Thiamethoxam 25 per cent WG 100 g/ha or Triazophos 40 per cent EC 1500-2000 ml/ha.

(D) Biological Measures

1. *Verticillium lecanii* 1.15 per cent WP @ 2500 g/ha.
2. *Eretrmocerus mundus* (Mercet) and *Enacarsia* sp. are special parasites of *B. tabaci*. Also, the predators *Amblyseius* sp. (predatary mite), the green lacewing bug (*Chrysoperala* sp.) the coccinellids *Brumus* sp., *Scymnus* sp. and *Menochilus* sp. play important role in reducing whitefly population.
3. A *chalcid parasite* attacks the older nymphs and the parasitisation is at times more than 30 per cent. Also, there are a few predators like some species of *Chrysopa* and *coccinellids*, which feed on the whitefly stages.

7. Cotton Aphids

Scientific Name, Order and Family

Aphis gossypii

(Hemiptera; Aphididae)

Taxonomical Position/Scientific Classification

Kingdom:	Animalia
Phylum:	Arthropoda
Class:	Insecta
Order:	Hemiptera
Family:	Aphididae
Genus:	Aphis
Species:	***A. gossypii***

Biology and Life Cycle

1. **Adult**: Adult is small, soft, yellowish, green or greenish brown in colour. It is found in colonies of hundreds on the tender shoot and the undersurface of tender leaves. They are characterized by the presence of two tubes like structures called cornicles, on the abdomen. They are wingless normally but winged forms are often found mostly in the beginning and towards the end of season. Wings are thin, transparent and are held like a roof over the body. They reproduce both sexually and parthenogenetically. Parthenogenetic females give rise to female's ovo - viviparously.
2. **Nymph**: Nymphs are light yellowish green or brownish or greenish black in colour. They colonise growing points, lower surface of leaves and tender shoots. There are four instars. Nymphal period is about 7-9 days.

Marks of Identification

Adult – The adult is an oblong insect, about 2 mm long, soft and greenish brown found in colonies the tender/growing part of the plants. The aphids are characterized by the presence of two tube like structures called cornicles on the abdomen. They are usually wingless but winged forms are often noticed mostly in the beginning and towards the end of the season. Wings are thin, transparent and held like a roof over the body.

Host Crops

It is a polyphagous species. Besides cotton, it also infests brinjal, chilies, okra, guava, amaranthus *etc.*

Nature and Symptoms of Damage

- Both nymphs and adults suck the sap by remaining on the lower surface of the leaves.

- In case of severe infestation, the plant become week, leaves curl up and wither. There is stunted growth, gradually drying and death of the plant. The young plants are more susceptable than the older once.
- The honeydew secreted by the aphid encourages shooty mould growth on the leaf.
- The dry condition favours rapid increase in pest population.
- The basic symptoms of damage are as follows:
  - Curled, faded and dried leaves.
  - Development of black sooty mould due to honeydew excretion on infested parts.

Economic Threshold Level (ETL)

Spray on visible damage on top terminals.

Management

(A) Cultural Measures

1. Planting resistant or tolerant varieties which are densely hairy and with stiff leaves are reliable control technique.
2. Seed treatment with Trichoderma viride or Imidacloprid 70 WS or Chlothianidin at 9 ml/kg can be effective in reducing population buildup.

(B) Mechanical Measures

1. Monitor adults and nymphs on the underside of main stem leaves, 3 - 4 nodes below the plant terminal. Presence of ants may indicate presence of aphids. Early detection of aphids is important as they can multiply rapidly. If a high proportion of plants have only the winged form of aphids, recheck within a few days to see if they have settled and young are being produced. Yellow traps are useful for monitoring winged aphids. The presence and abundance of natural enemies should also be recorded.

(C) Chemical Measures

1. Spraying of Oxydemeton methyl 25 EC or Dimethoate 30 EC @ 625 ml/ha.
2. Spray any one of the following insecticides (with 500 lt of water/ha): Methyl demeton 25 EC 500ml/ha or Acetamiprid 20 per cent SP @ 50 g/ha or Azadirachtin 0.03 per cent EC @ 500 ml/ha or Buprofezin 25 per cent SC @ 1000 ml/ha or Carbosulfan 25 per cent DS @ 60g/kg of seed or Chlorpyrifos 20 per cent EC @ 1250 ml/ha or Diafenthiuron 50 per cent WP @ 600 ml/ha or Dimethoate 30 per cent EC @ 660 ml/ha or Fipronil 5 per cent SC @ 1500-2000 ml/ha or Imidacloprid 70 per cent WG @ 30-35 kg/ha or Imidacloprid 17.8 per cent SL @ 100 -125 ml/ha or Malathion 50 per cent EC @ 1000 ml/ha or Methyl demeton 25 per cent EC @ 1200

ml/ha or Profenofos 50 per cent EC @ 1000 ml/ha or Thiacloprid 21.7 per cent SC @ 100-125 ml/ha or Thiamethoxam 25 per cent WG @ 100 g/ha.

(D) Biological Measures

1. *Coccinella* beetle feding on aphids (both nymph and adult).
2. Aphids are biologically controlled by several species of *Coccinellids, Chrysopids, Syrphids* and the parasitoid *Aphelinus gossypii.*

(B) PESTS OF SUNHEMP

1. Hairy Caterpillar

Scientific Name, Order and Family

Utethesia pulchella (Linn.)

(Lepidoptera; Arctiidae)

Taxonomical Position/Scientific Classification

Kingdom:	Animalia
Phylum:	Arthropoda
Class:	Insecta
Order:	Lepidoptera
Family:	Arctiidae
Genus:	Utetheisa
Species:	***U. pulchella***

Biology and Life Cycle

1. **Egg**: The female moths lay about 300-400 round, smooth whitish yellow eggs singly on tender leaves and shoots. Hatching/Incubation period is 3 to 4 days.
2. **Larvae**: Larval period is 10-20 days.
3. **Pupa**: Pupal period is 5 to 6 days. Pupation takes place either in leaf fold or in the soil. Pest hibernates as pupa in soil.
4. **Adult**: The caterpillar has yellow lines dorsally and dorsolaterally and black stripes and orange patches laterally and a brown head with long brown hairs on its body. The total life cycle takes 19-30 days.

Marks of Identification

1. Moth of *U. pulchella* has pale white fore wing with red and black dots on its upper wings. *A.cribraria* has orange yellow fore wing with black spots ringed with yellow. The hind wings are orange with black marginal blotches/spots on them.

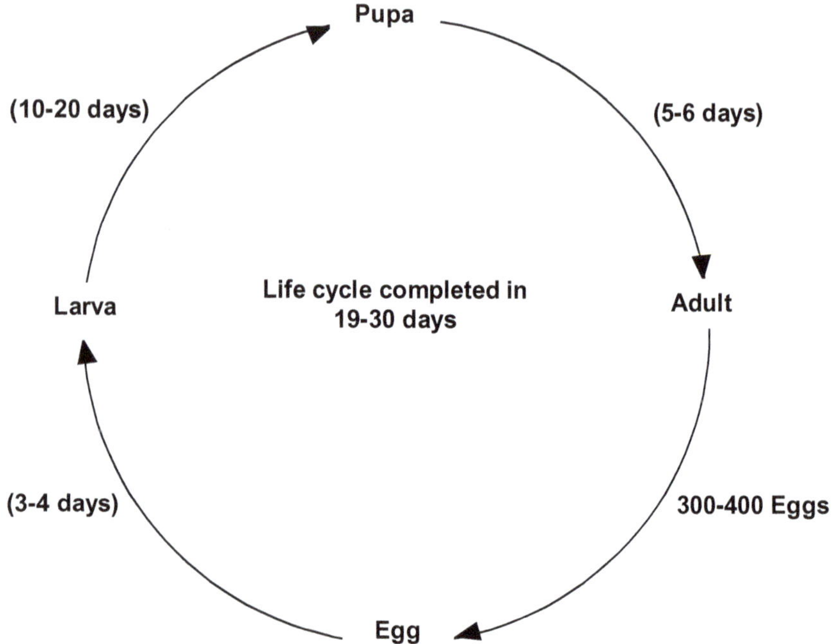

Figure 40: Life Cycle of *Utethesia pulchella.*

2. Full grown larva is hairy, 38 mm long having red and white markings on its body. Head of larva is brownish.

Host Crops

Sunnhemp and some leguminous crops.

Nature and Symptoms of Damage

- ☆ The just hatched larva either defoliates (vegetative stage) or bores into pods (in heading stage). The larvae feed on sunhemp and defoliate the plant.
- ☆ Caterpillars feed voraciously on tender foliage of host. When crop matures and seed capsule appears, larvae are seen boring and feeding on them by thrusting their head within leaving rest of body exposed. As a result, crop gets defoliated and there is substantial decline in yield of fibre and seed production.
- ☆ The damage will be more (serious in later stages), when they bore into pods and feed on the seeds. The symptoms are defoliation of plant, bored seed capsules.

Management

(A) Cultural Measures

1. Clean cultivation assists control.
2. Deep ploughing at summer to expose the pupa.

(B) Mechanical Measures

1. Hand collection and destruction of caterpillars.
2. Collection of moth by net or through light traps and killing them.

(C) Chemical Measures

1. Foliar spray with Carboryl 50 WP @ 2 kg/ha is effective measure.

(D) Biological Measures

1. Larval parasitoid *Bracon brevicornis* suppress the population naturally.
2. Entomophagous pathogen: *Aspergillus fumigates.*

(C) PESTS OF MESTA

1. Spiral Borer

Scientific Name, Order and Family

Agrilus acutus

(Coleoptera; Buprestidae)

Taxonomical Position/Scientific Classification

Kingdom:	Animalia
Phylum:	Arthropoda
Class:	Insecta
Order:	Coleoptera
Family:	Buprestidae
Genus:	Agrilus
Species:	***A. acutus***

Biology and Life Cycle

1. **Egg**: The eggs are ovate and scale like, and are laid singly on the stem, usually near a leaf scar. Egg period 10 -12 days.
2. **Larvae**: The larva bores through the lower surface of the egg directly into the stem and feeds under the bark forming a sprial tunnel. There are normally three moults in the active feeding stages, but exceptionally four, and a prepupal moult. The larval stage lasts 26 days.
3. **Pupa**: The fully grown larva is 21 mm. long and bores into the wood to pupate, making a pupal chamber 11 mm. long into which it fits itself by adopting an asymmetrical U-shaped posture. After a prepupal stage of 2.5 days it pupates. The pupal period occupies about 11 days and a further seven days are spent by the adult in the pupal chamber.

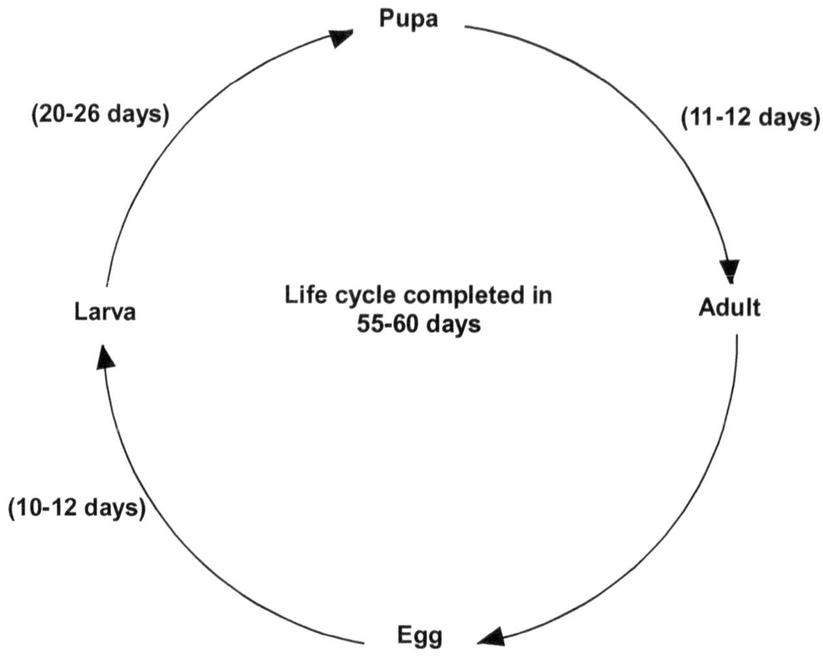

Figure 41: Life Cycle of *Agrilus acutus.*

4. **Adult**: The total life-cycle completes in about 60 days.

Marks of Identification

- An adult of A. acutus is always metallic green, blue in colour. The head is short more than twice as wide as long. The eyes are located laterally. The elytra have lateral parts of the abdomen uncovered.

Host Crops

Mesta only.

Nature and Symptoms of Damage

- The pest lays its eggs on the stem most frequently at the nodal region below the leaf base and the larva on hatching burrows its way beneath the cambium layer and starts feeding upon the woody tissues, making a spiral around the stem beneath the bark.
- During feeding, the larva travels spirally throughout the entire length of the stem, however, no influence of geotropism has been observed, "The larva at times may damage the inner fibre layers and may interrupt the flow of nutrient sap of the plant.
- The course of the spiral inside the stem as made by the larva can be recognised because of callus formation which results in some slight

swelling above its path. A portion of the infested region swells up considerably to form an elongated gall.

- ✰ The lignification of the cell wall of the sclerenchymatous tissue is affected by which stem at the level of gall becomes very weak and breaks by a strong gust of wind. The portion of the stem above the region of the gall dies and dries up after the break. The galls are usually 9-15 cm long.
- ✰ The fibre obtained from the infected plants become useless.

Management

(A) Chemical Measures

1. Best result was obtained when 20 ml per plant of 2.5 per cent Metasystox was applied by cotton swab method on the stem at a height of 37.5 cm from base of the affected plants containing borer inside.
2. Application of Triazophos 20 EC @ 2 ml/litre or Fipronil 5 EC @ 1 ml/litre or Acephate 75 SP @ 0.75 g/litre are effective to control borer.

(B) Biological Measures

1. Predatory birds: woodpickers (Picidae)

6

Pest Management of Cash Crops

(A) PESTS OF SUGARCANE

1. Top Shoot Borer

Also known as: Sugarcane Top borer

Scientific Name, Order and Family

Scirpophaga/Tryporza novella Fab.

(Lepidoptera; Pyralidae)

Taxonomical Position/Scientific Classification

Kingdom:	Animalia
Phylum:	Arthropoda
Class:	Insecta
Order:	Lepidoptera
Family:	Pyralidae
Genus:	Tryporza
Species:	***T. inovella***

Biology and Life Cycle

1. **Eggs**: About 32 to 216 dull white elongate overlapping eggs are laid in clusters, each cluster having 9-79 eggs laid mostly on underside of the leaves of young cane and near the midribs in older canes. The eggs are

covered by buff coloured hairs from female anal tuft. Incubation period is 6 to 11 days.

2. **Larvae**: The larval period lasts for 25 -42 days.
3. **Pupa**: Pupation takes place inside the larval tunnel. Pupal period lasts for 12 to 21 days.
4. **Adult**: Moth is medium sized, creamy white in colour with yellow head, slightly bigger than early shoot borer moth. Female has tuft of crimson coloured hairs at the tip of the abdomen. In case of certain males, each of the forewings has a black spot. One generation occupies 5 to 9 weeks. As many as, 5-7 generations are completed in a year.

Marks of Identification

- ☆ Adult is whitish moth with a wing expanse of 25 mm. The females are usually bigger than males. The female possesses orange coloured tuft of hairs at the tip of abdomen.
- ☆ Eggs are elongate, oval and covered with buff coloured hairs.
- ☆ Caterpillars are yellowish white measure about 25 to 35 mm in length when full-grown.
- ☆ Pupae are shiny brown in colour.

Host Crops

Sugarcane, jowar, maize, rice, wheat and some grasses.

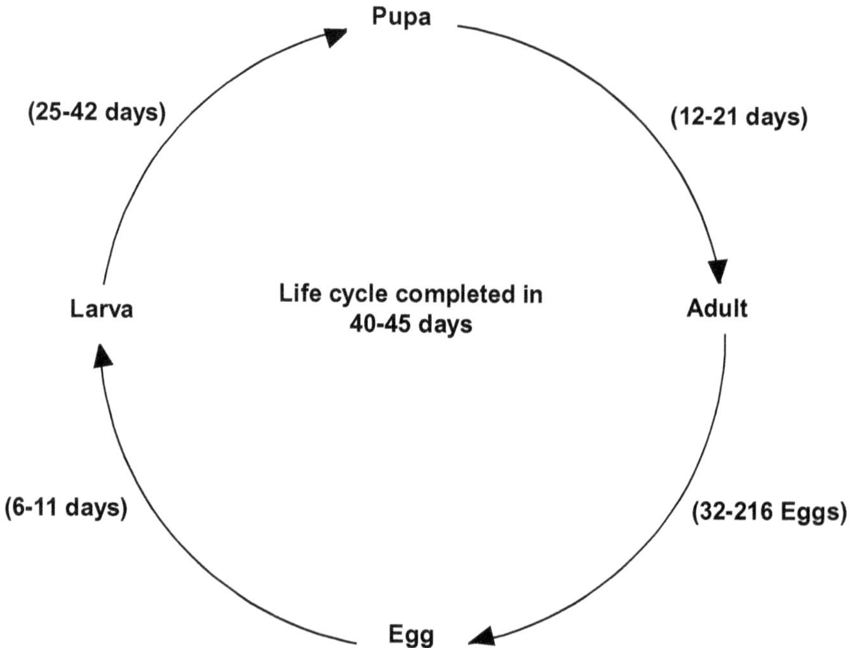

Figure 42: Life Cycle of *Tryporza novella*.

Nature and Symptoms of Damage

- ☆ The pest is injurious to sugarcane in all stages of its growth.
- ☆ Damage is caused by **caterpillars** which are generally found in the top portion of a cane.
- ☆ The newly hatched caterpillars enter first into the midrib of the leaf and bore downwards into the shoot from the top. From the midrib, it tunnels towards the central core of leaves and enters the shoot to feed. As a result of biting across the spindle, a row of shoot holes are formed.
- ☆ As a result of such feeding, the central shoot dries up in a characteristic way which results into secondary tillering in young plants and bunchy tops in the older ones.
- ☆ Young plants attack by this pest show characteristic reddish streak on the midrib in the leaves which altimatly cause dead heart.
- ☆ The punctures on the leaves, death of central shoot (form dead heart) and the bunchy top are the characteristic effects of this pest.
- ☆ Activity of the insect starts with the onset of monsoon rains in the north. Low humidity is unfavourable for the pest.

Economic Threshold Level (ETL)

15-22 per cent Attacked Plants or Incidence.

Management

(A) Cultural Measures

1. Collection and destruction of infested plant parts *i.e.* dead hearts.
2. Early earthing up of canes, usually one month after planting helps to minimize the damage of root borer and early shoot borer.
3. After harvest, all trashes and plant debris should be destroyed by burning to prevent carryover of internode borer infestation to the succeeding crop.
4. After harvest, deep ploughing is done to destroy stubbles and hibernated larvae of root borers, early shoot borer and internode borer.
5. Avoid late planting of cane. Correct time of planting adsali is in July-August, pre-seasonal in October and eksali in December-January.
6. Timely irrigation and mulching with straw or trash at planting or germination to reduce damage from early shoot borer.
7. Grow less infested varieties of canes such as Co-419 (less infested by top shoot borer) and Other resistant varieties are CoS 767, CoJ 67 and Co 1158.
8. At harvest, cut the canes below soil surface to minimize incidence of root borer.

(B) Mechanical Measures

1. Plants infested were rouged from April to September during first week of each month. Infested plants along with borers larvae and egg clusters

of all borers along with leaves were collected from April to September during first week of each month and fed to the livestock.

2. Collection and destruction of adult moths.
3. Collection and destruction of egg masses.
4. Collection and destruction of dead hearts.
5. Use of pheromone traps @ 4-5/acre for monitoring coinciding with brood emergence.
6. Installation of light trap with exit option for natural enemies @ 1/acre.

(C) Chemical Measures

1. Apply Sevidol® (Carbaryl + Lindane 4:4) @ 25 kg/ha in furrows at the time of planting for early shoot borer and internode borer.
2. Soil application of Phorate 10G @ 30 kg/ha or Carbofuran 3 G @ 33 kg/ha at the base of the shoot during last week of june or first week of July for top shoot borer is recommended.

(D) Biological Measures

1. Release egg parasite, *Trichogramma chilonis* @ 50,000 adults/ha/week in 6 to 8 instalments for internode borer. Trichocards are available commercially (20,000 eggs/card).
2. Prepupal parasite, *Isotima javensis* for top shoot borer.
3. The egg parasite, *Trichogramma australicum* for early shoot borer and internode borer.
4. Granulosis virus (GV) for internode borer.

2. Stem Borer

Also known as: Early shoot/Internode borer

Scientific Name, Order and Family

Chilo sacchariphagus indicus (Kapur)

(Lepidoptera; Pyralidae)

Taxonomical Position/Scientific Classification

Kingdom:	Animalia
Phylum:	Arthropoda
Class:	Insecta
Order:	Lepidoptera
Family:	Pyralidae
Genus:	Chilo
Species:	*C. sacchariphagus*

Biology and Life Cycle

1. **Eggs**: About 400 scale like white eggs are laid in masses, each batch containing 9-11 eggs in two rows on the leaf sheath or near the midrib of leaves or on stems. Incubation period is 5 to 8 days.
2. **Larvae**: Caterpillar has a white body with dark spots and a brown head. The larva becomes full grown in 37 to 53 days and pupates in the leaf sheath. Larval period is 28 to 38 days.
3. **Pupa**: Pupation takes place inside the cane within the tunnel. Caterpillar before pupating makes a large exit hole in the stem and blocks the opening with silken discs. Pupal period is 6 to 9 days.
4. **Adult**: Moth is small, straw coloured. Forewings have a marginal dark line and the hind wings are whitish. One life cycle is completed in 50 to 70 days. There is 6 generation in a year.

Marks of Identification

The adult is small straw coloured moth measuring 10 to 12 mm in length and 18 to 20 mm in wing expanse. The forewings have a marginal dark line while the hindwings are whitish. The caterpillar is pinkish in colour and 20 to 23 mm in length. The pupa is dark brown in colour and measures about 14 mm in length.

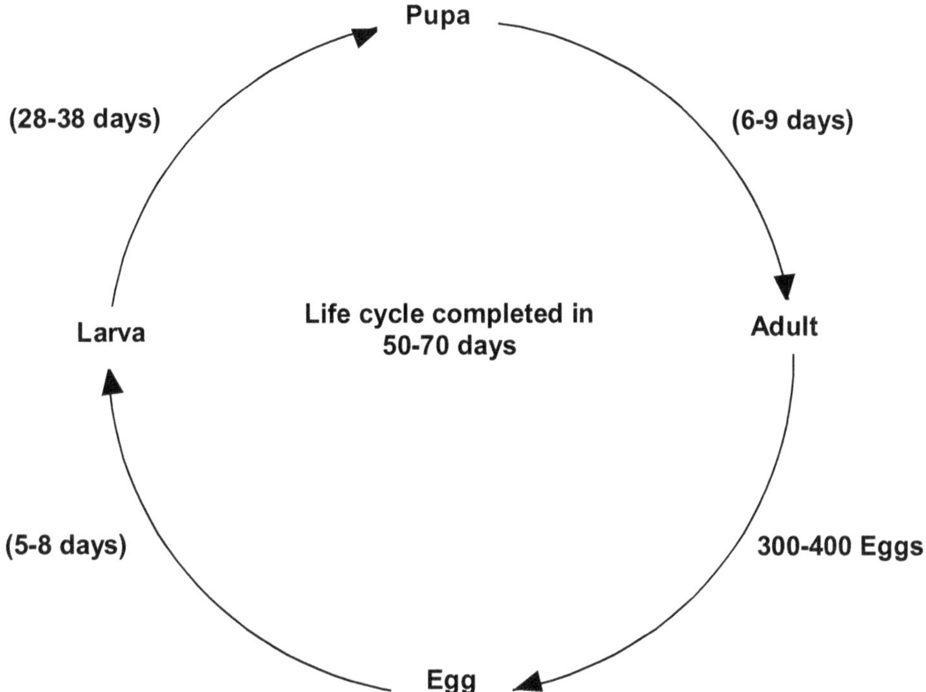

Figure 43: Life Cycle of *Tryporza novella.*

Host Crops

Sugarcane, bajra, wheat, maize, rice, kans, sorghum and some other grasses.

Nature and Symptoms of Damage

1. This pest attacks the crop after the third month of planting and continues to do so till harvest.
2. The caterpillar bores into the canes near the nodes. The entry holes are plugged with excreta. The larva migrates within the stem damaging many nodes.
3. Dead heart in 1-3 month old crop, which can be easily pulled out, rotten portion of the straw coloured dead – heart emites an offensive odour.
4. Larva has habit of boring one internode after another.
5. The loss caused by the pest to sugarcane is more to its tonnage than to its quality. According to one estimate, this pest causes on an average 16 per cent reduction in cane yield and a loss of 2.16 units in sugar recovery.

Economic Threshold Level (ETL)

16.15 to 28.39 bored canes1row of 6 m. length

Management

(A) Cultural Measures

1. Collection of egg masses and their destruction.
2. The infested tillers (canes) may be collected and destroyed.
3. High seed rate and earthing up the crop usually a month after planting helps in minimizing the damage.
4. Use resistant varities like CO 312, CO 421, CO 661, CO 917 and CO 853
5. Early planting during December – January escapes the shoot borer incidence.
6. Daincha intercropped sugarcane record the lowest early shoot borer incidence.
7. Trash mulching along the ridges to a thickness of 10-15 cm 3 days after planting.
8. Ensure adequate moisture to bring down the soil temperature and increase humidity (unfavourable condition for the multiplication of early shoots borer).

(B) Mechanical Measures

1. Setting pheromone traps at spindle level on 5th month of the crop at the rate of 6 traps per acre in a 15 metre grid. The pheromone septa need to be changed twice at 75 days interval.
2. Plants infested were rouged from April to September during first week of each month. Infested plants along with borers larvae and egg clusters

of all borers along with leaves were collected from April to September during first week of each month and fed to the livestock.

(C) Chemical Measures

1. Application of Chloropyriphos 20 EC @ 5 litres in 1500 litres of water over the setts in furrow and at the top.
2. Soil application of Cartap hydrachloride 4 G @ 1 kg ai/ha or Fiproni 0.3 G @ 25 kg/ha at planting reduces the shoot damage in endemic area.
3. Chlorpyriphos 1000 ml a sticker like Teepol (250 ml/500 l of water) can also be added to make the solution stick on to the surface of the crop and it is preferable to use high volume sprayer to be most effective.

(D) Biological Measures

1. Inundative release of the egg parasite *Trichogramma chilonis* and *Trichogramma australicum* at 50,000 parasites/ha/week.
2. Larval parasite, *Cotesia flavipes* and *Bracon chinensis* for stem borer.
3. Apply granulosis virus 1.1 x 10^5 IBS/ml (750 diseased larvae/ha) twice on 35 and 50 DAP.
4. Release 125 gravid females of sturmiopsis inferens a tachinid parasite per ac.

3. Leaf Hopper

Also known as: Pyrilla

Scientific Name, Order and Family

Pyrilla perpusilla Wlk.

(Hemiptera; Fulgoridae)

Taxonomical Position/Scientific Classification

Kingdom:	Animalia
Phylum:	Arthropoda
Class:	Insecta
Order:	Hemiptera
Family:	Fulgoridae
Genus:	Pyrilla
Species:	***P. perpusilla***

Biology and Life Cycle

1. **Eggs:** A single female lays about 600-800 yellowish eggs in groups of 20 to 25 on the under surface of the leaves and covers them with a white

filamentous waxy material secreted by the female. Hatching period is 7 to 14 days in summer and 30 to 40 days in winter.

2. **Nymph**: Newly hatched nymphs are milky white in color with a pair of characteristic processes or filaments covered by wax. They are very active and are found in very large numbers on sugarcane. The nymphal period lasts for 50 to 60 days. There are five nymphal instars.
3. **Adult**: Adults are straw coloured with two pairs of wings folded like a roof on the back and the head prominently drawn forward as a sort of rostrum. Life cycle is completed in about 60-70 days. Depending upon climatic conditions, 3 to 5 generations are completed in a year.

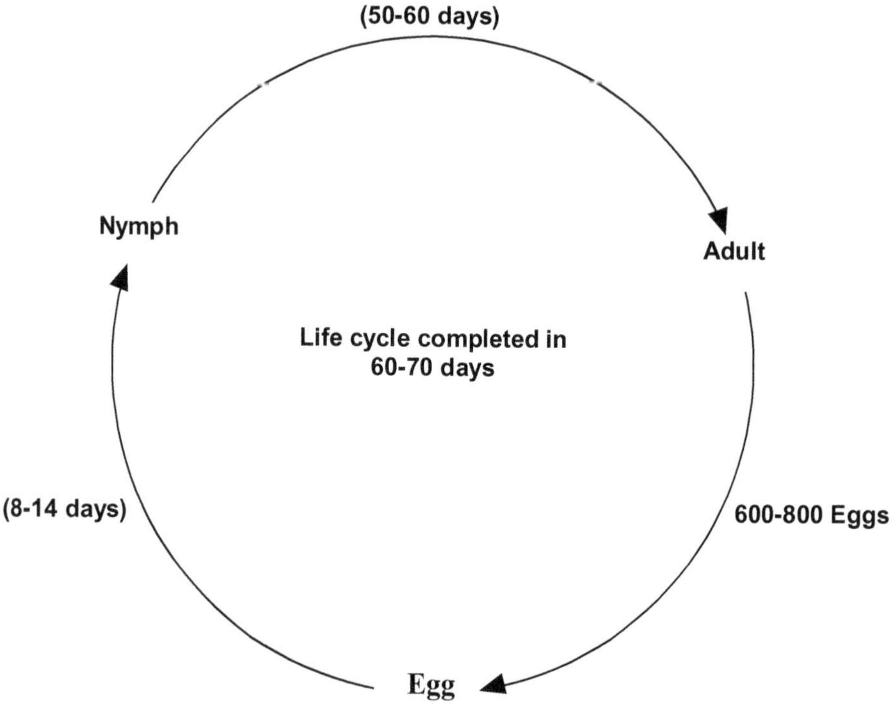

Figure 44: Life Cycle of *Pyrilla perpusilla*.

Marks of Identification

The adults is 12 mm long straw coloured adult with the head projecting forward in the form of rostrum. There are two pairs of transparent wings. Wings are held like a roof over the body when adult is at rest. The nymph is very active, pale brown or whitish with forkes tail and a pair of long wax covered anal processes. They are found in large numbers on canes. Nymphs are smaller than adults and wingless.

Host Crops

Sugarcane, wheat, jowar, rice, maize, oat, barley and bajra.

Nature and Symptoms of Damage

- ✰ Both nymphs and adults of pyrilla suck the cell sap usually from under surface of leaves. The attacked leaves become pale and wilted.
- ✰ The honey dew excreted by the bugs gets accumulated on the leaves causing profuse growth on them of the sooty mould. It interferes with proper functioning of the leaves and renders them unfit even as a cattle feed.
- ✰ In grown up crops, pyrilla causes deterioration in juice quality as a result of which sugar recovery in the mills is adversely affected, quality and yield of 'gur' also suffers. The sucrose content of the juice is reduced by 35 per cent.
- ✰ The factors which favour the multiplication of this pest are easterly winds, high atmospheric humidity during May and June, intermittent period of drought during July to September, heavy manuring and irrigation.

Economic Threshold Level (ETL)

-5 individuals/leaf or one egg mass per 1 leaf

Management

(A) Cultural Measures

1. Burn all trash immediately after harvest of sugarcane crop in an infested field.
2. Remove all the leaves bearing egg clusters in October planted crop and from ratoon sprouts during March-April.
3. Grow resistant varieties *i.e.* CO-365, CO-975 *etc.*

(B) Mechanical Measures

1. Mechanical methods of collecting and destroying egg masses in the initial stage *i.e.,* during April-May.

(C) Chemical Measures

1. As soon as, infestation is noticed or during May or June, spray the crop with following insecticides: Dimethoate 30 EC @ 0.03 per cent or Fenitrothion 50 EC @ 1 lt/ha or Quinalphos 25 EC 0.03 per cent or Malathion 50 EC @ 2.5 lt/ha. Give second application 15 days after first, if infestation continues

(D) Biological Measures

1. Mass release of nymphal parasitoid, *Epiricania melanoleuca* @ 3000 to 5000 cocoons/ha or 4 to 5 lakh eggs/ha during July onwards.
2. Egg parasite – *Tetrastichus pyrillae.*
3. Predator – *Coccinella* spp.
4. The green muscardine fungi – *Metarrhizium anisopliae*

4. White Fly

Also known as: Mealy wings

Scientific Name, Order and Family

Aleurolobus barodensis (Signoret)

(Hemiptera; Aleurodidae)

Taxonomical Position/Scientific Classification

Kingdom:	Animalia
Phylum:	Arthropoda
Class:	Insecta
Order:	Hemiptera
Family:	Aleurodidae
Genus:	Aleurolobus
Species:	***A. barodensis***

Biology and Life Cycle

1. **Egg**: A female lays about 60-65 eggs in group of 15-20 in linear rows on the under surface of the lower half leaves. Eggs are creamy yellow with

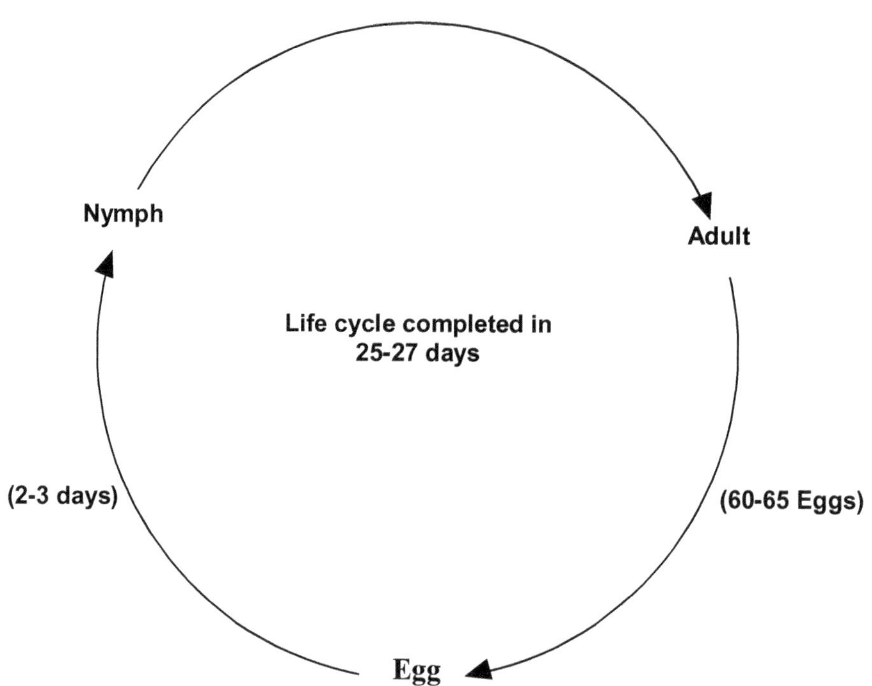

Figure 45: Life Cycle of *Aleurolobus barodensis*.

a small curved stalk when flushy laid, but soon it turns black. The egg period lasts for 2-3 days.

2. **Nymph**: The freshly hatched nymphs are creamy pale and slowly change colour to a shiny black hue. They develop fringes of wax and a waxy deposit on the body and the three nymphal instars last for respectively 4 - 7, 3 - 7 and 3 - 8 days. After each moult, the nymphs change colour from pale to blackish. Most 2nd and 3rd instar nymphs have an oval or elongate-oval body, and have shallow breathing folds in the ventral body wall, two thoracic and one caudal. These form a passage to the spiracles and may assist in the conduction of air. The nymphal period is completed in 7 days having 3-4 instars.
3. **Adult**: Pale yellow with white mealy wings. The life cycle is completed in 27 days and pest completes 9 generations a year.

Marks of Identification

The adult are active tiny insects with a pale body, black eyes and two pairs of mealy white wings. They measure about 3 mm in length. Adults are small, fly-like and often dull white in colour. The powdery wax on their membranous wings give the alternative name 'mealy wing'. Adults measure 1 - 3 mm in body length.

Host Crops

Sugarcane (Monophagous).

Nature and Symptoms of Damage

- ☆ Nymphs suck cell sap from leaves which dry up and characteristic yellow streaks appear along the length. Nymphs are stationary. Severely attacked plants become stunted.
- ☆ The sugarcane crop raised in low lying, water logged areas and in semi dry alkaline soils suffers more due to whitefly.
- ☆ Infestation is seen from August – October. Whiteflies prefer broad leaved succulent varieties.
- ☆ Due to attack by this pest, cane juice becomes more watery and the jaggery (gur) quality is adversely affected.
- ☆ A loss of 30-40 per cent in sucrose and 20-25 per cent in total solids was estimated due to its attack. It is reported that the loss to be of 15-20 per cent in yield and 1-2 units in sugar recovery due to the pest attack on crop.

Economic Threshold Level (ETL)

10 nymphs/Leaf

Management

(A) Cultural Measures

1. Clipping off and destroying the early infested leaves, prevents further spread of the pest.

2. Remove and destroy lower leaves containing pupae periodically.
3. Avoid ratooning in low lying areas, prompt clipping and destruction of affected parts.
4. Avoid water logging.

(B) Mechanical Measures

1. Detrashing the puparia bearing leaves and immediately disposing by burning or burying to prevent emergence of adult white flies.

(C) Chemical Measures

1. Spray Malathion 50 EC @ 2.5 lt/ha or Imidacloprid 17.8 SL @ 125 ml/ha or Cypermethrin 25 EC @ 300 ml/ha.
2. Spray with Methyl demeton 25 EC @ 0.08 per cent or Dimethoate 30 EC @ 0.08 per cent or Monocrotophos 36 WSC @ 0.08 per cent.
3. Foliar sprays with Quinalphos 2 ml/l against young nymphs is effective measures.

(D) Biological Measures

1. Nymphal parasitoid – *Azotus delihiensis, Encarsia isaaci, Eretmocerus* spp., *Dicyphus hesperus, Chrysocharis* sp. (nymphal and pupal).
2. Predator – *Cocinella* spp., Spiders and lacewings.

5. Mealy Bug

Also known as: Cockerell

Scientific Name, Order and Family

Saccharicoccus sacchari (Ckll)

(Homoptera; Pseudococcidae)

Taxonomical Position/Scientific Classification

Kingdom:	Animalia
Phylum:	Arthropoda
Class:	Insecta
Order:	Homoptera
Family:	Pseudococcidae
Genus:	Saccharicoccus
Species:	***S. sacchari***

Biology and Life Cycle

1. **Egg:** The females reproduce parthenogenetically and deposit eggs covered with a white waxy and mealy mass generally in the nodal region. On an

average, a female lays 1000 eggs which hatch within a few hours. The incubation period is last for 8 hrs.

2. **Nymph**: The nymphs migrate all over the plant and may often be blown off by wind from plant to plant which helps in spread of the pest. The nymphal period is completed in 2-3 weeks with 6 stages.
3. **Adult**: Small pinkish oval insect attached to the lower nodes, protected by leaf sheaths and covered by a white waxy powder. Adults and nymphs of these bugs are found in large number near the nodes. The females are sac like with clearly segmented body. Males are winged but rare. The total life cycle is completed in about a month.

Marks of Identification

Adults and nymphs are soft bodied, light coloured oval creatures found in large numbers near nodes. Their body is covered by a mealy white secretion of wax powder. It measures about 5 mm in length and 2.5 mm in breadth. Nymphs and female adults are wingless. Males are winged and sluggish in nature. Females are sac like with clear body segmentation.

- ☆ **Eggs:** Eggs are retained in the female reproductive organs untils almost fully mature. Incubation period is short. The females may bring forth hundreds of young ones parthenogenetically. Egg is yellowish, smooth, cylindrical and rounded at both ends.

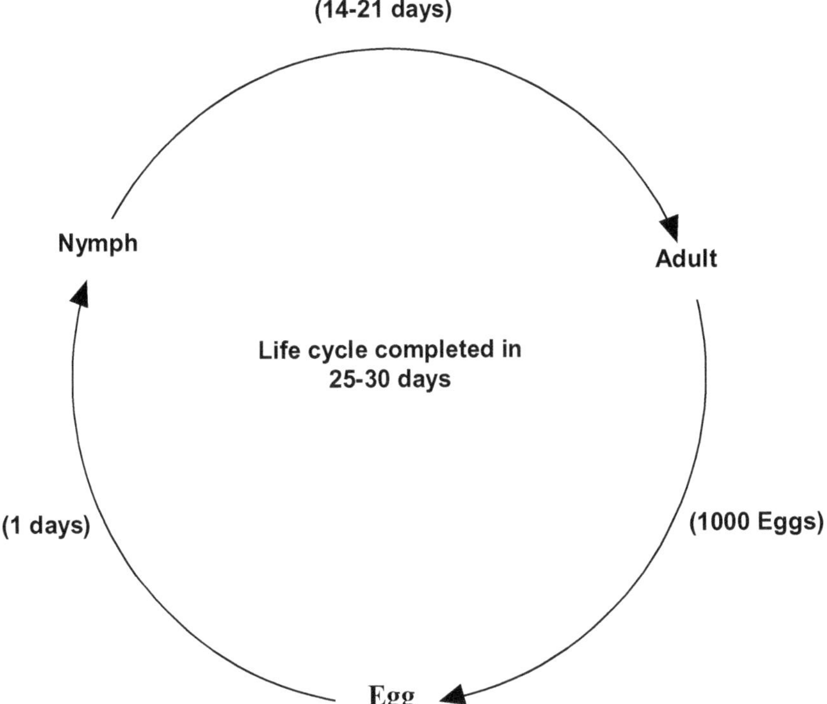

Figure 46: Life Cycle of *Saccharicoccus sacchari*.

- ✰ **Nymph:** Newly emerged nymphs are quite active with a pinkish transparent body.
- ✰ **Adult:** White with mealy coating, sessile.

Host Crops

Sugarcane, sorghum, mango, cashew, croton, *Saccharum spontaneum* (kans).

Nature and Symptoms of Damage

- ✰ Both nymphs and adults found in large number on nodal region and in the leaf sheath.
- ✰ They suck the sap, weaken the plant and reduce the sucrose content of juice.
- ✰ They also excrete honey dew, develop sooty mould, and affect photosynthesis.
- ✰ The pest population builds up under drought conditions.
- ✰ Pest acts as vector of mottling and spike disease of sugarcane.
- ✰ The infestation can be identified by the presence of mealy bugs at the nodes within the leaf sheath, reduced plant vigour and growth, movement of ants and mould on infested area.
- ✰ Yield is reduced by 55 per cent.

Management

(A) Cultural Measures

1. Selection and use of setts free from the infestation of mealy bugs.
2. Burning of the infested canes.
3. Before planting, sets are rubbed with muslin cloth soaked in 0.1 per cent Malathion.
4. Use resistant varieties like CO 439, CO 443, CO 720, CO 730 and CO 7704.
5. Drain excess water from the field.

(B) Mechanical Measures

1. Pruning of infested branches and burning them.
2. Crop residues should be removed and burnt.
3. Field borders should be free from weeds.
4. Remove alternate hosts plants.
5. Equipment should be sanitized before moving to uninfested portion in a crop.
6. Manual picking of bugs can be done in plants that are not severely infested.

7. Apply strong jet of water to remove bugs (avoid the damage to plants).
8. Destroy ant colonies during land preparation because their nests are located near the soil surface.

(C) Chemical Measures

1. Sett treatment with 50 EC Malathion or 200 ml Chloropyriphos 50 SC in 100 litre of water.
2. Apply malathion 50 EC 1000 ml/ha, when the incidence is noticed spray on the stem only.
3. Remove lower most 2-3 dry leaves and spray crop with Chlorpyriphos 20 EC @ 5 lit in 1000 lit water/ha.
4. Apply following insecticides: Diazinon 0.04 per cent or Malathion 50 EC @ 1.5 lt/ha or Quinolphos 25 ec @ 2 lt/ha.

(D) Biological Measures

1. Biological control by the parasitoid wasp Epidinocarsis (Apoanagyrus) lopezi.
2. Use of parasitoid wasp, *Pseudaphycus mundus.*
3. Among the biological control agents, introduction of the predatory Australian ladybird, *Cryptolaemus montrouzieri*; the parasitoid wasps *Anagyrus pseudococci* (Girault) and *Leptomastix dactylopii*; the predatory mite *Hypoaspis* sp., and the entomophagous fungi *Lecanicillium lecanii* (Zimmerman) and *Beauveria bassiana* have been effective in managing mealybug infestations.

(B) PESTS OF POTATO

1. Tuber Moth

Scientific Name, Order and Family

Phthorimaea operculella Zell.

(Lepidoptera; Gelechidae)

Taxonomical Position/Scientific Classification

Kingdom:	Animalia
Phylum:	Arthropoda
Class:	Insecta
Order:	Lepidoptera
Family:	Gelechiidae
Genus:	Phthorimaea
Species:	***P. operculella***

Biology and Life Cycle

1. **Eggs**: About 100-150 eggs are laid singly by each female on the underside of leaves or on exposed tubers. They are creamy white later change into yellow and finally black just before hatching. Hatching period is 3 to 6 days.
2. **Larva**: Larvae are pinkish white or greenish with dark brown head. Larval development is completed in 15-25 days.
3. **Pupa**: Pupation takes place in a silken cocoon among dried leaves, clod of earth and other trash on the ground. In storage, they are seen on the seams of storage bag and in the cracks on the wall and floors. Pupal period is 7 to 10 days.
4. **Adult**: Moth is narrow having silvery-gray body. Their forewings are gray brown with minute dark spot and a narrow fringe of hairs. The hindwings are dirty white in colour. One generation completed in 4 weeks and 8 to 9 generations completed in a year.

Marks of Identification

Adult is 6 mm long grayish brown moth with narrow wings having wing expanse of 12 mm. The posterior margin of hindwings has fringe of hairs. The full grown caterpillars are 18 mm long, pinkish white or greenish in colour with dark brown head.

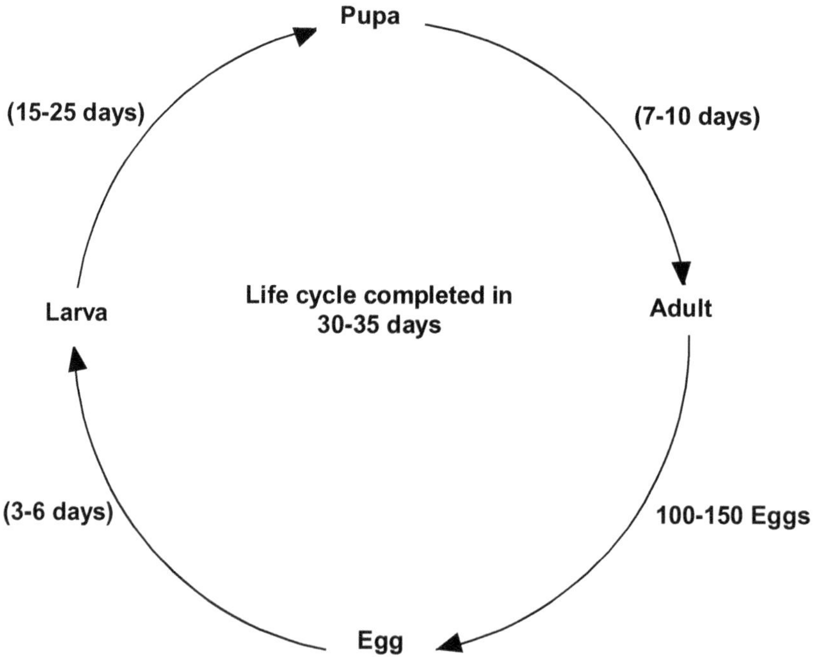

Figure 47: Life Cycle of *Phthorimaea operculella*.

Host Crops

Potato, brinjal, tomato, tobacco and several solanaceous weeds.

Nature and Symptoms of Damage

- ✰ The pests occur specially in hot and dry climate.
- ✰ Pest causes damage to potato crop both in the field as well as in the stores.
- ✰ In the field, caterpillars feed on plant by mining into the leaves, by boring into petioles, terminal shoots and potato tubers underneath the soil. The infested leaves show blisters on them, while the shoots wither and wilt.
- ✰ In storage, caterpillars are found boring into the stored potato through minute hole made near eyebud and feeding on the pulp. The infested potatoes ultimately rot and are made unfit for seed purpose as well as human consumption.
- ✰ The secondary infection of bacteria sets on such infested tubers. The presence of black spot and faecal matter around eyebuds helps to detect the pest infestation in storage. A single potato may have several caterpillars.
- ✰ As a result of pest infestation, the pulp and the eye buds of potatoes are destroyed.
- ✰ Under indigenous storage methods, pest causes losses to the extent of 30 to 70 per cent.

Economic Threshold Level (ETL)

15-20 moths/trap consecutive for 3 nights.

Management

(A) Cultural Measures

1. Use healthy potatoes free from the attack of potato tuber moth for planting.
2. Timely earthing up of crop to cover the exposed tubers in the field.
3. Store harvested potatoes immediately in well ventilated dry and cool place with temperature not exceeding 21°C.

(B) Mechanical Measures

1. Heaps of harvested potato should not be kept exposed in field.
2. Infested tubers should be rejected before storage.
3. As far as possible, clean pest free gunny bags should be used for storing the potatoes.

(C) Chemical Measures

1. Apply following insecticides in soil at 60 days after planting: Spraying Carboryl 50 WP @ 2.5 kg/ha or Quinolphos 25 EC @ 1.5 lt/ha.

2. Application of Carbaryl 10 D @ 20 kg/ha or Malathion 5 D @ 40 kg/ha in the infested field.
3. Infested tubers in storage can be treated with suitable fumigants like methyl bromide (@ 1 kg/27 m^3 area for 3 hours) or Carbon disulphide (@ 1 kg/27 m^3 space for 48 hours) or ED/CT mixture.
4. The walls of godown may be treated with suitable insecticide like Carbaryl once in every 3 months.
5. Bags used for storing the potatoes should be treated with 5 per cent Malathion or sun dried.

(D) Biological Measures

1. Release four times the egg larval parasite *Copidosoma koehleri* Blanchard @ 1250 mummies/ha/release at weekly interval starting from 45 days after planting or release egg larval parasite- *Chilonus blackburni* or egg parasite - *Argilus lepidus* @ 60,000 adults/ha, four times at weekly interval.

2. Aphid

Also known as: Green peach aphid or Cabbage aphid.

Scientific Name, Order and Family

Myzus persicae

(Hemiptera; Aphididae)

Taxonomical Position/Scientific Classification

Kingdom:	Animalia
Phylum:	Arthropoda
Class:	Insecta
Order:	Hemiptera
Family:	Aphididae
Genus:	Myzus
Species:	***M. persicae***

Biology and Life Cycle

Adults of *M. persicae* are usually of green colour but may be pale brown to pinkish, 1.5 - 2.5 mm long with long clavate siphunculi. *M. persicae* reproduces by parthenogenetical viviparity during summer, monsoon and autumn seasons but sexually in cooler regions during winter. Adults normally perish due to severe cold and eggs overwinter in cracks and crevices on the bark of various temperate fruit trees. When the temperature rises, the eggs hatch and nymphs start feeding on blossoms. They mature in 3 - 4 days and reproduce parthenogenetically producing young ones which develop into wingless adults. After 2 - 3 generations when the temperature rises further or when there is too much crowding of these aphids, the winged forms are produced and these migrate to other crops including brinjal; again with fall in temperature, they migrate back to temperate fruit trees.

Marks of Identification

They are known as peach potato aphid and cabbage aphid, respectively. The eggs are usually elliptical in shape, initially yellow or green, but soon turn black. Nymphs look like to the viviparous adults. Winged aphids of *M. persicae* have a black head and thorax, and a yellowish green abdomen with a large dark patch dorsally. They measure 1.8-2.1 mm in length. The wingless forms are yellowish or greenish in color and about 1.7-2.0 mm long. They may have a medial and lateral green stripes. The cornicles are moderately long, unevenly swollen along their length, and match the body in color. The appendages are pale. Wingless females of *B. brassicae* are about 2.1- 2.6 mm long oval shaped and the posterior end of the body tapers greatly. The pale green body is usually covered by white waxy powder. Underneath the wax there are eight dark brown or black spots located on the upper abdominal surface. They increase in size toward the posterior end. Winged females are a little smaller than the wingless forms and they are not covered with waxy powder. The head and thorax are dark brown to black. The yellowish green abdomen is with two dark spots on the dorsal anterior abdominal segments. Antennae are dark brown. Wings are short and stout with prominent veins.

Host Crops

Potato, tobacco, peach, beans, mustard and other cruciferous plants.

Nature and Symptoms of Damage

- Both nymphs and adults are found in large number sucking the cell sap from leaves and tender apical shoots.
- The under surface of the leaves get crinkled and slightly curled backwards.
- The vitality of the plant is diminished and the plants turn yellow, get deformed and dry away.
- Besides this direct damage, they also secrete copious quantity of honeydew on which sooty mould grows rapidly covering the affected parts with a thick black coating, which interferes with the photosynthetic activity of the plants.
- The infested plants become weak, pale and stunted in growth which consequently results in reduced fruit size. Young fruits are shriveled and drop pre maturely
- Fruit setting affected in severe infestation
- The insect also transmit various viral diseases like potato leaf roller virus, potato virus Y, PVX, PVS *etc.*
- This aphid remains active from December to March with peak activity during February.

Economic Threshold Level (ETL)

1-2 aphids/100 compound leaves of plant or at 25 per cent of plants infested.

Management

(A) Cultural Measures

1. Rouging the plants showing virus symptoms.
2. Remove and destroy the damaged plant parts along with nymphs and adults

(B) Mechanical Measures

1. Monitor weekly from flowering to grain fill of the crop. At least four sampling points of random 20 plants should be spread over the field and estimate the number of aphids per plant.
2. Also, yellow traps can use for monitoring aphids.

(C) Chemical Measures

1. Apply Dimethoate 30 EC or Odydemeton methyl 25 EC @ 0.02-0.3 per cent. 3-5 spraying should be given after 10-15 days interval for seed crop.
2. Two to three sprayings at 10 - 12 days interval with 0.05 per cent dimethoate or 0.01 per cent imidacloprid.
3. Spray dimethoate 0.03 per cent or methyl demeton 0.025 per cent.

(D) Biological Measures

1. Encourage parasitoid *i.e. Aphelinus mali* and predators *i.e. Coccinella septumpunctata* and *Bacillus eucharis.*
2. Conservation of the coccinellids and syrphids that are found to feed on the aphids will reduce the numbers considerably without any insecticidal spray.
3. The parasitoid, *Diaeretiella rapae* (McIntosh) normally provides good control of aphid.

3. Cut Worm

Also known as: Greasy cutworm or caterpillar.

Scientific Name, Order and Family

Agrotis ypsilon Hufn.

(Lepidoptera; Noctuidae)

Taxonomical Position/Scientific Classification

Kingdom:	Animalia
Phylum:	Arthropoda
Class:	Insecta
Order:	Lepidoptera
Family:	Noctuidae
Genus:	Agrotis
Species:	***A. ypsilon***

Biology and Life Cycle

1. **Egg**: Females lay about 300 creamy white, spherical eggs singly or in batches on the lower surface of leaves or on stem close to the ground surface or sometimes on ground itself. Hatching period is 2 to 13 days.
2. **Larva**: Larval period is 10 to 30 days.
3. **Pupa**: Pupation takes place in the soil after forming earthen cocoons. Pupal period varies from 11 to 18 days.
4. **Adult**: One generation is completed in 5 to 9 weeks.

Marks of Identification

Adult is medium sized moth, measuring about 25 mm in length and 42 to 56 mm in wing expanse. Each of the forewings has dark or blackish grey patches. Fullgrown caterpillar measures about 45 mm in length and smooth, dull gray in colour with small shining black spots on the body. The caterpillar has a habit of coiling at the slightest touch. Since the caterpillars are greasy in appearance, the pest is also known as 'Greasy caterpillar' or 'Greasy cutworm'.

Host Crops

Potato, tomato, brinjal, barley, oat, lucerne, tobacco, linseed, mustard, pea, gram and other leguminous crops, cotton, chilli (Polyphagous).

Nature and Symptoms of Damage

- ✰ The common name implies to the fact that the caterpillars cutoff the stalk of the young plant from the ground level.
- ✰ The caterpillars are nocturnal in habit. During day time, they hide under the clods of earth or cracks and crevices present in soil or debris around the plants.
- ✰ They come out and feed on the plants during night by cutting them close to the soil surface. Destruction is much more than actual feeding. Very occasionally, caterpillars are found nibbling the tubers.
- ✰ Damage to the extend of 37 per cent has been reported.

Economic Threshold Level (ETL)

3 per cent Attacked Plants.

Management

(A) Cultural Measures

1. Clean cultivation by regular interculturing operations helps to reduce the intensity of pest infestation.
2. Heaps of green grasses may be kept at suitable interval in infested field during evening to trap the caterpillars. These heaps of grasses may be removed next day early in the morning alongwith larvae and destroyed mechanically.
3. Irrigation brings the caterpillars to upper soil surface where they are destroyed by birds and other natural enemies.
4. Deep ploughing of field in summer season exposes the immature eggs of cutworm and other soil pest to sun and predatory fauna.
5. Breaking of the clods and use of woody stemmed varieties cal also be effective.
6. Use of Crop rotation.

(B) Mechanical Measures

1. Electric/pheromone light trap may be operated during night for mass collection of moth to reduce population of caterpillar of subsequent generation.

(C) Chemical Measures

1. Dusting the crop with 5 per cent carboryl @ 35 kg/ha.
2. Spraying the foliage and ridges with chloropyriphos 20 EC @ 2.5 lt/ha with 1000 litre of water.
3. Apply fenvalerate 2 per cent dust in soil once each at planting and earthing up operations @ 50 kg/ha.

(D) Biological Measures

1. Larval parasitoids – *Cotesia* spp. and *Microbracon* spp.

Pest Management of Vegetable Crops

(A) PESTS OF OKRA

1. Shoot and Fruit Borer

Scientific Name, Order and Family

Earias vitella Fab.

Earias insulana Boised.

(Lepidoptera; Noctuidae)

Taxonomical Position/Scientific Classification

Kingdom:	Animalia
Phylum:	Arthropoda
Class:	Insecta
Order:	Lepidoptera
Family:	Noctuidae
Genus:	Earias
Species:	***E. vitella***
	E. insulana

Biology and Life Cycle

1. **Egg**: The females lay bluish green eggs singly or in batches on tender shoots, flowers, flower buds and young fruits. Eggs are spherical in shape,

about half mm in diameter, light bluish-green in colour and beautifully sculptured with 26 - 32 longitudinal ridges; the alternate ridges project upwards to form a crown. Fecundity varies from 60 to 432. Incubation period is 3 to 7 days.

2. **Larva**: Larval development takes place within 9 to 20 days.
3. **Pupa**: Pupal period varies from 5 to 9 days. Pupation takes place in silken cocoon either on host or in the soil or in fallen leaves.
4. **Adult**: One generation completed in 19 to 21 days in summer and 30 to 31 days in winter. Total number of generations completed in a year is 12.

Marks of Identification

1. **Larva**: The caterpillars of both the species are brownish white with a dark head and prothoracic shield. Caterpillars have number of black and brown spots on the body. Full-grown larva measures about 19 mm in length.
2. **Moth**: The moths of *E. vitella* have pale white upper wings with broad greenish band in the middle. The moths of *E. insulana* possess completely greenish coloured forewings. The hind wings in both species are pale white. The moths of both species measure about 10 mm in length and 25 mm across the wings.

Host Crops

Cotton, Bhendi, Ambadi, Hollyhock and several other malvaceous plants, shoeflowers (Oligophagous).

Nature and Symptoms of Damage

- ☆ In the beginning of the crop growth (when the crop is only a few weeks old), larvae bore into the tender top shoot and feed internally causing shoots to droop, wither and dry up.
- ☆ When flowers, flower buds and fruits appear, the larvae are seen boring into them and feeding internally.
- ☆ The infested flowers and flower buds drop down prematurely. Damaged fruits present deformed appearance. The infested fruits show holes plugged with excreta.
- ☆ Infested fruits fetch less price in the market. The severely infested fruits are made unfit for human consumption.
- ☆ A single larvae may destroy several fruits.

Management

(A) Cultural Measures

1. All the alternate host plant from near the okra field should be destroyed.

2. Removal and destruction of infested shoots and fruits and shed materials helps in reducing intensity of infestation.
3. Deep ploughing after the harvest of crop helps to destroy hibernated pupae.
4. Infested fruits should be collected and destroyed promptly.
5. Grow "**Parbhani Kranti**" and "**Arka Anamika**" as they are resistant to yellow vein mosaic.
6. Trap crops like "**Lubia**" or "**Marigold**" may be planted around the border of main crop.

(B) Mechanical Measures

1. Hand picking of caterpillars and their mechanical destruction in the early stage of infestation can keep the population of this pest under check.
2. Use Trichocards of 1cc each @ 4-5 cards/ha.

(C) Chemical Measures

1. Give four sprays of following insecticides at 15 days interval starting from 15 days after germination: Cypermethrin 25 EC 0.01 per cent or Fenvalerate 20 EC 0.01 per cent or Deltamethrin 2.8 EC 0.01 per cent or Lambda cyhalothrin 5 EC 0.0015 per cent or Profanofos 50 EC 0.08 per cent.

(D) Biological Measures

1. Mass release *Trichogramma pretiosum* egg parasite 50,000 adults/ha five times at weekly interval from initiation of flowering.
2. Give five sprays of NPV 250 LE (larval equivalent) @ 500 ml/ha at weekly interval after initiation of 50 per cent flowering.
3. Spray 5 per cent Neem Seed Extract (NSE) @ 500 gm/500 litres of water.

(B) PESTS OF BRINJAL

1. Shoot and Fruit Borer

Scientific Name, Order and Family

Leucinodes orbonalis (Guen.)

(Lepidoptera; Crambidae)

Taxonomical Position/Scientific Classification

Kingdom:	Animalia
Phylum:	Arthropoda
Class:	Insecta

Order:	Lepidoptera
Family:	Crambidae
Genus:	Leucinodes
Species:	***L. orbonalis***

Biology and Life Cycle

1. **Egg**: Each female lays about 80-120 eggs singly on tender shoots, flower buds, leaves and young fruits. Eggs hatch out in 3 to 5 days.
2. **Larva**: Larval period is of about 8-28 days. White and last instar larvae is pink with sparsely distributed hairs on its body. The fully grown larvae measures about 10-20 mm in length. On hatching, caterpillars start boring into tende shoots and fruits when present. They become full grown in 7 to 13 days.
3. **Pupa**: Pupal period lasts for 7 to 11 days. Full grown caterpillars come out and pupate in cocoons either on follen leaves, shoots, fruits or in soil.
4. **Adult**: The moth is white but has pale brown or black spots on the dorsum of thorax and abdomen. Its wings are white with a pinkish or bluish tinge. Moths live for 2 to 5 days. The total life cycle is of about 17-50 days.

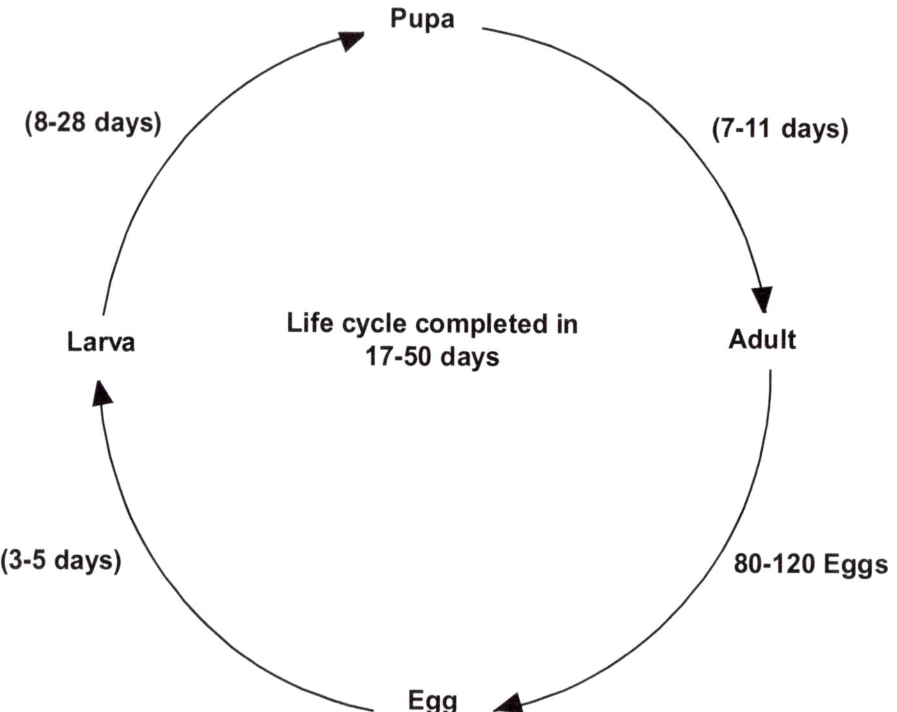

Figure 48: Life Cycle of *Leucinodes orbonalis*.

Marks of Identification

Adult is medium sized moth with whitish wings having large brown patches all over. Head and thorax are blackish brown. The moths measure about 10 mm in length and 20 mm in wing expanse. The caterpillars are pale white and measure about 15 mm in length when full grown.

Host Crops

Brinjal, Potato, Tomato, Bottle gourd, Pea pods.

Nature and Symptoms of Damage

- ☆ Infestation of pest starts from a few weeks after transplanting of crop and pest continues to damage the crop till the harvest of fruits.
- ☆ The larva bores into tender shoots in the early stage and causes "**dead hearts**".
- ☆ Caterpillars bore into the terminal shoots, petioles and midribs of large leaves. As a result, terminal portions wither and finally dry up. Later, when fruits develop, caterpillars are found boring into them and feeding internally.
- ☆ In case of young fruits, caterpillars make their entry below calyx. These entry holes get plugged with excreta leaving no visible sign of infestation. The larger holes seen on the fruits are usually the exit holes of the larva.
- ☆ The affected fruits are unfit for consumption.
- ☆ Under conditions of severe infestation, damage to the fruits is reported to be as high as 70 per cent.
- ☆ The varieties with spherical or round fruits are relatively more preferred than those with medium long and slender fruits, more compact vascular bundles, lower protein and sugar contents.

Economic Threshold Level (ETL)

1-5 per cent shoot/fruit infestation.

Management

(A) Cultural Measures

1. Remove and destroy the infested shoots and fruits along with larvae to avoid further build up of pest population.
2. Avoid continuous cropping of brinjal and potato in the same field as it may encourage the pest activity and hence, proper rotation should be followed.
3. Do not ratoon the brinjal crop.
4. Apply phorate 10 G in the seed bed @ 25 gm per seed bed of a size 2 m ×1 m.

5. Dip seedlings in the insecticidal suspension of imidacloprid 200 SL (10 ml/10 litres of water) before transplanting.
6. Grow resistant varieties *i.e.* Pusa purple long, Black beauty, H-165, Pusa purple cluster, Pusa Purple Round, Punjab Neelam, Kalyanpur- 2, Punjab Chamkila, PBR-129-5, Pant samrat, S_{519}, S_{520}, S_{521}, SM-202, SM-17, SM 17-4, PBr 129-5 Punjab Barsati, ARV 2-C, Gote-2, PBR-91, GB-1, GB-6.

(B) Mechanical Measures

1. Pheromone traps installed @ 12/ha for monitoring of borer.

(C) Chemical Measures

1. Apply 3 to 4 sprays of any one of the following insecticides at 15 days intervals with 500 liters of water/ha: Quinolphos 25 EC @ 2 lt/ha or Carboryl 50 WP @ 2kg/ha or Fenvelrate 20 EC @ 250 ml/ha or Permethrin 50 EC @ 100 ml/ha or Cypermethrin 10 EC @ 500 ml/ha.

(D) Biological Measures

1. 3- 4 releases of egg parasite, *Trichogramma chilonis* @1.0 lakh/ha.

2. Stem Borer

Scientific Name, Order and Family

Euzophera perticella

(Lepidoptera; Pyralidae)

Taxonomical Position/Scientific Classification

Kingdom:	Animalia
Phylum:	Arthropoda
Class:	Insecta
Order:	Lepidoptera
Family:	Pyralidae
Genus:	Euzophera
Species:	***E. perticella***

Biology and Life Cycle

1. **Egg**: Each female lays Creamy and scale-like,eggs singly or in batches on the under surface of young leaves or axil of branches. The fecundity is 104-363 eggs. The incubation period is of 10 days.
2. **Larva**: Fully grown larva is creamy white with few bristle-like hairs, 20 mm. Larval period is of about 26-58 days. Larva having 4 no. of instars.
3. **Pupa**: Pupation take place within a cocoon inside the larval tunnels or in the cracksin the soil. Pupal period lasts for 9-16 days.

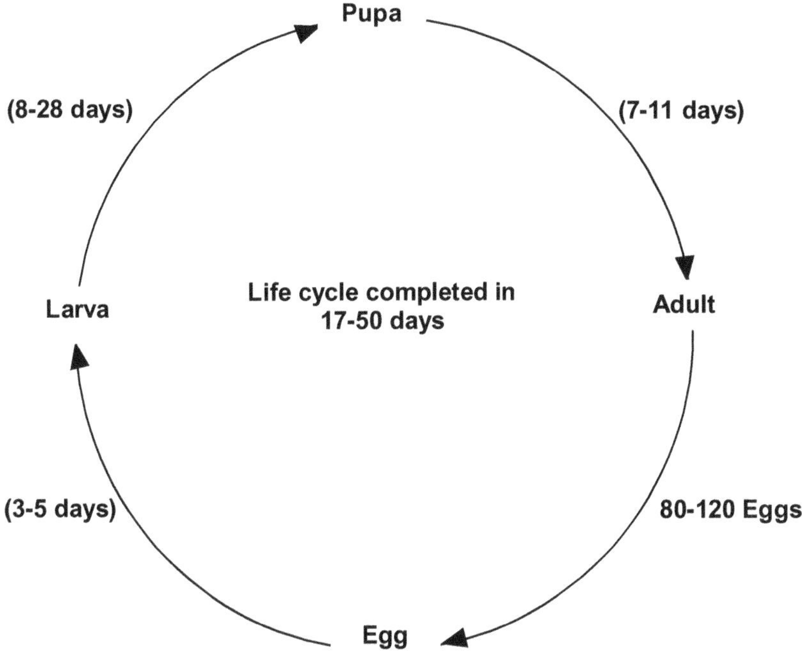

Figure 49: Life Cycle of *Euzophera perticella.*

4. **Adult**: Greyish brown, forewings with transverse line and white hindwings. Life cycle is completed in 35-76 days.

Marks of Identification

1. Full grown caterpillar is creamy white and has few hairs.
2. Moths are having pale yellow abdomen and head and thorax are grayish in colour. The forwings are pale straw yellow and hindwings ar whitish.

Host Crops

Brinjal, chilly, potato and Tomato *etc.*

Nature and Symptoms of Damage

- The newly hatched larva/caterpillars bore into the stem near the ground level, resulting in tunnels and filled with excreta and fross.
- Usually, boring is seems at the basal portion of main stem with abnormal swelling of the affected part.
- Gum like substance oozes out from the entry holes.
- The infected plant becomes stunted and later the intire plant withers.
- Both young and old plants are attacked.

Management

(A) Cultural Measures

1. Remove and destroy the infested plants along with larvae to avoid further build up of pest population.
2. Avoid continuous cropping of brinjal in the same field as it may encourage the pest activity and hence, proper rotation should be followed.
3. Do not ratoon the brinjal crop.
4. Apply phorate 10 G in the seed bed @ 25 gm per seed bed of a size 2 m ×1 m.
5. Dip seedlings in the insecticidal suspension of imidacloprid 200 SL (10 ml/10 litres of water) before transplanting.

(B) Mechanical Measures

1. Use light traps @ 1/ha to attract and kill the moths.

(C) Chemical Measures

1. Apply 3 to 4 sprays of any one of the following insecticides at 15 days intervals with 500 liters of water/ha: Quinolphos 25 EC @ 2 lt/ha or Carboryl 50 WP @ 2kg/ha or Fenvelrate 20 EC @ 250 ml/ha or Permethrin 50 EC @ 100 ml/ha or Cypermethrin 10 EC @ 500 ml/ha.

(D) Biological Measures

1. Conserve larval parasitoids *Pristomerus testaceus, P. euzopherae.*

3. Brinjal Mites

Also known as: Red vegetable mite.

Scientific Name, Order and Family

Tetranychus urticae

(Acarina; Tetranychidae)

Taxonomical Position/Scientific Classification

Kingdom:	Animalia
Phylum:	Arthropoda
Class:	Arachnida
Order:	Acarina
Family:	Tetranychidae
Genus:	Tetranychus
Species:	***T. urticae***

Biology and Life Cycle

1. **Egg**: Female lays 60-80 round eggs singly or in cluster on the lower surface of the leaf. Egg period lasts 2-7 days. The eggs hatch into the nymph named protonymph, and then turns into deutonymph, afterward adult stage form. *T. urticae* has worldwide distribution and infests to a wide range of plants.
2. **Nymph**: The completetly developed nymph is microscopic and measures about 0.33 mm in length. Nymph is light brown in colour and has two eye spots and four pairs of legs and quite active. Nymphs grow to maturity in 2 stages within 4-9 days and adults live for 9-11 days.
3. **Adult**: Adult male is of about 0.52 mm in length and 0.3 mm in breadth. The female is oval and variable in colour. It may be red, greenish, amber or rusty green. Two large pigmented spots are present on the body. Total life cycle in active period takes 9-19 days.

Marks of Identification

The size of the mite is less than 1 mm and varies in the color. The adults have two typical dark spots on the back so they are named two spotted mite. The female is 0.5 mm long; the male is smaller and slender and they have 4 pairs of legs. Adult female body is oval and is variable in colour *i.e.*, red, green, amber or rusty green and with two large pigmented spots on the body. Larva of reduced size has 3 pairs

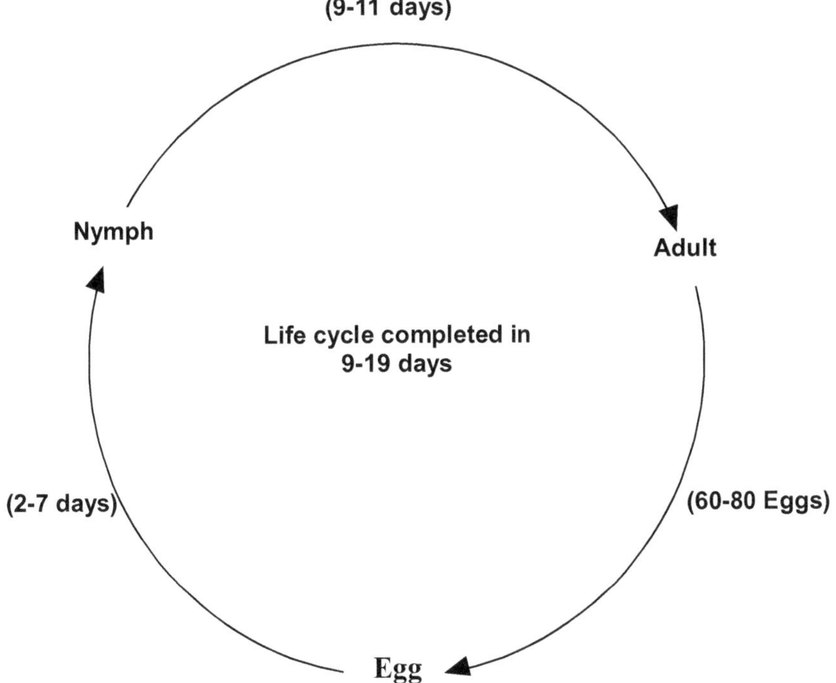

Figure 50: Life Cycle of *Tetranychus urticae*.

of legs. The common name, red spider mite, is because they spin silk webs to protect the eggs or colony from predators.

Host Crops

The mite is a polyphagous and is known to infest on 183 species of plants including cucurbits, brinjal and bhendi on which it is sometimes very serious.

Nature and Symptoms of Damage

All active stages (Both nymphs and adults) cause damage by feeding on the underside of the leaf or lower surface of the leaf by sucking cellsap. Gradually, the infested leaf dry up and the webbing interferes with plant growth.

- ✰ Mite infestation begins in the seedling stage and extends to harvest.
- ✰ Colonies of mites comprising of eggs, nymphs and adults are found feeding on ventral surface of leaves under protective cover of fine silken webs.
- ✰ As a result of their feeding innumerable yellow spots appear on the dorsal surface of leaves and the affected leaves gradually start curling and finally get wrinkled and crumpled.
- ✰ This in turn affects the growth and fruit formation capacity of the plants.
- ✰ Yield loss of the mites depends on when mite populations begin to increase and how quickly they increase. There is a poor setting of fruits and the yield is considerably reduced.

Economic Threshold Level (ETL)

10/Leaf or 30 per cent of plants infested through the bulk of the season.

Management

(A) Mechanical Measures

1. Monitor the oldest leaf when plants are very young. As plants grow, choose leaves that are from 3, 4 or 5 nodes below the plant terminal. Begin monitoring at seedling emergence and sample at least weekly. Sample more frequently if the weather conditions are hot and dry. Also, shake the leaves onto a white piece of paper and count the actual number of mites moving around with a hand lens. Eggs and immature stages are difficult to see with the naked eye, so a hand lens should be used.

(B) Chemical Measures

1. Foliar sprays with sulphur 50 WP 3 g/l or dicofol 5 ml/lt.
2. Spray Dimethoate 30 EC @ 625 ml/ha. Repeat the spray after 10 days, if necessary.

(C) Biological Measures

1. Biological control by ladybirds (*Hippodamia convergens* Guerin), big eyed bugs (*Geocoris* spp.), damsel bugs (*Nabis* spp.), lacewings (*Chrysopa* and

Micromus spp.) and *Phytoseiulus persimilis* (Athias-Henriot) can decrease the population of the mites in early season.

(C) PESTS OF CHILLI

1. Thrips

Scientific Name, Order and Family

Scirtothrips dorsalis Hood.

Thrips tabaci (Lind.)

(Thysanoptera; Thripidae)

Taxonomical Position/Scientific Classification

Kingdom:	Animalia
Phylum:	Arthropoda
Class:	Insecta
Order:	Thysanoptera
Family:	Thripidae
Genus:	Scirtothrips
Species:	***S. dorsalis***

Biology and Life Cycle

1. **Egg**: Females lay kidney shaped, yellow or white about 50-80 eggs singly in slits made by female in leaf or on flower tissue. Hatching period is 5-10 days.
2. **Nymph**: Nymph is wingless and slightly smaller than adult but similar in respect of shape and colour. The nymphal period is completed in 4-6 days. Pupation is takes place in soil at a depth of about 25 mm. the pupation period is completed in 3-6 days.
3. **Adult**: Slender, yellowish brown and measures about 1 mm in length. Males are wingless. Females are long, narrow, strape like wings.

Marks of Identification

1. *Thrips tabaci* adult is about 1 mm long; yellowish-gray to dark-gray and the head has post ocular setae of about the same length as the interocellar setae. The basal segments of the usually seven-segmented antennae are light brown, while the distal segments are dark brown. The prothorax has only two pairs of well developed setae on the posterior margin. The second segment of thoracic has convergent striations on the top surface, and lacks campaniform sensillae. Eggs are microscopic, white or yellow. The nymph ranging from 0.5 to 1.2 mm, white to pale yellow and look like

to the adult; just it has short antennas and does not have wings. Pupae are pale yellow to brown, they have short antennae and the wing buds are visible but short and not functional.

2. *Scirtothrips dorsalis* has small size less than 1 mm in length, yellow coloration, dark antennae with forked sense cones on antennal segments III and IV, antennomeres I-II are pale and III to IX are dark. Also, it has dark striping on the lower abdomen and three distal setae on the lateral margins of abdominal tergites, with pronotal posteromarginal seta II nearly one and a half times the length of I or III, a complete posteromarginal comb on tergite VII; and three ocellar setae with III between posterior ocelli.

Host Crops

Chilly, onion, cotton, brinjal, okra, cashew, mango, guava, bottle gourd, ridge gourd.

Nature and Symptoms of Damage

- ☆ Both nymphs and adults of this pest have mouth parts suitable for scrapping the epidermal tissues and sucking the oozing cell sap.
- ☆ Nymphs and adults suck sap from leaves and flower buds. Margins of affected leaves get slightly curled up and the leaf blades show uneven surface, when attack occurs in flowering stage, the affected flowers may wither away.
- ☆ The plant tissues damaged by thrips initially become whitish but later turn brown and ultimately dry.
- ☆ Besides causing direct damage to the plant, the pest acts as a vector transmitting leaf curl disease.
- ☆ Under conditions of severe infestation, 30 to 50 per cent crop may be lost.

Economic Threshold Level (ETL)

6 thrips/leaf or 10 per cent affected plants.

Management

(A) Cultural Measures

1. Avoid excess nitrogen and close spacing to avoid thrips.
2. Destruction and burning of severly infested plants.
3. Seed treatment with Imidacloprid 70 WS @ 3 g/kg of seed to prevent from early season infestation.

(B) Mechanical Measures

1. Use sticky traps @ 10/ha used for monitoring the thrips.
2. Monitoring should be started from seedling emergence of the crop and number of thrips on 20 - 30 plants counted weekly. Monitor by picking

and slapping a leaf on a white index card to see if the tiny specks move around. Open and microscopically examine the plant's growing point for thrips. Plucked growing points can also be dunked in alcohol to dislodge thrips.

(C) Chemical Measures

A. Nursery Treatment

1. Apply phorate 10 G @ 10kg/ha at the time of sowing in seed-bed.
2. If phorate is not available, then give two sprayings of Dimethoate 0.03 per cent at 15th and 30th days after seed germination in nursery.

B. Field Treatment

1. At the time of transplanting, dip the seedlings (leafy portion) in Monocrotophos 0.125 per cent solution.
2. On the main crop, spraying with Dimethoate 0.03 per cent or Methyl demeton 0.02 per cent should be undertaken 10 days to 15 days interval after transplanting.

(D) Biological Measures

1. Biological control of the thrips by Anthocoridae, Lygaidae and predator mites will be effective.
2. The *Franklinothrips vespiformis* (Crawford) and *Erythrothrips asiaticus* are predaceous in nature and their population may be encouraged by avoiding chemical sprays.

2. Fruit Borer

Scientific Name, Order and Family

Helicoverpa armigera (Hubn.)

(Lepidoptera; Noctuidae)

Taxonomical Position/Scientific Classification

Kingdom:	Animalia
Phylum:	Arthropoda
Class:	Insecta
Order:	Lepidoptera
Family:	Noctuidae
Genus:	Helicoverpa
Species:	*H. armigera*

Biology and Life Cycle

1. **Egg**: Females lay about 500 eggs singly on leaves, flowers and sometimes on fruits as well. Eggs are yellowish white, dome shaped, 0.4 to 0.5 mm in diameter. Hatching period is 6 to 7 days.
2. **Larva**: Larval development takes place within 14 to 38 days.
3. **Pupa**: Pupae are dark brown and measure about 11 to 14 mm in length. Pest pupates in the soil. Infestation is carried from one season to another through hibernated pupae in the soil. Pupal period varies from 6 to 12 days.
4. **Adult**: One generation is completed in 4 to 6 weeks. About 5 to 6 generations are completed in a year.

Marks of Identification

1. **Larva**: Caterpillars are green in colour with whitish and dark grey broken longitudinal strips. Caterpillars measure about 40 to 48 mm in length when full-grown.
2. **Adult**: Adult is medium sized, light yellowish brown coloured moth with wing expanse of 37 mm. Forewings are pale brown with a dark brown minute circular dots on it. Hindwings are smoky-white with a broad blackish outer border.

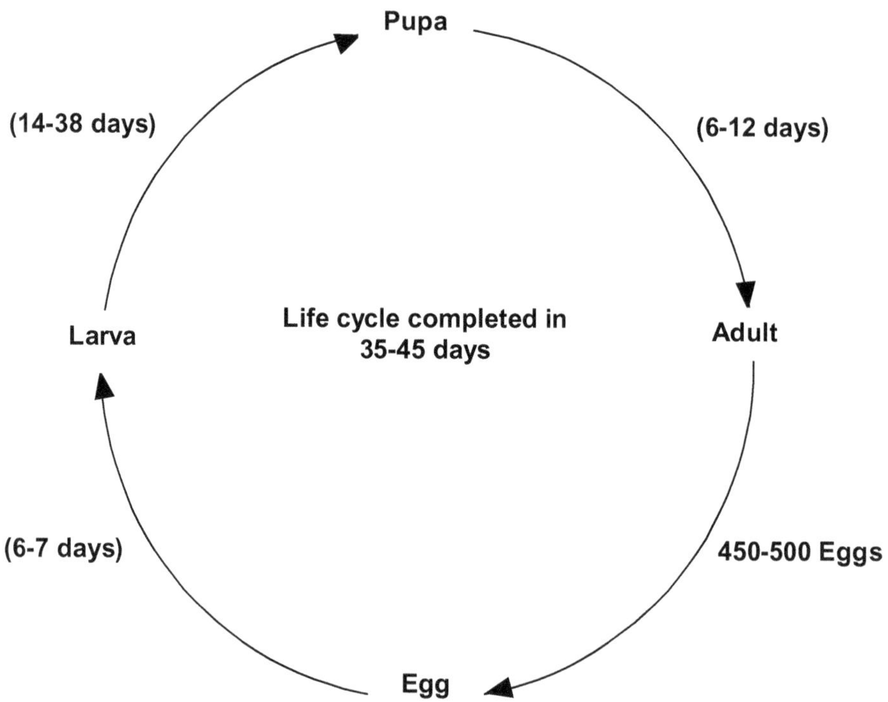

Figure 51: Life Cycle of *Helicoverpa armigera*.

Host Crops

Castor, cotton, citrus, indigo, groundnut, sorghum, millets, pulses, safflower, soyabean, sunflower, marigold, tobacco, linseed, okra. It is highly polyphagous pest.

Nature and Symptoms of Damage

The caterpillar causes damage initially by feeding on the foliage for couple of days and subsequently boring into fruits and feeding on the internal tissues by introducing it's head and anterior half part of the body alone within. Larva is migratory in habit. It moves from one fruit to another damaging as many as 2 to 8 fruits in it's life time. Young larvae feed on tender foliage while advanced stage larvae (4th instar onwards) attack the fruits.

Economic Threshold Level (ETL)

1 egg or 1 larva/plant or 1 damaged fruit/plant.

Management

(A) Cultural Measures

1. Deep ploughing after the harvest of crop helps to destroy hibernated pupae.
2. Infested fruits should be collected and destroyed promptly.

(B) Mechanical Measures

1. Hand picking of caterpillars and their mechanical destruction in the early stage of infestation can keep the population of this pest under check.
2. Install light traps @ 3 no./ha and pheromone traps @ 5 no./ha.
3. Growing of trap crop *i.e.* Marigold in border.

(C) Chemical Measures

1. Spraying the crop with Indoxacarb 14.5 EC or Novaluron 10 EC or Sponosad 45 EC are found effective.

(D) Biological Measures

1. Mass release *Trichogramma pretiosum* egg parasite 50,000 adults/ha five times at weekly interval from initiation of flowering.
2. Apply *Bacillus thuringinesis* var. *Kurstaki* @ 1 kg/ha.
3. Give five sprays of NPV 250 LE (larval equivalent) @ 500 ml/ha at weekly interval after initiation of 50 per cent flowering.

(D) PESTS OF TOMATO

1. Fruit Borer

Scientific Name, Order and Family

Helicoverpa armigera (Hubn.)

(Lepidoptera; Noctuidae)

Taxonomical Position/Scientific Classification

Kingdom:	Animalia
Phylum:	Arthropoda
Class:	Insecta
Order:	Lepidoptera
Family:	Noctuidae
Genus:	Helicoverpa
Species:	*H. armigera*

Biology and Life Cycle

1. **Egg**: Females lay about 500 eggs singly on leaves, flowers and sometimes on fruits as well. Eggs are yellowish white, dome shaped, 0.4 to 0.5 mm in diameter. Hatching period is 6 to 7 days.
2. **Larva**: Larval development takes place within 14 to 38 days.
3. **Pupa**: Pupae are dark brown and measure about 11 to 14 mm in length. Pest pupates in the soil. Infestation is carried from one season to another through hibernated pupae in the soil. Pupal period varies from 6 to 12 days.
4. **Adult**: One generation is completed in 4 to 6 weeks. About 5 to 6 generations are completed in a year.

Marks of Identification

1. **Larvae**: Caterpillars are apple green in colour with whitish and dark grey broken longitudinal strips. Caterpillars measure about 40 to 48 mm in length when full-grown.
2. **Adult**: Adult is medium sized, light yellowish brown coloured moth with wing expanse of 37 mm. Forewings are pale brown with a dark brown minute circular dots on it. Hindwings are smoky-white with a broad blackish outer border.

Host Crops

Castor, cotton, citrus, indigo, groundnut, sorghum, millets, pulses, safflower, soyabean, sunflower, marigold, tobacco, linseed, okra. It is highly polyphagous pest.

Nature and Symptoms of Damage

The caterpillar causes damage initially by feeding on the foliage for couple of days and subsequently boring into fruits and feeding on the internal tissues by introducing it's head and anterior half part of the body alone within. Larva is migratory in habit. It moves from one fruit to another damaging as many as 2 to 8 fruits in it's life time. Young larvae feed on tender foliage while advanced stage larvae (4th instar onwards) attack the fruits.

Economic Threshold Level (ETL)

1 egg or 1 larva/plant or 1-5 per cent fruit damage.

Management

(A) Cultural Measures

1. Deep ploughing after the harvest of crop helps to destroy hibernated pupae.
2. Infested fruits should be collected and destroyed promptly.
3. Grow fruit borer tolerant varieties *i.e.* Arka Vikash, Pusa Gaurav, Pusa Early Dwarf, Punjab Keshri, Punjab Chhuhara, Pant Bahar, Azad, Avinash -2, Hemsona, Krishna, Sartaj.

(B) Mechanical Measures

1. Hand picking of caterpillars and their mechanical destruction in the early stage of infestation can keep the population of this pest under check.
2. Install light traps @ 3 no./ha and pheromone traps @ 5 no./ha coinciding with the initiation of flowering and destruction of collected moths.

(C) Chemical Measures

1. Spray the crop with Carboryl 50 WP @ 0.1 per cent or Malathion 50 EC @ 0.05 per cent or Phosalon 50 EC @ 0.05 per cent twice at the interval of 2 weeks.
2. In case of severe attack, 5 per cent dust or 0.2 per cent spray of carbaryl or 0.005 per cent cypermethrin has been found to be effective.

(D) Biological Measures

1. Mass release of *Trichogramma pretiosum* egg parasitoid 50,000 adults/ha five times at weekly interval from initiation of flowering.
2. Apply *Baccillus thuringinesis var. Kurstaki* @ 1 kg/ha.
3. Give five sprays of HaNPV 250 LE (larval equivalent) @ 500 ml/ha at weekly interval after initiation of 50 per cent flowering.

2. Leaf Miner

Also known as: Serpentine leaf miner

Scientific Name, Order and Family

Liriomyza trifolii (Riley)

Diptera; Agromyzidae

Taxonomical Position/Scientific Classification

Kingdom:	Animalia
Phylum:	Arthropoda
Class:	Insecta
Order:	Diptera
Family:	Agromyzidae
Genus:	Liriomyza
Species:	***L. brassicae***

Biology and Life Cycle

1. **Egg**: Females lay about 329-358 eggs singly on the outer margin of leaves. The eggs are inserted just below the leaf surface. Hatching period is 2-5 days.

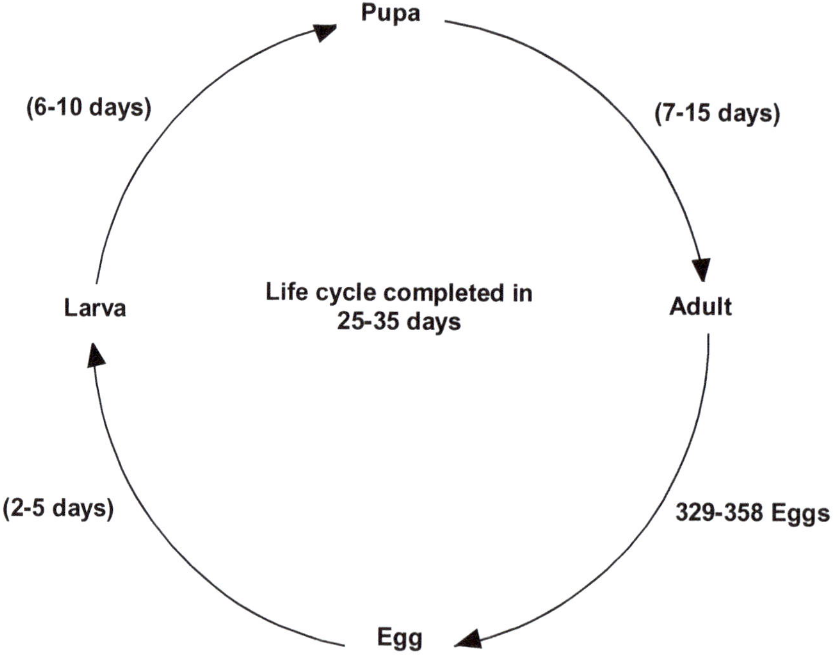

Fig.52: Life Cycle of *Liriomyza trifolii*.

2. **Larva**: Transparent and turn to yellow orange in later instar. Larval development completed within 6-10 days.
3. **Pupa**: Pest pupates in the soil and sometime on the leaf surface. Pupal period varies from 7-15 days.
4. **Adult**: Small flies are yellowish gray coloured. Adults are small, measuring less than 2 mm in length, with a wing length of 1.25 to 1.9 mm. The head is yellow with red eyes. The thorax and abdomen are mostly gray and black although the ventral surface and legs are yellow. The wings are transparent. Adults live about 13 to 18 days.

Marks of Identification

The egg is oval-shaped, white or yellowish. The larva is apodous, cylindrical and tapering at both ends. The length of the last instar larva is, as a rule, in the order of 2–3 mm. The tracheal system is metapneustic in the first instar early age and amphipneustic in the subsequent stages. The pupa is variable, from barrel shaped, to a more elongated shape. The outer surface can segmentation and is more or less smooth or wrinkled. The color varies from black to brown to yellowish white.

Host Crops

Tomato, beans, cowpea, cucurbits, okra *etc.*

Nature and Symptoms of Damage

- ✰ The major form of damage is the mining of leaves by larvae, which results in destruction of leaf mesophyll. The mine becomes noticeable about three to four days after oviposition and becomes larger in size as the larva matures. The pattern of mining is irregular. Both leaf mining and stippling can greatly depress the level of photosynthesis in the plant. Extensive mining also causes premature leaf drop, which can result in lack of shading and sun scalding of fruit. Wounding of the foliage also allows entry of bacterial and fungal diseases.

Economic Threshold Level (ETL)

2-5 miner/leaf

Management

(A) Cultural Measures

1. Avoidance of hybrids and judicious application of nitrogen fertilizer in the pest.
2. Destruction of weeds and deep ploughing of crop residues are recommended.
3. Avoidance of mulching and stalking of vegetables which may influence natural enemies.

(B) Chemical Measures

1. Spraying of Imidacloprid 17.8 SL @ 0.3 ml/ha or Dichlorvas 76 SL @ 0.3 per cent in case of severe infestation during fruiting stage.
2. Two Spraying of Cypermethrin 0.01 per cent or Dimethoate 0.03 per cent starting from 20 days after transplanting at 15 days interval.

(C) Biological Measures

1. Parasitic wasps (parasitoids) of the families Braconidae, Eulophidae and Pteromalidae are important in natural control.
2. At least 14 parasitoid species are known from Florida alone. Species of Eulophidae such as:

   Diglyphus begina (Ashmead), *D. intermedius* (Girault), *D. pulchripes*, and *Chrysocharis parksi* are generally found to be most important.
3. *Steinernema* nematodes have also been evaluated for suppression of leaf mining activity.

3. Stem Borer

Scientific Name, Order and Family

Euzophera perticella

(Lepidoptera; Pyralidae)

Taxonomical Position/Scientific Classification

Kingdom:	Animalia
Phylum:	Arthropoda
Class:	Insecta
Order:	Lepidoptera
Family:	Pyralidae
Genus:	Euzophera
Species:	***E. perticella***

Biology and Life Cycle

1. **Egg**: Each female lays eggs singly or in batches on the under surface of young leaves or axil of branches. The fecundity is 104-363 eggs. The incubation period is of 3-10 days.
2. **Larva**: Larval period is of about 26-58 days. Larva having 4 no. of instars.
3. **Pupa**: Pupation take place within a cocoon inside the larval tunnels or in the cracksin the soil. Pupal period lasts for 6-8 days.
4. **Adult**: It is creamy white in colour.

Marks of Identification

1. Full grown caterpillar is creamy white and has few hairs.
2. Moths are having pale yellow abdomen and head and thorax are grayish in colour. The forwings are pale straw yellow and hindwings are whitish.

Host Crops

Brinjal, chilly, potato and tomato *etc.*

Nature and Symptoms of Damage

- ☆ The newly hatched larva/caterpillars bore into the stem near root portion, resulting in tunnels and filled with excreta and fross.
- ☆ Usually, boring is seems at the basal portion of main stem with abnormal swelling of the affected part.
- ☆ Gum like substance oozes out from the entry holes.
- ☆ The fross and excreta may be seed at the collar region.
- ☆ The infected plant becomes stunted and later the intire plant withers.
- ☆ Both young and old plants are attacked.

Economic Threshold Level (ETL)

Management

(A) Cultural Measures

1. Remove and destroy the infested plants along with larvae to avoid further build up of pest population.
2. Avoid continuous cropping of tomato in the same field as it may encourage the pest activity and hence, proper rotation should be followed.
3. Apply phorate 10 G in the seed bed @ 25 g per seed bed of a size 2 m ×1 m.
4. Dip seedlings in the insecticidal suspension of imidacloprid 200 SL (10 ml/10 litres of water) before transplanting.

(B) Chemical Measures

1. Apply 3 to 4 sprays of any one of the following insecticides at 15 days intervals with 500 liters of water/ha: Quinolphos 25 EC @ 2 lt/ha or Carboryl 50 WP @ 2kg/ha or Fenvelrate 20 EC @ 250 ml/ha or Permethrin 50 EC @ 100 ml/ha or Cypermethrin 10 EC @ 500 ml/ha.

(E) PESTS OF ONION and GARLIC

1. Thrips

Scientific Name, Order and Family

Thrips tabaci (Lind.)

(Thysanoptera; Thripidae)

Taxonomical Position/Scientific Classification

Kingdom:	Animalia
Phylum:	Arthropoda
Class:	Insecta
Order:	Thysanoptera
Family:	Thripidae
Genus:	Thrips
Species:	***T. tabaci***

Biology and Life Cycle

1. **Egg**: Females lay kidney shaped, yellow or white about 50-80 eggs singly in slits made by female in leaf or on flower tissue. Hatching period is 5-10 days.
2. **Nymph**: Nymph is wingless and slightly smaller than adult but similar in respect of shape and colour. The nymphal period is completed in 4-6 days. Pupation is takes place in soil at a depth of about 25 mm. the pupation period is completed in 3-6 days.
3. **Adult**: Slender, yellowish brown and measures about 1 mm in length. Males are wingless. Females are long, narrow, strape like wings.

Host Crops

Onion, garlic, potato, cotton, cabbage, cauliflower, tobacco, tomato, cucumber *etc.*

Nature and Symptoms of Damage

Adults as well as by nymphs lacerate the leaf tissue and feed on the plant juice. The insects are just visible to the unaided eye and are seen moving briskly on the flowers and leaves of onion and garlic plants. They usually congregate at the base of a leaf or in the flower. Leaves of attacked plants turn silvery white, curl, wrinkle and gradually dry from tip downwards. The plants do not form bulbs nor do the flowers set seed. Leaf tip discoloration and drying is the main symptom.

Economic Threshold Level (ETL)

5 thrips/leaf

Management

(A) Cultural Measures

1. Grow tolerant varieties *i.e.* PBR-2, PBR-6, Arka Niketan, Pusa Ratnar, PBR-4, PBR-5, PBR-6 and resistant varieties *viz.*, White Persian, Grano, Sweet Spanish and Crystal Wax.
2. Grow mustard crop as border crop.
3. Avoid excess nitrogen and close spacing to avoid thrips outbreak in garlic.
4. Destruction and burning of severly infested plants.
5. Seed treatment with Imidacloprid 70 WS @ 3 g/kg of seed.
6. Use neem coated urea to reduce the infestation of the pest.

(B) Mechanical Measures

1. Install yellow blue colour sticky traps@ 25/ha.

(C) Chemical Measures

1. Spraying of Fipronil 5 SC @ 1 ml/lt or Acephate 75 SP @ 1 g/litre of water.
2. Apply any one of following insecticides at 3 to 4 weeks interval starting from 2 to 3 weeks after transplanting: Quinolphos 25 EC @ 0.03 per cent or Malathion 50 EC @ 0.05 per cent or Fenthion @ 0.05 per cent or Fenitrothion @ 0.05 per cent.

(D) Biological Measures

1. Release parasitoid – *Ceronisus* sp.
2. Conserve predators like *Scymnus nubilis, Orius albidipenis, Chrysopa* sp. and predatory thrips *viz., Aelothrips collartris.*

2. Onion Caterpillar

Also known as: Tobacco leaf eating caterpillar/Armyworm.

Scientific Name, Order and Family

Spodoptera litura

(Lepidoptera; Noctuidae)

Taxonomical Position/Scientific Classification

Kingdom:	Animalia
Phylum:	Arthropoda
Class:	Insecta
Order:	Lepidoptera

Family:	Noctuidae
Genus:	Spodoptera
Species:	***S. litura***

Biology and Life Cycle

1. **Egg**: Female lays about 500 to 2000 eggs in cluster in parallel rows on tender/central leaves. The eggs are covered over by brown hairs. The eggs hatch in about 4-5 days.
2. **Larva**: Caterpillar measures 35-40 mm in length, when full grown. It is velvety, black with yellowish – green dorsal stripes and lateral white bands with incomplete ring – like dark band on anterior and posterior end of the body. It passes through 6 instars. Larval stage lasts for 15-30 days. Full-grown larvae drop down and pupate in soil at a depth of 0.5 to 6 cm.
3. **Pupa**: The pupae are reddish brown and are found in the soil close to the plant. Pest usually hibernates as pupa in the soil. Pupal period is 8 to 10 days.
4. **Adult**: Tobacco bud worm moths are light brown with a wing span of about 30 mm and mottled forewings. Adults live for 7-10 days. The total period of their life-cycle is 30 to 40 days. There are several generations in one year.

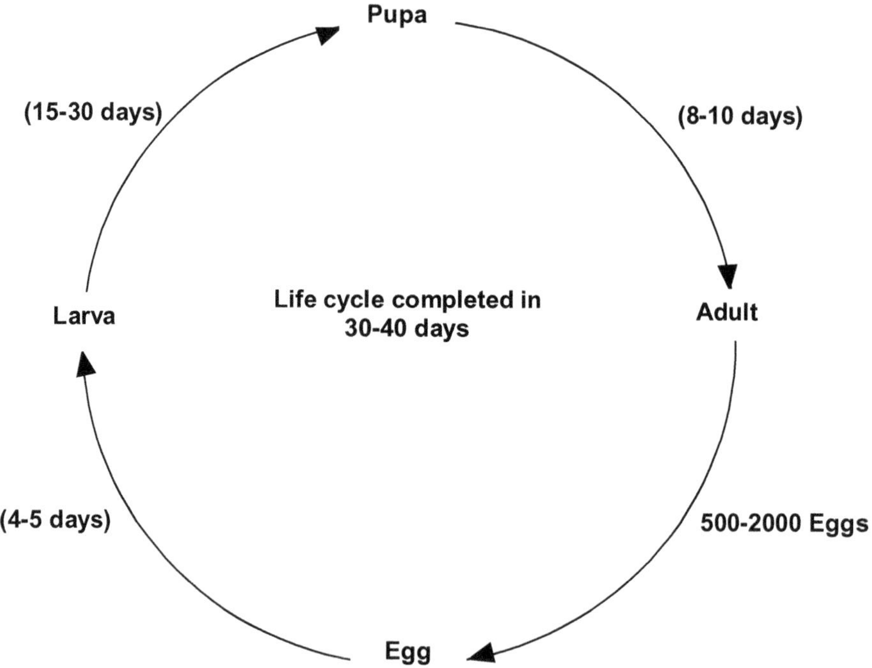

Figure 53: Life Cycle of *Spodoptera litura*.

Marks of Identification

Adult is medium sized moth with pale grey or dark brown forewings having weavy white markings. Hind wings are whitish with brown patches along the margin of wing. The caterpillars are pale greenish brown with dark head and prothoracic shield. The full-grown caterpillars measure about 37.5 mm in length. The pupae are dark brown and eggs are round and greenish white in colour. Moths are active at night.

Host Crops

Tobacco, onion, pea, brinjal, castor, banana are the main hosts.

Nature and Symptoms of Damage

- ☆ Pest is nocturnal in habit, during daytime caterpillars hide in leaf whorl or in soil.
- ☆ In early stages, the caterpillars are gregarious on tender leaves and scrape the chlorophyll content of leaf lamina at night, giving it a papery white appearance.
- ☆ Later they become voracious feeders making irregular holes on the leaves and finally leaving only veins and petioles.
- ☆ In case of severe infestation, only mid-ribs and leafstalks are left over. The infested field looks like as if it is grazed by cattle.
- ☆ During flowering and boll formation stage, the caterpillars also feed on the internal contents of bolls causing irregular holes.
- ☆ The pest usually breaks out in serious proportion particularly when there is good monsoon rain followed by long dry spell.
- ☆ The pest remains active from June to November.

Economic Threshold Level (ETL)

1 caterpillar/hill

Management

(A) Cultural Measures

1. Crop rotation for two or more years.
2. Collection and destruction of the infested material from the field.
3. Deep ploughing of infested fields after harvest for exposing and killing hibernating pupae.
4. Clean cultivation in field and border and optimum plant population per unit area help in reducing pest infestation.
5. Grow maize crop as border crop around the main crop.

(B) Mechanical Measures

1. Hand collection and destruction of egg masses and caterpillars in early stage of infestation helps in reducing pest infestation.
2. Light traps may be set up to destroy adult moths.
3. Setting up of pheromone traps @ 12/ha.
4. Survey the field thoroughly and remove plant parts or plants having gregarious stage of caterpillars.

(C) Chemical Measures

1. Apply Carbofuron 3G @ 25 kg/ha or Phorate 10G @ 10 kg/ha to soil followed by light irrigation at initial stage.
2. At the early stage of infestation, dusting with 10 per cent Carbaryl @ 20 to 25 kg/ha or spraying with 0.0075 per cent Cypermethrin controls the pest effectively.
3. As soon as pest appears apply: Cypermethrin 0.006 per cent or Quinalphos 0.05 per cent.
4. Foliar application of novaluron 10 EC @ 0.75 ml/l, chitin synthesis inhibitor against eggs of *S. litura*.
5. Baiting with rice bran 10kg + jaggery 2 kg+ Chlorpyriphos 750 ml or Thiodicarb 300g in sufficient quantity of water in form of small balls and broadcasting in evening hours in one acre.
6. Spray extract of custard apple as feeding deterrent against the pest.

(D) Biological Measures

1. Spraying NPV @ 200LE/ac in combination with jaggery 1 kg, sandovit 100 ml or Robin Blue 50 g thrice at 10-15 days interval on observing the eggs or first instar larvae in the evening hours.
2. Release of egg parasitoid *Trichogramma* @ 50,000/ha/week four times.

3. Onion Fly

Also known as: Onion Maggot.

Scientific Name, Order and Family

Delia (Hylemya) antique

Diptera, Anthomyiidae

Taxonomical Position/Scientific Classification

Kingdom:	Animalia
Phylum:	Arthropoda
Class:	Insecta

Order:	Diptera
Family:	Anthomyiidae
Genus:	Delia
Species:	***D. antiqua***

Biology and Life Cycle

1. **Egg**: The female lays elongate, white eggs near the base of the plant, in cracks in the soil. The eggs hatch in 2-7 days.
2. **Larva**: After hatching the maggots crawl up, enter the leaf sheath and reach the bulb. They feed there and become full-grown in 2-3 weeks.
3. **Pupa**: The maggots then crawl out of the bulb and pupate in the soil.
4. **Adult**: The flies are slender, greyish, large-winged. The maggots are small, white and about 8 m. in length. After 2-3 weeks, the adults emerge and start the new generation.

Marks of Identification

The onion fly has an ash-grey body and resembles a housefly. The male has a longitudinal stripe on the abdomen which is lacking in the female. The legs are black, the wings transparent, and the compound eyes brown. The eggs are white and elongated and are laid in groups on the shoots, leaves and bulbs of host plants and on the ground nearby. The larvae are white and cylindrical. The larvae moult three times, feed for about twenty days, and grow to about one centimetre long. The pupa is brown, ringed and ovoid and measures 7 millimetres (0.28 in) long.

Host Crops

The larvae or maggots feed on onions, garlic and other bulbous plants.

Nature and Symptoms of Damage

The larvae damage bulb onions, garlic, chives, shallots, leeks and the bulbs of flowering plants. The first generation of larvae is the most harmful because it extends over a long period owing to the females' longevity and occurs when the host plants are small. Seedlings of onion and leek can be severely affected as can thinned-out onions and shallots. Less damage occurs in wet and cold springs as this delays the development of the larvae. When plants are attacked, the leaves start to turn yellow and the bulbs rot quickly, especially in damp conditions.

Economic Threshold Level (ETL)

1 maggot/hill.

Management

(A) Cultural Measures

1. Sanitation is a important component of onion larva (maggot) management. Damaged bulb left in the field are an important food source for overwintering population therefore all onions are removed from the field.

2. Avoid mechanical injury, bulb should be made to minimize damage at harvest as these injured bulbs are very suitable for onion maggot development.
3. Crop density affects onion maggot damage, high density cropping onion maggot of injury over more.
4. Adopt proper crop rotation, which reduce onion maggot population and the resulting damage by 'escaping the infestation in space'.
5. Avoiding planting in soils that are high in undecomposed organic matter such as weedy situation.
6. Delay planting time, early spring planted crops is more likely to be damaged when the soil is too cool for rapid germination and emergence.

(B) Mechanical Measures

1. Collection and destruction of infested onion bulbs.
2. Using insecticide (spinosad) bait devices (Delia lure), to attract and kill the adult flies.

(C) Chemical Measures

1. Apply 10 kg of carbaryl 4G or phorate lOG to the soil followed by light irrigation.
2. Spray methyl demeton 25 EC or dimethaote 30 EC 1.0 L in 500 – 750 L of water per ha.

(D) Biological Measures

1. Entomophagous pathogen: fungus, *Entomophthora muscae* can infect and kill large number if adults.
2. Predators: carabid beetle; *Bembidion quadrimaculatum*, staphylinid beetle; *Alecochara bilineata.*
3. Parasitoid: *Bracon* sp., *Aphaereta pollipes.*
4. Nematode: *Heterorhabditis bacteriophora.*

(F) PESTS OF SWEET POTATO

1. Weevil

Scientific Name, Order and Family

Cylas formicarius Fab.

(Coleoptera; Curculionidae)

Taxonomical Position/Scientific Classification

Kingdom:	Animalia
Phylum:	Arthropoda
Class:	Insecta
Order:	Coleoptera
Family:	Curculionidae
Genus:	Cylas
Species:	*C. formicarius*

Biology and Life Cycle

1. **Egg**: Female bite small cavities on vines and tubers, make holes and deposit about 100-200 eggs singly in holes (one per hole). The holes are covered by a grayish mass that hardens to form protective caps over developing eggs. Hatching period is 5 to 7 days.
2. **Larva**: Grubs are apodus, pale-yellowish white in colour. Grub becomes fully developed for 2 to 3 weeks.
3. **Pupa**: Pupation takes place inside infested vines and tubers. Pupal period lasts for 7 days.
4. **Adult**: Adult weevils are ant-like, slender bodied having elongated snout-like bluish-brown head with non-geniculate antenna, bright red thorax

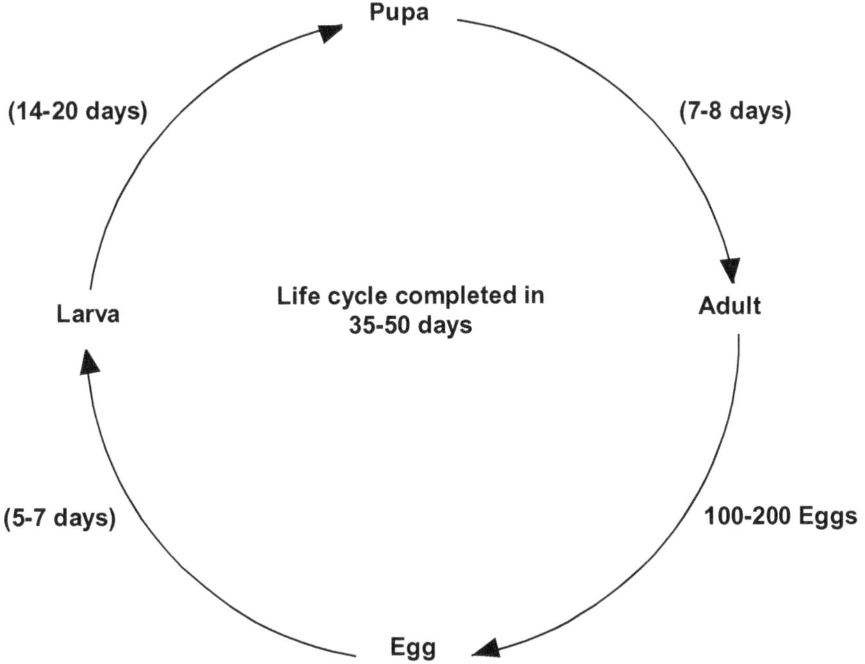

Figure 54: Life Cycle of *Cylas formicarius*.

and legs and brownish-red abdomen. Infestation of pest is carried from one field to another field through infested vines and from one season to another season by breeding in the tubers left over in the field after harvest. One life cycle occupies 5 to 7 weeks.

Marks of Identification

Adult is small, 6 to 8 mm long, wingless, steel black coloured weevil with a brown elongated snout like head. The thorax and legs are bright and abdomen is red. The grubs are 8 to 10 mm long, pale yellow in colour and apodous. Adults look like ant, hence the species name "*formicarius*".

Host Crops

Sweet potato only (monophagous).

Nature and Symptoms of Damage

- ☆ This is important pest of sweet potato both in the field and in storage.
- ☆ The grubs bore into the tubers and feed on the pulp making them unfit for human consumption. The grubs may induce terpenoid production inside the tuber.
- ☆ The tunnels frequently contain larvae, pupae, or newly transformed adults. Adult exit holes in the root are about the size of a match stick.
- ☆ The grubs also bore into the vines by making a hole at collar region. As a result, the thickness of vine is enlarged forming gall like structure at the point of pest infestation.
- ☆ The adults feed on leaves, vines and tubers but cause negligible damage.
- ☆ Loss in yield to the extent of 60 to100 per cent has been reported.

Management

When sweetpotato weevil populations are high, no single control method provides adequate protection. The integration of different techniques, with emphasis on the prevention of infestation, provides sustainable protection.

(A) Cultural Measures

1. Collect and destroy all plant debris or all remains of the harvested crop.
2. Use healthy cuttings for planting, free from the attack of sweet potato weevil.
3. Use transplants or slips that are cut an inch above the soil line, or use vine cuttings rather than pulled plants.
4. Destroy excess plants and seed potatoes in the bed once transplanting has been completed.
5. Follow proper crop rotation.
6. Dip the cuttings in insecticidal suspension of 0.1 per cent carbaryl for 10 minutes at the time of planting.

7. Mulch the field with paddy straw or black plastic cover.
8. Timely earthing up of crop at 30 and 60 days after planting, irrigate the field regularly to fill up soil cracks.
9. Store sweet potatoes by mixing them with sand.

(B) Mechanical Measures

1. Install sex pheromone trap in the field @ 1 trap/100 m^2 area.

(C) Chemical Measures

1. Give three applications of 0.05 per cent Malathion or 0.1 per cent Fenthion or 0.1 per cent Carbaryl at triweekly interval starting from 1½ month after planting.
2. Apply Carboryl 10 D @ 25 kg/ha in soil on the vine.
3. Apply Phorate 10 G @ 10 kg/ha in soil at the time of planting and 45 days thereafter.

(D) Biological Measures

1. **Microbial control:** Promising biological control agents for sweetpotato weevils appear to be the fungi *B. bassiana* and *Metarrhizium anisopliae* and the nematodes *Heterorhabditis* spp. and *Steinernema* spp.The fungi attack and kill adult weevils, whereas the nematodes kill the larvae.
2. **Predators:** Ants, spiders, carabids and earwigs are important generalist predators that attack weevils.They are described more fully in the section on "Natural Enemies."

(G) PESTS OF GINGER

1. Shoot Borer

Also known as: Rhizome borer

Scientific Name, Order and Family

Dichocrocis punctiferalis (Guenee)

(Lepidoptera; Pyralidae)

Taxonomical Position/Scientific Classification

Kingdom:	Animalia
Phylum:	Arthropoda
Class:	Insecta
Order:	Lepidoptera
Family:	Pyralidae
Genus:	Dichocrocis
Species:	***D. punctiferalis***

Biology and Life Cycle

1. **Egg**: The female pink, oval, flat eggs laid singly or in groups on leaves and other soft and tender parts of the plants. Hatching period is 2 to 6 days.
2. **Larva**: The larvae are light brown with sparse hairs. Grub becomes fully developed for 12-16 days with 5 instars.
3. **Pupa**: Pupation takes place inside the seed or sometimes in grass with loose silken cocoon. Pupal period lasts for 7-10 days.
4. **Adult**: The adult is medium sized moth, the wings are orange yellow with minute black spots. There are 6-7 generations in a year.

Marks of Identification

The adult is a medium sized moth with a wingspan of about 20 mm; the wings are orange-yellow with minute black spots. Fully-grown larvae are light brown with sparse hairs. The pest population is higher in the field during September-October.

Host Crops

Ginger

Nature and Symptoms of Damage

The larvae bore into pseudostems and feed on internal tissues resulting in yellowing and drying of leaves of infested pseudostems. The presence of a bore-hole on the pseudostem through which frass is extruded and the withered and yellow central shoot is a characteristic symptom of pest infestation.

Economic Threshold Level (ETL)

1 egg mass/m^2

Management

(A) Cultural Measures

1. Pruning and destroy freshly infested pseudostems or shoots.
2. Mulching with green leaves @ 4-4.5 ton/acre at the time of sowing. Use of *Lantana camera* and *Vitex negundo* as mulch at the time of planting may reduce the infection of shoot borer.
3. Use of attractant plants for natural enemy's conservation like carrot family, sunflower family, buckwheat, corn, shrubs *etc.*

(B) Mechanical Measures

1. Install light trap for collecting and monitoring shoot borer adults if such infestation is observed.
2. The shoots infested by borer are cut open and the caterpillar are handpicked and destroy.

(C) Chemical Measures

1. The shoot borer can be managed by spraying malathion (0.1 per cent) at 21 day intervals during July to October. The spraying is to be initiated when the first symptom of pest attack is seen on the top most leaf in the form of feeding marks on the margins on the pseudostem. An integrated strategy involving pruning and destroying freshly infested pseudostems during July-August (at fortnightly intervals) and spraying malathion (0.1 per cent) during September-October (at monthly intervals) is also effective against the pest.

(D) Biological Measures

1. Parasitoids: *Xanthomonas autralis, Apenteles taragamae, Trichogramma* sp., *Myosoma* sp., *Bracon* sp.
2. Entomophagous nematodes: *Rhabditis/Oscheius* as biopesticides for management of the shoot borer.

2. Fly Maggot

Also known as: Rhizome fly

Scientific Name, Order and Family

Mimegralla coeruleifrons Macq.

(Diptera; Micropisidae)

Taxonomical Position/Scientific Classification

Kingdom:	Animalia
Phylum:	Arthropoda
Class:	Insecta
Order:	Diptera
Family:	Micropisidae
Genus:	Mimegralla
Species:	***M. coeruleifrons***

Biology and Life Cycle

1. **Egg**: The whitish eggs are laid on the lower surface of leaves or on exposed rhizomes. The hatching period is completed in 4 days.
2. **Larva**: The maggots pupate in rotten rhizomes.
3. **Pupa**: Pest hibernates in pupal stage.
4. **Adult**: Flies are noticed in the field in the month of August and September. The adult has longevity of 18 days. The fly completes its life cycle in 38 to 62 days. The total number of generations completed in a season is 2 to 3.

Marks of Identification

Adults are fairly large flies with slender body and long legs. The body is black and wings are transparent with ashy spots. Wing expanse of adult is 13 to 15 mm. Maggots are creamy white, apodous, 9.5 mm in length and 1.95 mm in breadth. Eggs are white cigar-shaped tapering at either side.

Host Crops

Ginger and turmeric.

Nature and Symptoms of Damage

The maggots on hatching bore into the rhizomes of ginger and turmeric and damage them. The yellowing of plants and rotting of rhizomes takes place due to severe infestation of pest. 'Dead hearts' are also seen due to primary injury of the maggots. The maggots pave way for the invasion by the fungus (*Pythium* spp., *Sclerotium* spp.) and the nematodes (*Meloidogyne* spp.).

Management

(A) Cultural Measures

1. Select healthy rhizomes for planting.
2. Apply insecticides three times at monthly interval by broadcasting them at the base of the plants followed by light irrigation.
3. Remove and destroy the rotten rhizomes from the field along with maggots after harvest of crop to check the breeding of the pest.
4. Destroy stray plants in off season.

(B) Chemical Measures

1. Treat the rhizome sets by dipping in carbaryl 50 WP (0.4 per cent) suspension for 10 minutes before planting.
2. Apply following insecticides three times at monthly interval by broadcasting them at the base of the plants followed by light irrigation: Phorate 10 G @ 20 kg/ha or Carbaryl 10 D @ 20 kg/ha.
3. Spray dimethoate 500 ml in 500 -750 L water per ha.
4. Soak seed rhizomes, in insecticide solution of either dimethoate 30 EC or phosalone 1.5 ml/L or dichlorvos 0.5 ml/L for 15 min. for storing

(C) Biological Measures

1. **Parasitoids:** Trichogrammatids, *Aracerus fasciculatus*, *Trichopria* sp.

3. Rhizome Scale

Also known as: Yam Rhizome scale

Scientific Name, Order and Family

Aspidiella hartii

(Hemiptera; Coccidae)

Taxonomical Position/Scientific Classification

Kingdom:	Animalia
Phylum:	Arthropoda
Class:	Insecta
Order:	Hemiptera
Family:	Coccidae
Genus:	Aspidiella
Species:	***A. hartii***

Biology and Life Cycle

Scales are minute, circular, light brownish to grey with a thin pale membrane. It reproduces either ovoviparously or parthenogenetically. Female lays about 100 oval, yellowish eggs under the scale. Egg period one day, nymphal period 30 days. Adult is yellow to deep yellow in colour.

Marks of Identification

Adult (female) scales are circular (about 1 mm diameter) and light brown to grey and appear as encrustations on the rhizomes

Host Crops

Ginger and turmeric.

Nature and Symptoms of Damage

The rhizome scale infests rhizomes in the field (at later stages) and in storage. The infested plants become weak, pale and withered in the field that results in shrivelling of rhizomes and buds. They feed on sap and when the rhizomes are severely infested, they become shriveled and desiccated affecting its germination.

Management

1. Apply well rotten sheep manure/poultry manure in two splits @ 10 tons/ha, first before planting and the second at the time of earthing up.
2. Drench soil with dimethoate 30 EC or phosalone 35 EC @ 2 ml/L of water.
3. The rhizome scale can be managed by timely harvest, discarding severely infested rhizomes, and treating the seed rhizomes with quinalphos (0.075 per cent) (for 20-30 minutes) before storage and also before sowing in case the infestation persists. The seed rhizome may be stored in sawdust + Strychnos nuxvomica leaves (dried) after seed treatment.

4. Soak seed rhizomes, in insecticide solution of either dimethoate 30 EC or phosalone 1.5 ml/L or monocrotophos 36 WSC 1.5 ml/L or dichlorvos 0.5 ml/L for 15 min. for storing.

(H) PESTS OF CORIANDER

1. Aphid

Scientific Name, Order and Family

Hyadaphis coriandri (Das)

(Hemiptera, Aphididae)

Taxonomical Position/Scientific Classification

Kingdom:	Animalia
Phylum:	Arthropoda
Class:	Insecta
Order:	Hemiptera
Family:	Aphididae
Genus:	Hyadaphis
Species:	***H. coriandri***

Biology and Life Cycle

Aphids breed throughout the year by parthenogenetic and viviparous reproduction. About 40-50 nymphs are produce by a single female. All young ones are viviparous female. Winged form are seen. Nymphal period is of about 14-21 days in summer and of 45 days in winter.

Marks of Identification

Nymphs and adults are pear shaped, light green in colour but looks like bluish white due to powdery substance. They have short, dusky, slightly swollen, siphunculi (or cornicles) that are about twice as long as wide. They form dense and often damaging colonies on leaves, heads, and stems of their host plants. Nymphs are upto 11 mm in length where as adults are 12-14 mm in length.

Host Crops

It is a pest of Umbelliferous plant *viz.*, coriander, celery, parsley and carrot.

Nature and Symptoms of Damage

Both nymph and adult suck the cell sap from leaves, flowers and fruits and excuding copious quality of honeydew which favours the growth of shooty mould. In case of severe infestation at early stage of crop, leaves and growing point withers and the plant become dries up. Due to their fast multiplication in few days, aphids cover the entire surface of apical shoots and as a result of continuous feeding by such a large population, yellowing, curling and subsequent drying of leaves takes place, resulting in poor and shriveled seed formation.

Economic Threshold Level (ETL)

1 aphid index/plant

Management

(A) Cultural Measures

1. Grow tolerant varieties of coriander *i.e.* Panipat local, *CS-193 etc.*
2. Avoid late sowing and adopt early flowering types.
3. Clip off the infested plant parts bearing aphid colonies and destroy them.
4. Dust the canopy with cowdung ash early in the morning.

(B) Chemical Measures

1. Spraying any one of insecticide at flowering stage: Oxydemeton methyl 25 EC @ 0.05 per cent or Chloropyriphos 20 EC @ 0.02 per cent or Dichlorvas 76 EC @ 0.05 per cent or Methyl demeton 0.02 per cent.
2. Spray foliage with 2 per cent neem oil suspension. Apply fungal pathogen, *Beauveria bassiana* under humid condition.
3. Avoid use of toxic pesticides on coriander grown under kitchen garden.
4. Harvest the plants at least seven days after the application of insecticides.

(C) Biological Measures

1. Release predator – *Menochilus sexmacullatus* and *Coccinella septempunctata.*

2. White Fly

Scientific Name, Order and Family

Bemisia tabaci

(Hemiptera; Aleyrodidae)

Taxonomical Position/Scientific Classification

Kingdom:	Animalia
Phylum:	Arthropoda
Class:	Insecta
Order:	Hemiptera
Family:	Aleyrodidae
Genus:	Bemisia
Species:	***B. tabaci***

Biology and Life Cycle

1. **Egg:** The females mostly lay eggs near the veins on the underside of leaves. They prefer hairy leaf surfaces to lay more eggs. Each female can lay about 300 eggs in its lifetime. Egg period is about 3-5 days during summer and 5 to 33 days in winter.
2. **Nymph:** Upon hatching, the first instar larva (nymph) moves on the leaf surface to locate a suitable feeding site. Hence, it is commonly known as a "crawler." It then inserts its piercing and sucking mouthpart and begins sucking the plant sap from the phloem. Nymphal period is about 9 to 14 days during summer and 17 to 73 days in winter. Adults emerge from puparia through a T-shaped slit, leaving behind empty pupal cases or exuviae.
3. **Adult:** The adult males are slightly smaller in size than the females. Adults live from 15-30 days.

   Egg to Nymph – 3-30 days

   Nymph to Pre-pupa – 9-19 days

   Pupa to Adult – 2-8 days

Marks of Identification

1. **Egg:** Eggs are small (about 0.25 mm), pear-shaped, and vertically attached to the leaf surface through a pedicel. Newly laid eggs are white and later turn brown. The eggs are not visible to the naked eye, and must be observed under a magnifying lens or microscope.
2. **Nymph:** The first instar nymph has antennae, eyes, and three pairs of well-developed legs. The nymphs are flattened, oval-shaped, and greenish-yellow in color. The legs and antennae are atrophied during the next three instars and they are immobile during the remaining nymphal stages. The last nymphal stage has red eyes. This stage is sometimes referred to puparium, although insects of this order (Hemiptera) do not have a perfect pupal stage (incomplete metamorphosis).
3. **Adult:** The whitefly adult is a soft-bodied, moth-like fly. The wings are covered with powdery wax and the body is light yellow in color. The wings are held over the body like a tent.

Host Crops

Coriander, cotton, okra *etc.*

Nature and Symptoms of Damage

Both nymphs and adults suck sap from lower side of leaves resulting in:

- ✰ Chlorotic spots which later coalesce forming irregular yellowing of leaves which extends from veins to outer edges.

- ☆ Leaves mottled and yellowish in colour
- ☆ The vegetative growth retarded and seed formation seriously hampered.

Management

(A) Cultural Measures

1. Avoid the alternate or cultivated host crops of the white fly.
2. Adopting crop rotation with non-preferred hosts such as sorghum, ragi, maize *etc.*, for the white fly to check the build up of the pest.
3. Timely sowing with recommended spacing.
4. Field sanitation may be given proper attention.

(B) Mechanical Measures

1. Monitoring the activities of the adult white flies by setting up yellow pan traps and sticky traps.

(C) Chemical Measures

Chemical control same as under cotton leaf hopper.

1. Spray of any one of the following insecticides: Quinolphos 25 EC @ 0.06 per cent or Chloropyriphos 20 EC @ 0.06 per cent or Methyl demeton 25 EC 500 ml/ha or Dimethoate 30 EC 500 ml/ha or Malathion 50 EC @ 0.06 per cent or Fenvelrate 20 EC @ 0.016 per cent.
2. Spraying of Imidacloprid 17.8 SL @ 0.3 ml/litre of water.
3. Acetamiprid(20 per cent SP 100 g/ha or Azadirachtin 0.15 per cent 500-1000 ml/ha or Buprofezin 25 per cent SC 1000 ml/ha or Clothianidin 50 per cent WDG 40-50 g/ha or Diafenthiuron 50 per cent WP 600 g/ha or Fipronil 5 per cent SC 250-340 ml/ha or Profenofos 50 per cent EC 1000 ml/ha or Thiacloprid 21.7 per cent SC 500-600 ml/ha or Thiamethoxam 25 per cent WG 100 g/ha or Triazophos 40 per cent EC 1500-2000 ml/ha.

(D) Biological Measures

1. *Verticillium lecanii* 1.15 per cent WP @ 2500 g/ha.
2. *Eretrmocerus mundus* (Mercet) and *Enacarsia* sp. are special parasites of *B. tabaci*. Also, the predators *Amblyseius* sp. (predatary mite), the green lacewing bug (*Chrysoperala* sp.) the coccinellids *Brumus* sp., *Scymnus* sp. and *Menochilus* sp. play important role in reducing whitefly population.
3. A chalcid parasite attacks the older nymphs and the parasitisation is at times more than 30 per cent. Also, there are a few predators like some species of *Chrysopa* and *coccinellids*, which feed on the whitefly stages.

3. Flower Stink Bug

Also known as: Green stink bug

Scientific Name, Order and Family

Nezara viridula (Linnaeus),

(Hemiptera, Pentatomidae)

Taxonomical Position/Scientific Classification

Kingdom:	Animalia
Phylum:	Arthropoda
Class:	Insecta
Order:	Hemiptera
Family:	Pentatomidae
Genus:	Nezara
Species:	***N. viridula***

Biology and Life Cycle

1. **Egg**: Females laid 200-300 barrel-shaped, yellow to green coloured eggs in clusters (about 36 eggs/cluster) primarily on leaves and stems but also on pods. Eggs measure 1.4 x 1.2 mm. the egg period lasts for 5-7 days.
2. **Nymph**: Green stink bug nymphs are predominantly black when small, but as they mature, they become green with orange and black markings. Nymphs hatch from these eggs and pass through five instars before becoming adults. Nymphal period lasts for 28-32 days.
3. **Adult.** All adult stink bugs are shield-shaped, bright green and measure 14.0 to 19.0 mm long. The major body regions of the green stink bug are bordered by a narrow, orange-yellow line.

Marks of Identification

The first instar nymphs are yellowish orange and slightly bigger than the eggs, shortly after emergence, they darken, remaining compactly clustered and motionless on the empty egg mass or adjacent to it. The second instar nymph has a bright red on the head, large orange thorax, black abdomen and black eyes. The colour, spots, and other markings remain essentially the same in the third instar. The colour pattern varies markedly during the fourth instar can be separated into light and dark.

The adult is a large green bug. It has the shield-shaped form characteristic of a pentatomid, and it has the appearance typical of a stink bug. The female is larger than the male, and the males can be differentiated from females by a notch and two brown spots on the ventral surface of the terminal end of the abdomen.

Host Crops

Polyphagous pest damaging so many crops like coriander, soybean, beans, tomato, cucmber and other host belonging to family cruciferae.

Nature and Symptoms of Damage

- ✰ In general, they prefer to suck on developing seeds within green pods.
- ✰ The plant compensates for lost seeds by setting new ones but these seeds remain small.
- ✰ In damaged seeds, germination is adversely affected.

Economic Threshold Level (ETL)

2 per cent infestation or 1 bug/m^2 or the maximum bug damage permitted being only 2 per cent or for *N. viridula* when the adults number ranges from 0.3-0.8/m^2.

Management

(A) Mechanical Measures

1. The pheromone methyl (E,Z,Z)-2,4,6-decatrienoate may be used to attract the bugs away from fields.

(B) Chemical Measures

1. Spraying of Dimethoate @ 0.03 per cent or Quinolphos 25 EC @ 0.05 per cent.

(C) Biological Measures

1. Two scelionid parasites, *Asolcus mitsukurii* and *Telenonus makagawai,* heavily infested eggs and were decisive in keeping the population of *N. viridula* from causing major losses.
2. A proctotrapid egg parasite, *Microphanurus megacephalus* (Ashmead), was the only biotic agent preventing stink bug from becoming a pest.
3. A reduvid bug, *Sycanus collaris* (F.) keeps *N. viridula* under control.
4. One hymenopterous parasite, *Ooencyrtus* sp., preying on the eggs of *N. viridula.*
5. *Trissolcus basalis* (Wollaston), and *Trichopoda giacomellii* (Blanchard) may control stink bugs.

(I) PESTS OF CUCURBITACEOUS VEGETABLES

1. Red Pumpkin Beetle

Scientific Name, Order and Family

Raphidopalpa foveicollis (Lucas)

Aulacophora foveicollis (Lucas)

(Coleoptera; Chrysomelidae)

Taxonomical Position/Scientific Classification

Kingdom:	Animalia
Phylum:	Arthropoda
Class:	Insecta
Order:	Coleoptera
Family:	Chrysomelidae
Genus:	Raphidopalpa
Species:	*R. foveicollis*

Biology and Life Cycle

1. **Egg:** A female lays 150 to 300 eggs singly or in clusters of 8-9 in moist soil around the host plants. Eggs are brownish and elongated. Hatching period is 5 to 8 days.
2. **Larva**: Grub is whitish creamy in colour. Grub development is completed in 13 to 25 days.
3. **Pupa**: Pupation takes place in soil at a depth of 150 to 250 mm. Pupal period is 9 to 22 days.
4. **Adult**: Beetle is reddish brown in colour. Adult longevity is about a month. 5 to 8 overlapping generations are completed in a year. Overwintering takes place in adult stage. Total life cycle is completed in 32 to 65 days.

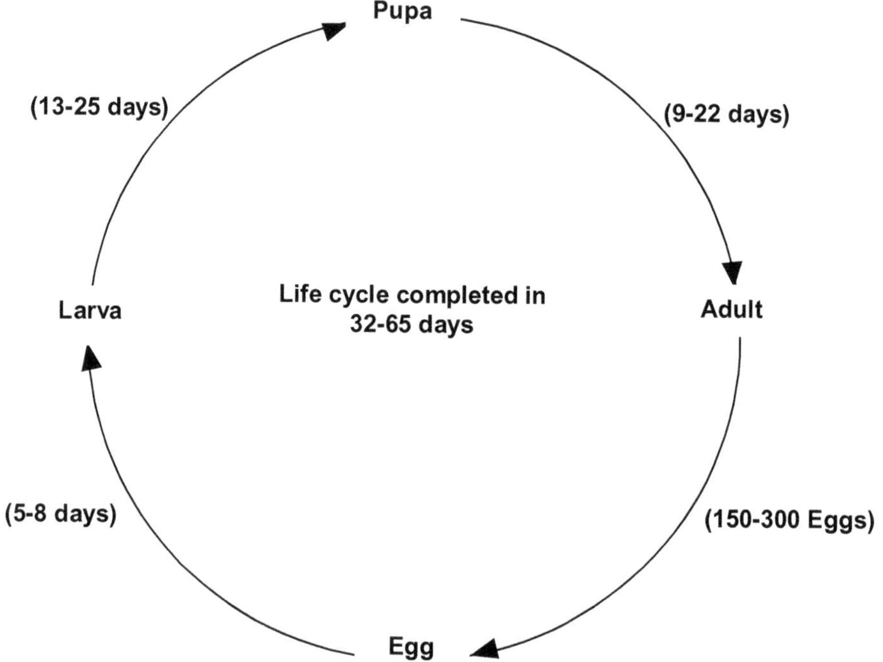

Figure 55: Life Cycle of *Raphidopalpa foveicollis.*

Marks of Identification

Adults are small beetles, 6 to 9 mm long, brilliant red coloured dorsally and black ventrally. The grubs are creamy yellow with brown head and 10 to 14 mm long. Eggs are elongated, yellowish-brown initially but soon become orange in colour. Pupae are pale white and are found in earthen cells, 150 to 250 mm deep in the soil.

Host Crops

Cucurbitaceous vegetables *i.e.* bittuergaurd and pumpkin, melons, leguminous crops.

Nature and Symptoms of Damage

On hatching, grubs feed on the roots and underground portion of host plants as also fruits touching the soil. The damaged roots and infested underground portion of the stems start rotting due to secondary infection by saprophytic fungi and the unripe fruits of such vines dry up. Infested fruits become unfit for human consumption.

Adult beetles are mainly responsible for the damage of the plant above ground attacking on the flowers, leaves and fruits. Adult beetles feed voraciously on leaf lamina making irregular holes. They prefer young seedlings and tender leaves. Damage at that stage may even kill the seedlings. The beetles are active from March to October, though the peak period of activity is April to June.

Economic Threshold Level (ETL)

1/10 Plants (at seedling stage) or 1/Plants (at crop stage)

Management

(A) Cultural Measures

1. Collection and destruction of infested leaves and fruits.
2. Prunning of old creepers.
3. Early planting of pumpkin during October – November to avoid damage by this pest
4. After harvesting, deep ploughing of infested field to kill the grub in the soil.
5. Frequent raking of soil beneath the crop to expose and kill the eggs and grubs.
6. Dusting the plant with ash temporarily repel the beetle.

(B) Mechanical Measures

1. Collection and destruction of beetles in early stage of infestation.

(C) Chemical Measures

1. Spraying with Dichlorvos (DDVP) or 0.2 per cent Carbaryl or 0.05 per cent Malathion has been suggested.

2. Spray malathion 50 EC 750 ml, dimethoate 30 EC 500 ml, methyl demeton 25 EC 500 ml, 500 g of carbaryl 50WP in 500-750 L of water per ha or apply 7.0 kg of carbofuran. 3G per ha 3-4 cm deep in the soil near the base of the plants just after germination and irrigate.

2. Fruit Fly

Scientific Name, Order and Family

Bactrocera cucurbitae (Cog.)

(Diptera, Tephritidae)

Taxonomical Position/Scientific Classification

Kingdom:	Animalia
Phylum:	Arthropoda
Class:	Insecta
Order:	Diptera
Family:	Tephritidae
Genus:	Bactrocera
Species:	***B. cucurbitae***

Biology and Life Cycle

Maggots legless and appear as headless, dirty-white wriggling creatures, thicker at one end and tapering to a point at the other. The adult flies are reddish brown with lemon-yellow markings on the thorax. Adult flies emerge from pupae in the morning hours and mate at dusk. The female, on an average, lays 58-95 eggs in 14-54 days. Egg period 1-9 days, larval period 3 - 21 days. It pupates deep in the soil. The pupae are barrel-shaped, light brown, pupal period 6-30 days. There are several generations in a year.

Marks of Identification

Adult fly (5 mm) resembles common housefly but has yellow legs, conical brown abdomen and transparent outstretched wings having grey spots on them. Maggot is dirty white, 10 mm long and tapering anteriorly. Hyaline wings with costal band broad and prominent, anal stripes well developed and hind cross veins thickly margined with brown and grey spots at the apex

Host Crops

Cucurbits, papaya, mango, guava, pulm, peach, pear, fig, ber, apple, banana, pomegranate, citrus, *etc.*

Nature and Symptoms of Damage

☆ Adult fruit fly is not harmful. Maggots/Larval stage is are the damaging ones.

☆ The female flies puncture the soft and tender fruits with their stout and hard ovipositor and lay eggs below the epidermis. A puncture made by one female is often used by others also for ovipositing and a single fruit may have more than one puncture made by one or more females. On hatching the maggots feed inside on the pulp of fruits and the infested fruits can be identified by the presence of brown resinous juice which oozes out of the punctures made by the flies for oviposition. These punctures also serve as an entry for various bacteria and fungi; as a result, the infested fruits start rotting, get distorted and malformed in shape and fall off from the plants pre-maturely.

Management

(A) Cultural Measures

1. All the fallen and infested fruits should be collected and destroyed to prevent the carry over of the pest.
2. Fruits should be harvested before they start ripening.
3. Raking soil or ploughing after harvest to destroy hibernated pupae.
4. Mulching beneath the plant helps to kill the pupae inside the soil, if infestation is of severe nature.
5. To avoid infestation by fruit flies, growing of resistant or early maturing varieties has been recommended.

(B) Mechanical Measures

1. Use fly trap: Keep 5 g of wet fishmeal in plastic container with six holes (3 mm dia), 2 cm from the bottom of the bag. Add a drop (0.1 ml) of dichlorvos in cotton plug and keep it inside the bag. Dichlorvos should be added every week and fishmeal renewed once in 20 days (20 traps/ acre) or, Use fly traps having methyl eugenol soaked plywood piecies (2″ x 2″). Collect and destroy the flies.

(C) Chemical Measures

1. Spray Malathion 0.05 per cent or 0.2 per cent Carbaryl during flowering to reduce intensity of infestation.
2. The flies when they congregate and rest on the under surface of large leaves of ribbed gourd may be controlled by spray application of cypermethin 0.025 per cent.
3. Spray application of three to five rounds of profenofos 0.05 per cent or fenthion 0.1 per cent or carbaryl 0.1 per cent at intervals of 15 days commencing from flowering may be useful.
4. Citronella oil or Clensel or liquid ammonia or Vanilla or Ammonium sulphate or Ammonium carbonate or Ammonium nitrate or Molasses or Eugenol or Methyl eugenelor isoeugenol and fermented fruit juice mixed

with any poison will act as a chemical trap (10 ml of the chemical + 45 ml of water + Poison).

5. Spot spraying with liquid bait tartar emetic 28 ml + jaggery 700 g + water 9 litres.
6. Use of 0.4 ml methyl engenol with 1ml of dichlorvos in bait traps was also found effective.

(D) Biological Measures

1. Conserve pupal parasitoids *viz., Opius fletcheri, O.compensatus* and *O. insisus, Spalangia philippinensis* and *Pachycepoideus debrius, Dirhinus giffardi* and *D. lzonensis.*

3. Cucurbits Vine Borer

Also known as: Squash Vine borer or Cucurbit stem borer.

Scientific Name, Order and Family

Melitta curcurbitae (Harris)

(Lepidoptera; Sesiidae)

Taxonomical Position/Scientific Classification

Kingdom:	Animalia
Phylum:	Arthropoda
Class:	Insecta
Order:	Lepidoptera
Family:	Sesiidae
Genus:	Melittia
Species:	***M. cucurbitae***

Biology and Life Cycle

1. **Egg:** The eggs of the squash vine borer are laid singly on the lower part of the main stem of the host plant, as well as on the leaf stalks, leaves, and fruit buds. Eggs are dark to reddish brown, ovoid but slightly flattened, and small; they measure approximately 1 mm in length and 0.85 mm in width. Eggs hatch in 9 to 14 days.
2. **Larva:** The larvae are white, with a darkened head capsule, three pairs of thoracic legs, and five pairs of prolegs (peg-like legs on the abdomen of a larva). Newly emerged larvae are 1.5 to 2 mm long, tapered on the posterior end, and covered with numerous large hairs. As the larva matures, it develops a dark thoracic shield, loses its tapered shape and hairy appearance, and grows to a length of approximately 25 mm. Late instar larvae burrow into the soil before pupation. Larvae feed inside the stem for 4 to 6 weeks.

3. **Pupa:** The pupae of the squash vine borer are mahogany brown, approximately 14 mm long, and enclosed in silk cocoons.
4. **Adult:** The adult squash vine borer is thought to resemble a wasp. They are approximately 16 mm in length, with a wingspan range of 25 to 38 mm. The front wings are covered with scales that give them a metallic green to black sheen. Large portions of the hind wings lack scales, making their look clear. The abdomen is covered with conspicuous orange to reddish hairs, punctuated dorsally with black dots. The hind legs are covered with long black hairs inside and orange hairs outside. Females are larger and less colorful than males, with wider abdomens. Moths are active for about 30 days. There is usually one generation per year.

Marks of Identification

The adult borer resembles a wasp. It is about 1/2 inch long with an orange abdomen with black dots The first pair of wings is metallic green while the back pair of wings is clear, although that may be hard to see as the wings are folded behind them when they at rest. Eggs are flat, brown, and about 1/25 inch long. The larvae are white or cream-colored with brown heads, growing to almost an inch in length

Host Crops

It is a serious pest of vine crops, commonly attacking summer squash, winter squash, and pumpkins. Cucumbers and melons are less frequently affected.

Nature and Symptoms of Damage

- Often the first symptom of a borer attack is wilting of affected plants.
- Larvae bore within stems, usually in the lower 30 cm (one foot) of the stem. When gourds are grown on a trellis, borers can be found at any stem node, even several feet above ground. Stems can be girdled by borers, which prevent water and nutrients from circulating in the plant.
- The point where a borer enters a stem is marked by a hole with yellow granular or sawdust-like frass exuding from it.
- Injured vines often decay and become wet and shiny.
- Over 100 larvae have been found in a single vining plant in plants with multiple stems, but one larva per plant is most common.

Management

(A) Cultural Measures

1. Destroy vines soon after harvest to destroy any larvae still inside stems.
2. Till or disk the soil in fall or spring to destroy overwintering cocoons.
3. Cover vines at leaf joints with moist soil to promote formation of secondary roots that will support the plant if the main root and stem are injured.

4. A trap crop of very early-planted Hubbard squash can be used to alleviate pest pressure on other cucurbits.
5. Promptly pull and destroy any plants killed by squash vine borers.
6. Don't use row covers if cucurbits were planted in the same area the previous year. This is because squash vine borers overwinter in the soil near their host plants. When the adults emerge the following summer, they may end being trapped under the row cover instead of being kept out. Practice rotation to minimize this issue by planting cucurbits in different areas of your garden (if possible) or alternate seasons when you grow cucurbits.

(B) Mechanical Measures

1. Crush the eggs by hand before they hatch.
2. Stems can be covered with a barrier such as strips of nylon to prevent egg laying.
3. Borers can be removed from vines if detected before much damage is done. Examine stems in early summer; once holes are detected, slit the stem longitudinally with a fine, sharp knife, remove the borer, then cover the wounded stem with moist soil above the point of injury to promote additional root formation.
4. Catch and destroy the moths, especially at twilight or in early morning when they are resting on the upper side of leaf bases.

(C) Chemical Measures

1. If insecticides are needed, spray or dust the stems at their base. Start treatments when vines begin to run (or the last week of June or early July for non-vining varieties) or when the first adult borers are detected. Repeat in 7-10 days. Two applications help manage most squash vine borer adults. Spraying the crop with carboryl or permethrin or bifenthrin or esfenvalerate insecticide.

(D) Biological Measures

1. The stage most susceptible to natural enemies is the egg stage, which is attacked by parasitic wasps. Larval and adult ground beetles can attack larvae of squash vine borer, but do not appear to cause significant mortality.

(J) PESTS OF CRUCIFEROUS VEGETABLES

1. Cabbage Semilooper

Scientific Name, Order and Family

Trichoplusia ni

(Lepidoptera, Noctuidae)

Taxonomical Position/Scientific Classification

Kingdom:	Animalia
Phylum:	Arthropoda
Class:	Insecta
Order:	Lepidoptera
Family:	Noctuidae
Genus:	Trichoplusia
Species:	*T. ni*

Biology and Life Cycle

1. **Egg:** Eggs are greenish white, spherical and sculptured and are laid singly on ventral surface of leaves. They hatch in 3 to 5 days.
2. **Larva:** Mature larvae range in size from 1 to 1 1/2 inches long and eat ragged- edged holes through the leaves or in from the leaf margins. The full development of larvae completed in about 14-15 days.
3. **Pupa:** Pupation takes place in thintransparent cocoons on ventral surface of leaves.
4. **Adult:** Adults are stout moths. Head and thorax grey in colour, while abdomen is white with basal tuft of hairs. Life cycle occupies about 30 days. Moths are very active at dusk.

Marks of Identification

☆ **Larvae:** plump and pale green having three pairs of prolegs and are generally found mixed with the caterpillars of *P. brassicae*

☆ **Adults:** light brown with a golden patch on each fore wing and measures about 42 mm across the wings.

☆ Loopers are distinguished from cutworms and armyworms by their "looping" behavior when moving and by having only two pairs of abdominal prolegs (cutworms and armyworms have four pairs).

Host Crops

Its is a polyphagous pest infested on onion, groundnut, sugarbeet, cabbages, cauliflowers, pigeon pea, safflower, soybean, sunflower, raddish, tobacco, potato, urd, mung, cowpea *etc.*

Nature and Symptoms of Damage

☆ The pest is active from March to October.

☆ Larvae of semilooper act as a defoliator, cause the damage by biting round holes into the leaves.

- ☆ Newly hatched caterpillars tend to feed on the lower surfaces of leaves which they scarify being unable to eat through the lamina. This results in the appearance of 'windows', particularly on the top canopy leaves.
- ☆ Older caterpillars feed on the entire leaf tissue between the veins, giving a very lacey appearance.

Economic Threshold Level (ETL)

20 larvae/10 plants or 5 per cent defoliation.

Management

(A) Cultural Measures

1. Early maturity varieties are less subject to attach than late maturity varieties.
2. The use of transplants that are free of larval contamination is a key step in avoiding damage.
3. Floating row covers can provide a physical barrier to cabbage loopers from depositing eggs on plants.

(B) Mechanical Measures

1. Hand picking and destroy caterpillars.
2. Use light trap to attract and kill adults.

(C) Chemical Measures

1. Spray quinolphos 0.5 per cent or malathion 0.1 per cent.

(D) Biological Measures

1. Egg parasitoids: *Trichogramma* sp.
2. Larval parasitoids: *Voria ruralis, Microplitis brassicae, Copidosoma truncatellum.*
3. Predators: *Orius tristicolor* (Pirate bug), *Nabis americoferis* (Damsel bug).
4. Entomophagous fungus: *Bacillus thuringiensis* var. *Kurstaki* has long been used for effective suppression of cabbage semilooper.

2. Cabbage Diamond Back Moth

Scientific Name, Order and Family

Plutella xylostella (Linn.)

Lepidoptera; Plutellidae

Taxonomical Position/Scientific Classification

Kingdom:	Animalia
Phylum:	Arthropoda
Class:	Insecta
Order:	Lepidoptera
Family:	Plutellidae
Genus:	Plutella
Species:	***P. xylostella***

Biology and Life Cycle

1. **Egg**: Female lays yellowish pin head size 40 to 60 eggs singly on the lower surface of leaves. Incubation period is about 3-9 days.
2. **Larva**: Full grown larvae are pale yellowish green with fine black hairs scattered all over the body. Larval development is completed within 9-20 days. There are 4 no. of larval instars.
3. **Pupa**: Pupation takes place on the leaves in silken cocoons. Pupal period varies from 7 to 11 days.
4. **Adult**: Brown or grey with 3 white spots on the posterior margin of forewings, which appears like diamond pattern when the wings lie flat

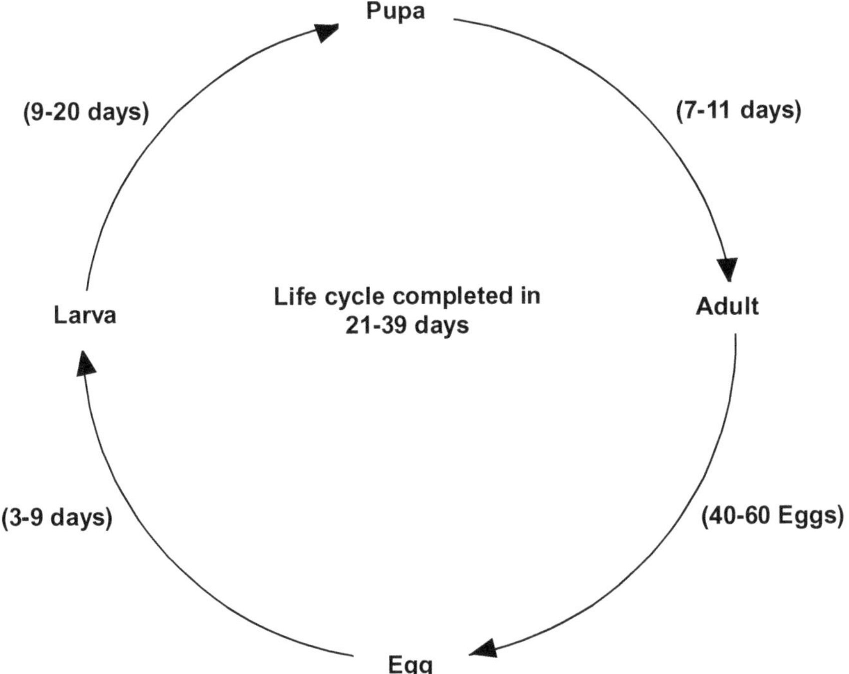

Figure 56: Life Cycle of *Plutella xylostella.*

over the body. Longevity of adult is 10 to 18 days. One generation occupies in 21 to 39 days. About 5 to 7 generations are completed in a year.

Marks of Identification

Adult is brownish gray coloured moth measuring about 8 to 10 mm in length and 15 mm across the wings. The hind wings have a fringe of long fine hairs. There are three pale white triangular markings on hind margins of forewings. When the wings are closed, three diamond shaped yellowish white markings are clearly visible by joining both the forewings. On account of this, the pest is commonly known as "*Diamond back moth*". The full-grown larvae are yellowish green in colour with fine erect black hairs scattered on body and they measure of about 12 mm in length.

Host Crops

Cabbage, Cauliflower, Knol-khol, Radish, Turnip, Mustard *etc.*

Nature and Symptoms of Damage

The newly hatched larvae feed on the epidermal tissues/lower surface of the leaves producing typical whitish patches. It eats away completely the green portion leaving a network of vein. The advanced stage larvae feed on the leaf lamina by biting small holes in the leaf lamina. Growth of plant is inhibated. There is no head formation because they eat away the central portion of cabbage and cauliflower. These larvae also bore into the heads of cabbage and cauliflower. In case of severe infestation, plant may be completely skeletonized.

Economic Threshold Level (ETL)

Economic threshold is when 25-30 larvae observed per square foot or 1-2 larvae per plant.

Management

(A) Cultural Measures

1. Grow cabbage intercropped with mustard (16:2) as trap crop to attract DBM.
2. Sprinkler irrigation unable to fly the adult of DBM.

(B) Mechanical Measures

1. Monitor the presence of *P. xylostella* by pheromone traps or yellow sticky traps. Traps should be checked weekly and count the number of moths.

(C) Chemical Measures

1. Spray the crop with Quinolphos @ 1 litre/ha when the pest is noticed and repeated it 2 weeks after.
2. Spraying cypermethrin @ 30 g a.i. or fenvalerate @ 50 g a.i. or deltamethrin @ 10 g a.i. or cartap hydrochloride @ 175 g a.i./ha once at primordial

initiation (22 days after planting) and repeated either thrice at 7 days interval or twice at 10 days interval.

(D) Biological Measures

1. Spray *Bacillus thuringienesis var. kurstaki* @ 1 gm/lt of water or 1 kg/ha.
2. Bio-control agents such as *Cotesia plutellae* (Kurdjumov), *Diadegma mollipla* (Holmgren), *Oomyzus sokolowski* (Kurdjumov) and *Apanteles* sp. can control diamond back moth.

3. Leaf Webber

Scientific Name, Order and Family

Crocidolomia binotalis (Zeller)

(Lepidoptera, Pyralidae)

Taxonomical Position/Scientific Classification

Kingdom:	Animalia
Phylum:	Arthropoda
Class:	Insecta
Order:	Lepidoptera
Family:	Crambidae
Genus:	Crocidolomia
Species:	***C. binotalis***

Biology and Life Cycle

1. **Egg**: Female lays 100-150 flat eggs in cluster of 40-100 in the lower surface of leaf. The incubation period is of 5-15 days depending on weather.
2. **Larvae**: The larval period lasts for 25-27 days in summer and 50 days in winter.
3. **Pupa**: Pupation takes place in soil after making earthen cocoons. Pupal period lasts for 14-20 days.
4. **Adult**: The small moth with light brownish fore wings lays the eggs in masses.

Marks of Identification

1. **Larvae**: The larva with red head has brown longitudinal stripes and rows of tubercles with short hairs on its pale violaceous body.
2. **Moth**: Small, light brown in colour with light reddish forwings having wavy lines.

Host Crops

Cabbage, cauliflower, raddish, mustard and other cruciferous plants.

Nature and Symptoms of Damage

The leaves are skeletonized by the larvae which remain on the under surface of leaves in webs and feed on them. Larvae webbed the leaves with silken thread and feed from lower surface of leaves. They also feed on flower buds and bore into the heads. Often it assumes serious proportions. It attacks cabbage, cauliflower, radish, mustard and other crucifers and the weed *Gynandropsis pentaphylla*.

Management

(A) Cultural Measures

1. Remove and destroy webbed leaves with larvae with in.

(B) Mechanical Measures

1. Monitor the presence of insect by pheromone traps or yellow sticky traps.

(C) Chemical Measures

1. Spraying of Malathion 50 EC or Dichlorvas 76 SL @ 320 ml/ha or Cypermethrin @ 30 gm a.i./ha or fenvalerate @ 50 g a.i./ha or Deltamethrin @ 10 g a.i./ha or cartap hydrochloride @ 175 g a.i./ha once at primordial initiation (22 days after planting) and repeated either thrice at 7 days interval or twice at 10 days interval.

(D) Biological Measures

1. Spraying of BT formulation @ 500 g/ha.
2. Larval parasitoid – *Microbeacon mellus* and *Apanteles crocidolmie.*

4. Cabbage Borer

Also known as: Cabbage head borer

Scientific Name, Order and Family

Hellula undalis (Fab.)

(Lepidoptera; Pyralidae)

Taxonomical Position/Scientific Classification

Kingdom:	Animalia
Phylum:	Arthropoda
Class:	Insecta
Order:	Lepidoptera
Family:	Crambidae
Genus:	Hellula
Species:	***H. undalis***

Biology and Life Cycle

1. **Egg**: Females lay eggs singly, or in groups or chains of 2 or 3 on the leaves near the bud and after 4 to 5 days, the eggs hatch. The eggs are small, oval and slightly flattened upon the plant surface. They are creamy white when freshly laid, become pinkish the next day and then turn brownish-red.
2. **Larva**: Larvae are pale whitish brown with 4 or 5 pinkish-brown longitudinal stripes. Larval period is 7-12 days.
3. **Pupa**: The larva pupates in the cocoon among the leaves in the ground or in the larval burrows. Pupation occurs in a silk cocoon and early stage pupae are soft and very pale yellowish-white in color with a bright red, dorsal blood. Pupal period is 5-7 days.
4. **Adult**: One life cycle is completed in about 3 weeks.

Marks of Identification

1. Caterpillar is 12-25 mm long and creamy yellow with a pinkish tinge and seven purple brown longitudinal stripes.
2. Moths of *H. undalis* are pale greyish-brown, suffused with reddish color. Their forewings have wavy gray markings, a curved pale patch sub terminally, and a kidney shaped mark one third length from the tip. The wingspan is about 18 mm. Hindwings are pale, with the tip being lightly colored.

Host Crops

Cabbage, Cauliflower, Radish, Knol khol, Turnip.

Nature and Symptoms of Damage

The caterpillars first mine into the leaves and make it white papery. Later they feed on leaves and bore into stems; entrance hole is covered with silk and excreta. later on they feed on the leaves surface and the shoots by sheltering within silken passages. At later stage, they bore into the heads resulted in to deformed in to heads. When the attack is severe, the plants are riddled out and the heads look deformed.

Economic Threshold Level (ETL)

ETL is when 15-25 per cent of plants infested in a random pattern.

Management

(A) Cultural Measures

1. Collection and destruction of infested plant parts (leaves and heads of plant).
2. Screening of seedling beds and clean culture are helpful in reducing damage caused by cabbage webworm.
3. Grow tolerant varieties *i.e.* Early Patna, EMS-3, KW-5, KW-8, Kathmandu Local.

(B) Mechanical Measures

1. Regular monitoring of young plants in the nursery and after transplant is important. Install pheromone traps and detect larvae.
2. Collect and destroy caterpillars in early stage of attack.

(C) Chemical Measures

1. Spray Malathion 50 EC @ 0.05 per cent at 10 days interval.
2. Spraying of Carboryl 50 WP @ 375 gm/ha or Cypermethrin 25 EC @ 200 ml/ha. Repeat the spray at after 15 days, if needed.

(D) Biological Measures

1. *B. hebetor* and *C. blackburni* can control *H. undalis*.
2. Biopesticides such as *Bacillus thuringiensis* and spinosad are useful for the management of caterpillars. If the threshold is exceeded.

Pest Management of Fruit Crops

(A) PESTS OF MANGO

1. Inflorescence Midge

Also known as: Mango Gall Midge

Scientific Name, Order and Family

Erosomyia indica (Grover)

Erosomyia mangiferae

Dasinuera amaramanjarae

Procystiphora mangiferae (Felt)

(Diptera, Cecidomyiidae)

Taxonomical Position/Scientific Classification

Kingdom:	Animalia
Phylum:	Arthropoda
Class:	Insecta
Order:	Diptera
Family:	Cecidomyiidae
Genus:	Erosomyia
Species:	***E. indica***

Biology and Life Cycle

1. **Egg**: Female lay elongated, oval shaped and creamy white 40-50 eggs at the base of developing fruits or in the dorsal surface of the leaf near the base or on the inflorescence peduncle. Eggs hatch out in 2-3 days.
2. **Maggot**: Maggot period is 8-10 days. A maggot light yellowish colour and moults three times.
3. **Adult/Fly**: Small, yellowish midge with graish back. Male being larger than female.
   (i) *Erosomyia indica* (Grover): The maggots attack the inflorescence stalk, flower buds and small developing fruits. The adult fly is yellowish and lays the eggs on the infloresence peduncle or at the base of' developing fruit. The maggots are yellowish and when full grown pupate in the soil.
   (ii) *Dasineura amaramanjarae* (Grover): The adult flies insert the eggs into unopened flower buds. The maggots feed inside the buds and they fail to open and drop down. The maggots hibernate in the soil and thus carry-over of the pest to the next year is accomplished. When favourable conditions set in they pupate and emerge as adults

(i) *Procystiphora mangiferae* (Felt): The light orange coloured fly lays the eggs inside immature blossoms. The maggots that hatch out from the eggs feed on stalks of stamens, anthers, ovary, *etc.* Only one maggot is found in each

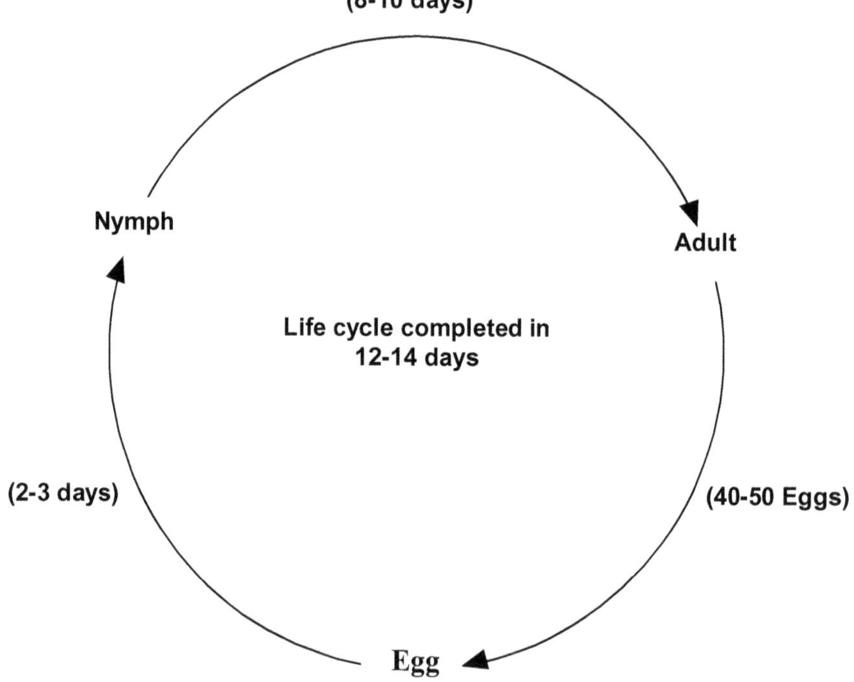

Figure 57: Life Cycle of *Erosomyla indica*.

bud and it pupates inside the bud itself. The life-cycle from egg to adult occupies 12-14 days).

Host Crops

Mango (Monophagous).

Nature and Symptoms of Damage

- ☆ Maggot affects the crop at 3 stages, *i.e.*, at floral bud burst, fruit set and tender leaves particularly encircling the inflorescence.
- ☆ The first phase is more damaging as the entire inflorescence destroyed before flowering and fruit set. The inflorescence show stunting growth and its axis has curve at the entry point of the larvae and ultimately die before fruit set. Its attack on inflorescence could be recognized by presence of tiny black spots.
- ☆ Infested mango buds, shoots and young fruits bear many small blister galls, about 3-4 mm long, each containing a yellow E. mangiferae larva.
- ☆ Small emergence holes may be detectedwhere larvae have left the galls to go to the soil and at this stage secondary fungal infections may develop.
- ☆ When young fruits are attacked, the exit holes are usually on the lower side of the fruit near its point of attachment to the axis of the inflorescence.
- ☆ In severe attacks, the affected plant parts shrivel and die.

Economic Threshold Level (ETL)

10 spots/twig or inflorescence

Management

(A) Cultural Measures

1. Remove and destroy affected flowers and tender shoots.
2. Deep ploughing of orchard exposing pupae and diapausing larvae to sun's heat kills them.
3. The soil below the conopy of plant is covered with the plastic sheets; it prevents the emergence of adults from soil and also prevents the dropping larvae to go into soil for pupation.

(B) Mechanical Measures

1. Remove and destroy affected and tender shoots.

(C) Chemical Measures

1. Stem injection by making 5 – 10 cm deep holes in the main branches with dimethoate @ 0.5 ml a.i./cm circumference gave effective control of the pest.

2. Single spray of 2, 4-D @150 mg/l in October resulted in opening of galls causing 90 per cent autocidal mortality of the nymphs.
3. Spray Dimethoate 30 EC @ 0.06 per cent or Methyl demeton 25 EC @ 0.05 per cent.
4. Spray dimethoate 30 EC or methyl demeton 25 EC 3.0 - 4.0 L in 1500-2000 L of water per ha (10-15 L of spray fluid per tree).

(D) Biological Measures

1. Parasitoids: *Platygaster* sp., *Tetrastichus gala, Aprostocetus* sp., *Systatis* sp.
2. Predators: *Formica* sp., *Componatus* sp., *Oecophila* sp.

2. Shoot Gall Psylla

Common name: Mango shoot gall maker

Scientific Name, Order and Family

Apsylla cistella

(Homoptera, Psyllidae)

Taxonomical Position/Scientific Classification

Kingdom:	Animalia
Phylum:	Arthropoda
Class:	Insecta
Order:	Homoptera
Family:	Psyllidae
Genus:	Apsylla
Species:	***A. cistella***

Biology and Life Cycle

1. **Egg**: Female lay about 150 eggs singly during end of March or early April. Eggs hatch out in 3-4 months. Eggs are partially embedded within the midrib and lateral veins of new leaf.
2. **Nymph**: Flat or round and pale yellow in colour. Nymphal mperiod is completed in 5-6 months.
3. **Adult**: Adult is 3-4 mm long with black thorax and head and light brown abdomen.

Marks of Identification

1. **Nymphs**: Freshly hatched nymphs are yellowish in colour, but change in size and colour with time
2. **Adults**: 3-4 mm long with black head and thorax and light brown abdomen. Membranous wings.

Host Crops

Mango (Monophagous).

Nature and Symptoms of Damage

- ✰ The pest is active from August onwards with the nymphs emerging from eggs during August- September and crawling to the adjacent buds to suck cell sap. As a result of feeding, the buds develop into hard conical green galls which are usually seen during September-October.
- ✰ Terminal shoots affected.
- ✰ Formation of green conical galls in leaf axis in response to egg-laying by adult insects or feeding by nymphs.
- ✰ Development of the green galls results in no flowering and fruit setting.

Management

(A) Cultural Measures

1. Grow least susceptible varieties like Alfanso, Sirohi *etc.*
2. New mango orchard in humid region need to be discouraged.

(B) Mechanical Measures

1. Galls with nymphs should be collected and destroyed.
2. Pruning, gall bearing branches is pruned by secateurs.
3. Use of yellow sticky traps.

(C) Chemical Measures

1. Spraying of Dimethoate 30 EC @ 0.06 per cent or Quinolphos 25 EC @ 0.05 per cent or Carboryl 50 WP @ 0.2 per cent. Three spray beginning from mid August at 15 days intervals may be given for effective control.
2. Spray thiomithoxam (0.05 per cent) or quinalphos (0.05 per cent) at fortnightly interval starting from the middle of August.

(D) Biological Measures

1. Predators: Coccinellids, pirate bug, brown lacewing, syrphids, anthocorids.
2. Parasitic wasps: *Inostemma apsyllae*

3. Leaf Hopper

Also known as: Mango Jassids

Scientific Name, Order and Family

Amritodus atkinsoni (Lethiery)

Idioscopus clypealis

I. niveosparsus

(Hemiptera; Cicadellidae)

Taxonomical Position/Scientific Classification

Kingdom:	Animalia
Phylum:	Arthropoda
Class:	Insecta
Order:	Hemiptera
Family:	Cicadellidae
Genus:	Amritodus
Species:	***A. atkinsoni***

Biology and Life Cycle

1. **Egg**: Females lay 100-200 eggs, singly inside the leaf tissues near midrib and also on flowering shoots, flower, buds and tender leaves. Hatching period is 4 to 7 days.
2. **Nymph**: There are 5 nymphal instars which together occupy 8 to 13 days. The nymphs develop faster during the flowering and fruiting period.
3. **Adult**: Adults overwinter by hinding the cracks and crevices in bark of tree. Hopper love damp and shady places. About 14-20 days are required to complete one generation. There are many generations in a year.

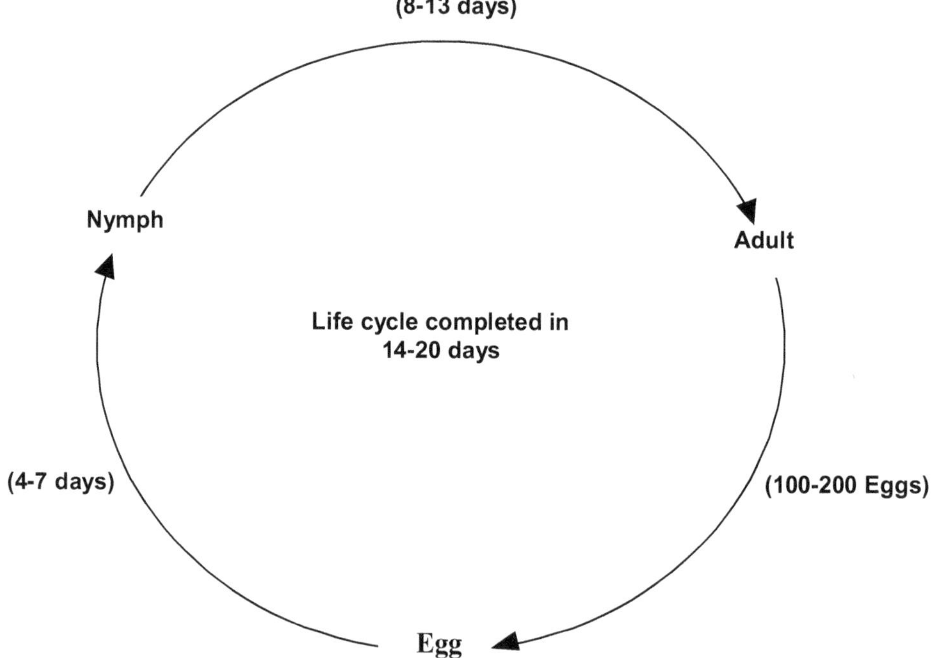

Figure 58: Life Cycle of *Amritodus atkinsoni.*

Marks of Identification

1. **Eggs** - Cigar-shaped, creamy-yellow in colour. Size: 0.9–1 mm in length.
2. **Nymph** - Nymphs greenish with black or brown markings, resemble small adults but without wings. They are very active and hide in flower rachis.
3. **Adult** – Golden brown or dark brown resembling to bark colour wedge-shaped. Size:4–5 mm in length.
   - ✰ *Idioscopus niveoparsus*: dark with wavy lines on wings and three spots on scutellum.
   - ✰ *Idioscopus clypealis*: small, light brown with dark spots on the vertex and two spots on scutellum.
   - ✰ *Amirtodus atkinsoni*: large, light brown with two spots on scutellum.

Host Crops

Mango (Monophagous).

Nature and Symptoms of Damage

1. Both nymphs and adults suck the cell sap from tender leaves of mango during non- flowering season.
2. The over-wintering hoppers become active with the advent of flowering season and suck the cell sap clustering together in large number on flowers and flower buds.
3. Due to continuous draining of sap from the inflorescence, the flowers and flower buds drop down and affect adversely fruit setting. Even if few fruits are set on plant, they remain undersized and drop down prematurely in due course.
4. The hoppers excrete large amount of sugary substance on leaves and inflorescence which adversely affects its fertilization. The saprophytic fungi *Capnodium mangiferum* and *Meliola mangiferae* develops on the sugary secretion giving complete blackish appearance to the plant. The photosynthetic activity of plant is hampered because of thick coating of black sooty mould on the leaves.
5. The hopper scause a loss of 20-100 per cent of the inflorescense.

Economic Threshold Level (ETL)

5/Leaf (In Summer) or 1/Leaf (In Winter), or 10/Inflorescence or Twig.

Management

(A) Cultural Measures

1. Grow hopper resistant variety *i.e.* Totapari and moderate resistance varieties *i.e.* Ratna, Mallika, Dashehri, Langra, Keshar *etc.*
2. Avoid close planting, as the incidence very severe in overcrowded orchards.

3. Orchards must be kept clean by ploughing and removal of weeds.
4. Pruning of dense canopy to facilitate aeration and sunlight.
5. Avoid excess use of nitrogenous fertilizers.

(B) Chemical Measures

1. Spray dimethoate 30 EC or methyldemeton 25 EC or malathion 50 EC 1.5 -2.0 L in 1500 – 2000 L of water per ha or acephate 75 SP @ 1 g/L, phosalone 35 EC @1.5 ml/L, or new molecules like buprofezin 25 SC 1-2ml/L of water or imidacloprid 17.8 SL 2-4ml/tree or lambda cyhalothrin 5 EC 0.5-1.0ml/L of water at 10 -15 L of water per tree
2. Spraying of Carboryl 50 WP @ 2 kg/ha or Malathion 50 EC @ 1.75 lt/ha or Bufrozin 25 SC @ 1-2 ml/lt with 5-15 litres of water per tree.
3. Spray Imidacloprid 17.8 SL @ 60 ml/ha at penicle emergence and 2nd spray at pea size of fruit stage reduce the hopper density.
4. Spray two rounds of acephate 75 SP@ 1g/lit or phosalone 35 EC@ 1.5 ml/litre of water or Spray two rounds of imidacloprid 0.2ml/lit or phosphamidon 40SL @ 2 ml/lit of water. First spray at the time of panicle emergence, second spray two weeks after first spray.
5. Wettable sulphur @ 2 g/lit may be sprayed after spraying carbaryl to avoid mite resurgence.
6. The mixture toxaphene with sulphur (1:1) have been reported to be effective against pest.
7. Neem oil 5 ml/lit of water can be mixed with any insecticides. Spray 3 per cent neem oil or neem seed kernel powder extract 5 per cent

(C) Biological Measures

1. Various species of spider *i.e. Rhene indicus, Marpiss* spp. and *Oxyopes shweta* were found to predate on hopper.
2. Predators: *Mallada boninensis, Chrysopa lacciperda.*
3. Egg parasite: *Polynema* spp. *Gonatocerus* sp. *Tetrastichus* sp.
4. Fungus: *Verticillium lecanii.*

4. Mealy Bug

Also known as: Gaint mealy bug.

Scientific Name, Order and Family

Drosicha mangiferae (Green.)

(Hemiptera; Pseudococcidae)

Taxonomical Position/Scientific Classification

Kingdom:	Animalia
Phylum:	Arthropoda
Class:	Insecta
Order:	Hemiptera
Family:	Pseudococcidae
Genus:	Drosicha
Species:	***D. mangiferae***

Biology and Life Cycle

1. **Egg**: The females crawl down the trees during end of March to end of May and enter the soil (80 to 150 mm deep) wherein they excrete whitish foam. This forms a pouch in which females go on depositing the eggs for 7 to 16 days and soon after completing the oviposition, the females die. Each female gives 400 to 500 eggs and these remain in the soil in a state of diapause. Winter chilling terminates diapause. The hatching of these eggs also depends upon the humidity of soil and normally starts from end of November and continues till January.
2. **Nymph**: There are three nymphal instars. The nymphal stage is completed in 69-119 days.
3. **Adult**: Total life cycle occupies 77 to 135 days in case of female and 67 to 119 days in case of male. Adult longevity is 22 to 27 days in April for female and about a week for the male.

Marks of Identification

Adult females are wingless while males are crimson coloured bugs with two dark brownish black wings and cause no damage except fertilizing the females. The nymphs and adult females are flat, oval, waxy whitish insects, sometimes mistaken for fungal growth.

Host Crops

Apple, apricot, ber, cherry, citrus, phalsa, fig, grapevine, fuava, jackfruit, jamun, litchi, mango, mulberry, papaya, pear, peach, plum, pomegranate *etc.*

Nature and Symptoms of Damage

1. Soon after hatching, the majorities of nymphs starts crawling up the tree trunks and congregates in large number on growing shoots and panicles and suck the cell sap.
2. As a result, the shoots become wilted and dry up and ultimately fruit setting is affected.

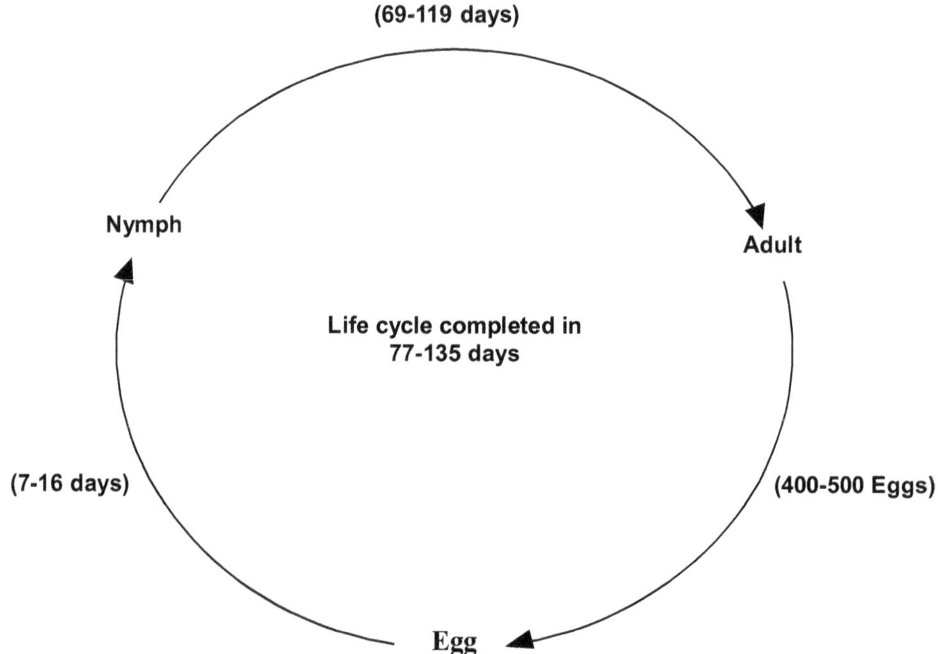

Figure 59: Life Cycle of *Drosicha mangiferae.*

3. The young fruits also become juiceless and immature fruits drop down.
4. The also excrete honeydew over which shooty mould develops as a result veaves and inflorescence become shining black and sticky affecting photosynthesis. The overall growth of tree is adversely affected.

Management

(A) Cultural Measures

1. Ploughing below the tree during summer to kill the eggs by exposing the same to sun's heat, ants, birds *etc.*
2. Remove weeds like *Clerodendrum inflortunatum* and *Parthenium* by ploughing during June-July to avoid build up of mealy bug population.
3. Mix thoroughly with upper 400 to 500 mm of soil around tree with 2 per cent Methyl parathion dust to kill the freshly hatched nymphs.

(B) Mechanical Measures

1. Band the trees with 20 cm wide alkalthene of polythene (400 gauge) in the middle of December (50 cm above the ground level and just below the junction of branching).

(C) Chemical Measures

1. Spray Chloropyriphos 20 EC @ 1 lt/ha or Malathion 50 EC @ 1 lt/ha if nymphs are observed on tree.
2. Early instar nymphs of the mealy bug can be controlled by spraying of 0.05 per cent carbaryl from January to March.

(D) Biological Measures

1. Release of predators, *Menochilus sexmaculatus, Rodolia fumida, Sumnius renardi* and Australian ladybird beetle, *Cryptolaemus montrouzieri* @ 10-15 No./tree are effective in controlling the nymphs of the mealy bug.

5. Stone Weevil

Also known as: Mango seed or nut weevil

Scientific Name, Order and Family

Sternochaetus mangiferae (Fab.)

(Coleoptera; Curculionidae)

Taxonomical Position/Scientific Classification

Kingdom:	Animalia
Phylum:	Arthropoda
Class:	Insecta
Order:	Coleoptera
Family:	Curculionidae
Genus:	Sternochetus
Species:	***S. mangiferae***

Biology and Life Cycle

1. **Egg**: Females lay tiny, elongated and creamy white 12-36 eggs which is covered by a protective brown coating at just beneath the rind of the ripe fruit. The egg period is lasts for 7 days.
2. **Larva**: The grubs feed on inner contents of the stone. The larval period is completed in 33-43 days.
3. **Pupa**: The pupation takes place inside the stone in a pupal shell. Pupal period lasts for 30-45 days.
4. **Adult**: The adults are not strong fliers and are not found far from fallen fruit. New areas are infested by human transport of infested fruit. The adults emerging out of the pupae remain inside the stone till they are thrown away after consumption of pulp. Adults can also emerge out by boring and cutting through the nut. The adult weevils remain dormant

within the cracks present in the bark of stem from July-August till next March-April. One life cycle occupies from 40 to 50 days. Only one generation completed per year.

Marks of Identification

Adult is short, stoutly built, oval, dark brown weevil measuring about 6 mm in length and 3 mm in breadth. Antennae are 10 segmented. Weevils are found inside the stone. Grub is white, thick, fleshy and legless. Eggs are minute in size and whitish in colour.

Host Crops

Mango only (monophagous).

Nature and Symptoms of Damage

- ✰ The grubs on hatching from eggs make their way through the pulp and damage the stone by boring and feeding on its inner content.
- ✰ The pest passes through all the stages of its growth inside the stone.
- ✰ The grubs and adults feed and damage the cotyledons. No signs of infestation are noticed outside the fruit. The discolouration of the pulp adjacent to the stone due to its excreta is often seen when the infested fruits are cut.

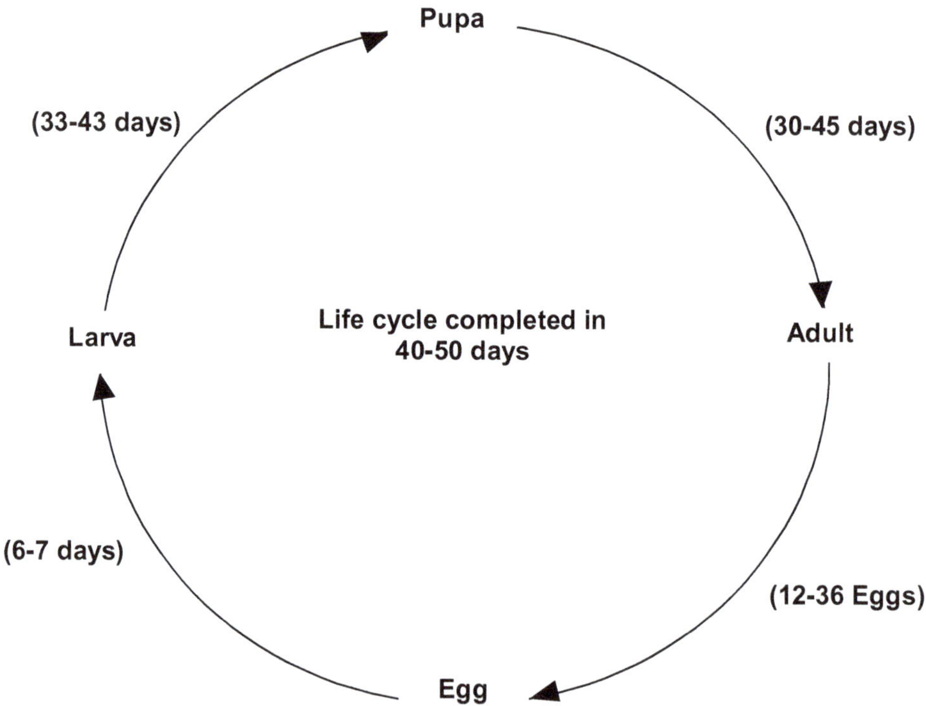

Figure 60: Life Cycle of *Sternochaetus mangiferae*.

☆ The pest is not highly injurious to mango fruits but their presence within fruits has rendered them unacceptable to foreign countries. The sweet varieties of mango *viz.*, Alphonso, Bangalora, Neelam, Totapuri *etc.*, are relatively more damaged.

Management

(A) Cultural Measures

1. Disposal of the stones, fallen and infested fruits and debries under the tree and digging of soil to destroy hibernating adult weevils.
2. Destruction of weevils on the bark during August helps in reducing infestation of pest.
3. In stone grafting programme, expose collected stones for five days at 12°C to kill grubs, pupae and adults present in the stone.

(B) Chemical Measures

1. Spray of Carboryl 50 WP @ 0.02 per cent or 0.1 per cent Malathion at monthly interval starting from 1½ month after fruit set.
2. Three sprays of Fenthion 0.05 per cent on young fruits at fortnightly interval as per necessity. Fruit exposed to sub-freezing temp. 12°C for 5 days kill all the grubs and pupae inside the fruits.
3. Spray kerosene oil emulsion after harvest to kill weevils, hinding under bark.

6. Stem Borer

Also known as: Mango tree borer.

Scientific Name, Order and Family

Batocera rufomaculata (DeG.)

Batocera rubus (Linn.)

(Coleoptera; Cerambycidae)

Taxonomical Position/Scientific Classification

Kingdom:	Animalia
Phylum:	Arthropoda
Class:	Insecta
Order:	Coleoptera
Family:	Cerambycidae
Genus:	Batocera
Species:	***B. rufomaculata***

Biology and Life Cycle

1. **Egg**: The female deposits 25 to 30 eggs singly under the loose bark or in the crevices of the stems. Eggs measure about 5.0 to 6.5 mm in length, shining white in colour and oval in shape. The eggs hatch out in about 14 to 17 days.
2. **Larva**: The grubs tunnel directly into the bark as they possess stout biting mouth parts. They feed for about 6 months.
3. **Pupa**: Pupae are yellowish brown to dark brown and measure about 50 to 55 mm in length. Pupate inside the tunnel itself.
4. **Adult**: The adults emerge from the pupae in about 1 to 3 months. The beetles are nocturnal in habit. Beetles live for about 3 months. There is only one generation in a year.

Marks of Identification

1. Adult is stoutly built, dark brown longicorn beetle measuring about 50 to 55 mm in length. Adults possess two large kidney shaped orange spots on the prothorax and a thick spine like projection on its either side. The elytra is decorated with small light orange spots.
2. The full-grown grubs are fleshy, yellowish ivory in colour and 85 to 95 mm in length with well defined segmentations.

Host Crops

Mango, apple, mulberry, cashew, jackfruit, rubber, eucalyptus, ber, citrus and fig.

Nature and Symptoms of Damage

☆ Grubs on hatching make zigzag burrows beneath bark of the tree trunk or branches. Grubs move upward by making tunnels and feed on the wood inside.

☆ The adult beetles feed on the bark of young twigs and petioles.

☆ Grub feeding the vascular tissues and interruption of nutrient and water transport on the tissue.

☆ Drying of terminal shoot in early stage.

☆ Frass comes out from several points and some times sap oozes out of the holes.

☆ Wilting of branches or entire tree.

Management

(A) Cultural Measures

1. Clean cultivation and avoid over crowding of trees.
2. Avoid injury at the base of trunk while pruning

3. Cut and destroy infested branches along with grubs and pupae inside.
4. Grow tolerant mango varieties *viz.*, Neelam, Humayudin.
5. Remove alternate host, silk cotton and other hosts.

(B) Mechanical Measures

1. Killing of grub by poking stiff wire inside the burrow.

(C) Chemical Measures

1. Remove the frass near the holes on main stem and inject 5 ml of kerosene or petrol into the whole and sealing it with cotton or mud that kill the larvae and give active control.
2. Swab Coal tar + Kerosene @ 1:2 or Carbaryl 50 WP 20 g/l (basal portion of the trunk - 3 feet height) after scraping the loose bark to prevent oviposition by adult beetles.
3. Inject borer solution (2 parts of carbon disulphide + 1 part of chloroform + 1 part of creosote oil) or carbon disulphide or ED/CT mixture into the larval burrows by using plastic syringe for controlling grubs feeding inside.
4. Injection of any one of insecticide either Carbofuran 3G @ 5 g per hole Chloropyriphos 20 EC or Quinolphos 25 EC @ 5 ml/lt of water into the hole and sealing it with cotton or mud.
5. One celphos tablet (3 g aluminum phosphide) per hole.
6. Padding with monocrotophos 36 WSC 10 ml in 2.5 cm/tree soaked in absorbent cotton.
7. Hook out the grub from the bore hole - apply monocrotophos 36 WSC 10 to 20 ml/hole.

(B) PESTS OF CITRUS

1. Lemon Butterfly

Also known as: Swallow tail butterfly or Citrus Caterpillar.

Scientific Name, Order and Family

Papilio demoleus (Linn.)

(Lepidoptera; Papilionidae)

Taxonomical Position/Scientific Classification

Kingdom:	Animalia
Phylum:	Arthropoda
Class:	Insecta
Order:	Lepidoptera

Family:	Papilionidae
Genus:	Papilio
Species:	***P. demoleus***

Biology and Life Cycle

1. **Egg**: Female deposits about 180 rounded, pale yellow eggs singly or in group of 2-5 on the tender leaves. Hatching period is 3 to 7 days.
2. **Larva**: Larvae are dirty white and oblique mark when fully grown. Larval development is completed in 8-22 days.
3. **Pupa**: Pupation takes place on leaves or on plant part itself, attached with silken girdle. Pupal period varies from 8-11 days.
4. **Adult**: Large beautiful butterfly, head and thorax are black and its wings are dull black ornamental with yellow markings. One generation is completed in 20 to 100 days depending upon climatic conditions. As many as 6 overlapping generations are completed per year.

Marks of Identification

1. **Adult**: Adult is a beautiful butterfly with yellow and black markings on it's large wings. Adult has wing expanse of about 50 to 60 mm. Hind wings have a brick red oval patch near anal margin and a tail like extension behind on account of which pest is commonly known as 'Swallow tail butterfly'.

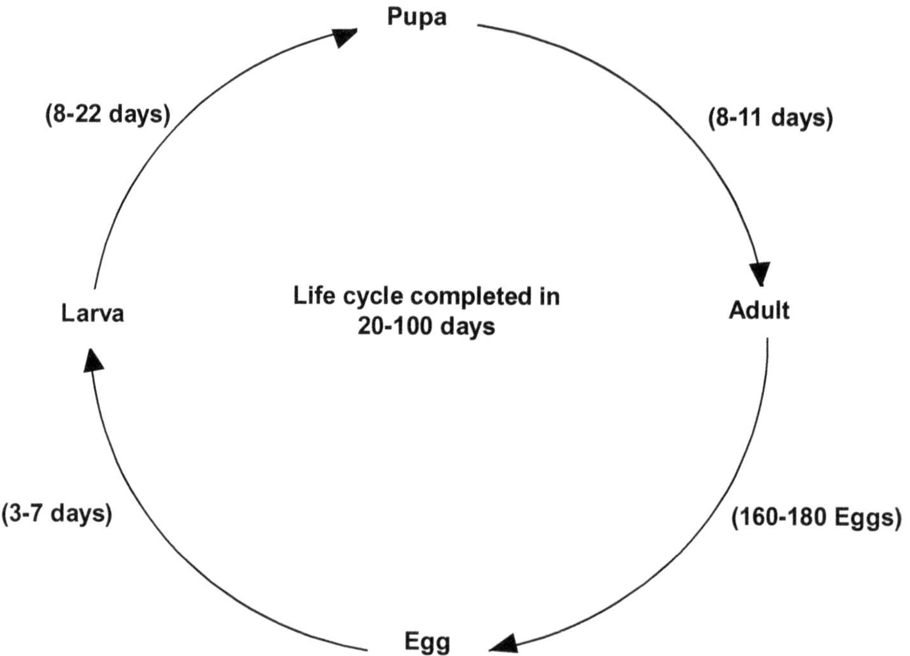

Figure 61: Life Cycle of *Papilio demoleus*.

2. **Larva/Caterpillar**: Newly hatched larva is dark brown with milky white markings just look like bird droppings. When full-grown, it turns deep green in colour and cylindrical in form and measures about 38 mm in length. The young caterpillar is often mistaken as bird excreta. The caterpillar when disturbed, pushes out from top of it's prothorax a bifid purple or red horn like structure called "osmeterium" which emits a distinct smell. After a few seconds of its extrusion, it is withdrawn.

Host Crops

All species of citrus and ber, wood apple, kadhilimb, ornamental and medicinal plants, bael.

Nature and Symptoms of Damage

Damage is caused by the larvae alone by feeding on tender leaves right upto the mid-rib and thus defoliate the plants. In case of severe infestation, only midribs are left over. The tender foliage and young shoots of plant are mainly damaged. The pest is not of regular occurrence but occasionally becomes serious, particularly on young plants and seedlings. Heavily infested plants bear no fruits.

Economic Threshold Level (ETL)

20-30 per cent foliar damage.

Management

(A) Cultural Measures

1. Eradication of wild hosts such as Bawchi and Bhira (Rutaceae family).

(B) Mechanical Measures

1. Hand picking of larvae and pupae and destroy them in kerosinized water.

(C) Chemical Measures

1. Spray the crop with Quinolphos 25 EC @ 2 lt/ha or Carbaryl 50 WP @ 2 kg/ha when attack is noticed. For an effective control, spraying should be undertaken when the larvae are in early stages.

(D) Biological Measures

1. Release of egg parasitoids – *Trichogramma evanscense* or *Telenamus* sp. @ 500 adults per tree.
2. Release of larval parasitoids – *Cotesia flavips* or *Charops* sp.

2. Citrus Psylla

Scientific Name, Order and Family

Diaphorina citri (Kuwayama)

(Hemiptera; Psyllidae)

Taxonomical Position/Scientific Classification

Kingdom:	Animalia
Phylum:	Arthropoda
Class:	Insecta
Order:	Hemiptera
Family:	Psyllidae
Genus:	Diaphorina
Species:	***D. citri***

Biology and Life Cycle

1. **Egg**: About 180 to 860 bright yellow eggs are laid singly in leaf axil or on terminal shoots. Eggs are bright yellow, tapering towards both the ends and 1 mm in length. Hatching period is 4-6 days in summer and 10-20 days in winter.
2. **Nymph**: Nymph moults four times before becoming adult. Nymphal period is 15-20 days.
3. **Adult**: Longevity of adult is 6 months. As many as 8 generations are completed in a year.

Marks of Identification

1. **Adult**: Adult is dark brown coloured insect measuring about 3 mm in length. Wings are folded like a roof over body when insect is at rest. Wings are membranous, semi-transperent with a brown band in the apical half of forewing. Hindwings are shorter and thinner than the forewings.
2. **Nymph**: Nymph is flat, louse like and orange yellow in colour. The nymphs prior to becoming adults are light brown in colour and 2 mm in length.

Host Crops

All citrus spp.

Nature and Symptoms of Damage

- ☆ Damage to the plant is caused by both nymphs and adults.
- ☆ Both nymph and adult aggregate on just emerged shoot buds and suck the cell sap from tender leaves, terminal shoots and flower buds. As a result, curling of leaves, defoliation and even death of shoots may result.
- ☆ This insect excretes honeydew on which black shooty mould develops.
- ☆ The infested fruits remain undersized, poor in juice content and insipid in taste.
- ☆ Pest is responsible for transmitting greening virus of citrus.

Economic Threshold Level (ETL)

6 no./leaf.

Management

(A) Cultural Measures

1. Prune the affected trees and dried shoots.
2. Propogation of clean nursery stock.
3. To avoid more nitrogen fertilizers.
4. Minimize shearing or clipping of terminal, shearing stimulates new growth which is preferred for feeding and egg laying.
5. Alternate hosts of *D. citri* in close proximity to orchards should be remove.
6. Citrus grown in intercropped with guava are preferred from psylla and some relative released from the guava leaves give repellant effects.

(B) Mechanical Measures

1. Use of yellow sticky traps.

(C) Chemical Measures

1. Spraying of Dimethoate 30 EC @ 2 lt/ha or Oxydemeton methyl 25 EC @ 2 lt/ha or Imidacloprid 17.8 SL @ 120 ml/ha, or
2. 1st spray is given as soon as new sprouts appear in June/January with Malathion 0.05 per cent or Carbaryl 0.1 per cent or Dimethoate 0.025 per cent or Oxydemeton methyl 0.025 per cent or Fenthion 0.05 per cent or fenvalerate 0.01 per cent. 2nd spray may be given two weeks after first application.

(D) Biological Measures

1. Release Nymphal parasitoid: *Tetrastichus radiates.*
2. Release Predator: *Coccinella septumpunctata, Menochilus sexmaculatus, Mallada boninensis, Cheilommenes sexmaculata, Horminia axyridis.*
3. Pathogen: *Hirsutella citriformis, Isaria fumosorosea.*
4. Parasitoids: *Tamarixia dryi* and *T. radiate.*

3. Fruit Sucking Moth

Scientific Name, Order and Family

Ophideres conjuncta

Eudocima (Othreis) fullonica (Linn.)

E. materna

(Lepidoptera; Noctuidae)

Taxonomical Position/Scientific Classification

Kingdom:	Animalia
Phylum:	Arthropoda
Class:	Insecta
Order:	Lepidoptere
Family:	Noctuidae
Genus:	*Ophideres*
Species:	***O. conjuncta***

Biology and Life Cycle

1. **Egg**: Female moth lays about 200 to 300 egg, singlys on lower side of leaves of certain weeds like Gulwel and Vasanwel. Eggs are transparent, spherical and about 1 mm in diameter. Hatching period is 3 to 4 days.
2. **Larva**: The larvae passes through 5 instars in 4-5 weeks. The larvae feed on foliage of above weeds. Larval period is 13 to 17 days.
3. **Pupa**: Full-grown larva pupates in soil by forming a cell of soil particles by webbing leaves. Pupa is thick and reddish brown. Pupal period lasts for about 12 to 18 days.
4. **Adult**: The moths on emergence swarm towards the odour released by ripened fruits. The body colour is faint orange brown. Its forwings are dark gray and the hindwings are orange red having 2 black patches. Moth lives fairly for a long time.

Marks of Identification

1. **Adult**: Adult is large sized moth measuring about 25 mm in length and 75 mm in wing expanse. The forewings are usually grey or reddish brown while hindwings are orange or yellow with circular or semicircular black spots or marginal black bands. The proboscis of the adult is strongly built to pierce the rind of fruits.
2. **Larvae**: The larva is a cylindrical, stout bodied semilooper having dark brown colour. These caterpillars are usually found feeding on certain creepers/weeds like gulwel. *Tinospora cardifolia, Coculus pendulus., Convolvulus hirsutus etc.,* belonging to natural order Menispermaceae and Anacardiaceae.

Host Crops

Banana, peach, pear, plum, pomegranate, citrus, mango, grape, cashewnut.

Nature and Symptoms of Damage

☆ Only moths of this pest are harmful.

- ☆ The adult moths after emerging from pupae migrate to citrus orchard and attack the fruits only during night at dusk.
- ☆ The adults with their strongly developed proboscis suck the juice from the fruits after piercing the rind of fruits.
- ☆ It inserts its proboscis through the rind of fruit and sucks the juice. It may be about 50 punctured places on fruits. The punctured area become soft and pale white turns brownish. Later on, fruit turns yellow.
- ☆ The secondary infection of bacteria and fungi sets up on injured/punctured portion which make the fruits to rot quickly. Such fruits drop down prematurely. When squeezed, such fruits ooze out fermenting juice from the puncture.
- ☆ Such fruits are unfit for human consumption.

Management

(A) Cultural Measures

1. Collect and destroy the fallen and decaying fruits which attract the moths.
2. As the pest breeds on certain wild plants, eradication of alternate and wild host plants in the vicinity of citrus orchard helps in checking pest infestation.
3. Growing tomato as a trap crop in the orchards to attract the moths.

(B) Mechanical Measures

1. Bagging of fruits with polythene or paper to prevent the damage of adult moth.
2. When fruit ripen, make smoke in the garden at dusk to repel the moths at 6 to 7 p.m..
3. Set up light traps in the field to destroy adult moths.

(C) Chemical Measures

1. Spray the tree with Carboryl 50 WP @ 2 kg/ha at the time of maturity of fruits.
2. Poison baiting with 20 ml of Malathion + 200 g of jaggery + 2 lit. of water or mixture of gur one kg + vinegar 60 gm + 20 ml of Malathion may be filled in wide mouth bottle and hanged on tree for killing moths.

4. White Fly

Scientific Name, Order and Family

Dialeurodes citri (Ril. and How.)

(Hemiptera; Aleurodidae)

Taxonomical Position/Scientific Classification

Kingdom:	Animalia
Phylum:	Arthropoda
Class:	Insecta
Order:	Hemiptera
Family:	Aleurodidae
Genus:	Dialeurodes
Species:	***D. citri***

Biology and Life Cycle

1. **Egg**: Females lay about 150 to 200 pale yellowish eggs singly on under surface of tender leaves, which hatch out in a period of 10 to 14 days.
2. **Nymph**: Nymphal period varies from 25 to 71 days. Pupation takes place on under surface of leaves. Pupal period varies from 114 to 159 days.
3. **Adult**: Longevity of adult is about 2 to 8 days. The pest has reported to complete two generations in a year.

Marks of Identification

1. **Adult**: Adult is minute sized insect measuring about 0.5 mm in length having white or greyish wings, yellowish body and red medially constricted eyes.
2. **Nymph**: Nymphs and pupae are oval in shape, scale like blackish with marginal bristle like fringes. Nymphs remain stationary once they have settled down. The black dots present on lower surface of leaves are generally pupae of the pest.

Host Crops

Citrus, coffee, cotton, castor, banana and some ornamental plants.

Nature and Symptoms of Damage

- ☆ Damage is caused by both nymphs and adults by sucking the cell sap from lower surface of leaves through it's piercing stylets, resulting into leaf curl.
- ☆ Infested leaves curl upward. The plant growth remains stunted.
- ☆ The nymphs excrete honey dew in large amount which spreads all over the leaf surface, attracting saprophytic fungi, *Capnodium citri* and *Meliola camelliae*.
- ☆ Due to thick coating of black sooty mould through fungi, the photosynthetic activities of plant are hampered and the roots fail to function properly.

Management

(A) Cultural Measures

1. Close planting, water logging, excessive application of nitrogen or any other stress condition should be avoided.
2. In case of localized infection, the affected shoot should be clipped off and destroy.

(B) Chemical Measures

1. Give two applications of following insecticides at 15 days interval starting from as soon as 50 per cent eggs are hatched: Dimethoate 30 EC @ 2 lt/ha or Malathion 50 EC @ 1 lt/ha or Imidacloprid 17.8 SL @ 120 ml/ha or Acephate 75 SP @ 1.25 g or or Quinalphos 0.05 per cent or Triazophos 0.03 per cent.

(C) Biological Measures

1. Release predator – *Malada boninensis* @ 200 eggs/tree, *Brumoides suturalis, Verania cardoni.*

5. Leaf Minor

Scientific Name, Order and Family

Phyllocnistis citrella (Stainton)

(Lepidoptera; Phyllocnistidae)

Taxonomical Position/Scientific Classification

Kingdom:	Animalia
Phylum:	Arthropoda
Class:	Insecta
Order:	Lepidoptera
Family:	Phyllocnistidae
Genus:	Phyllocnistis
Species:	***P. citrella***

Biology and Life Cycle

1. **Egg**: About 36 to 76 eggs are deposited singly by female on undersurface of leaves near midrib. Eggs hatch out in 2 to 10 days.
2. **Larva**: On hatching, the caterpillar enters tissue and starts mining between the two layers. It is thin and yellowish green in colour. In about one or two weeks caterpillars become full grown and pupate inside the mine. Larval period is 5 to 10 days.

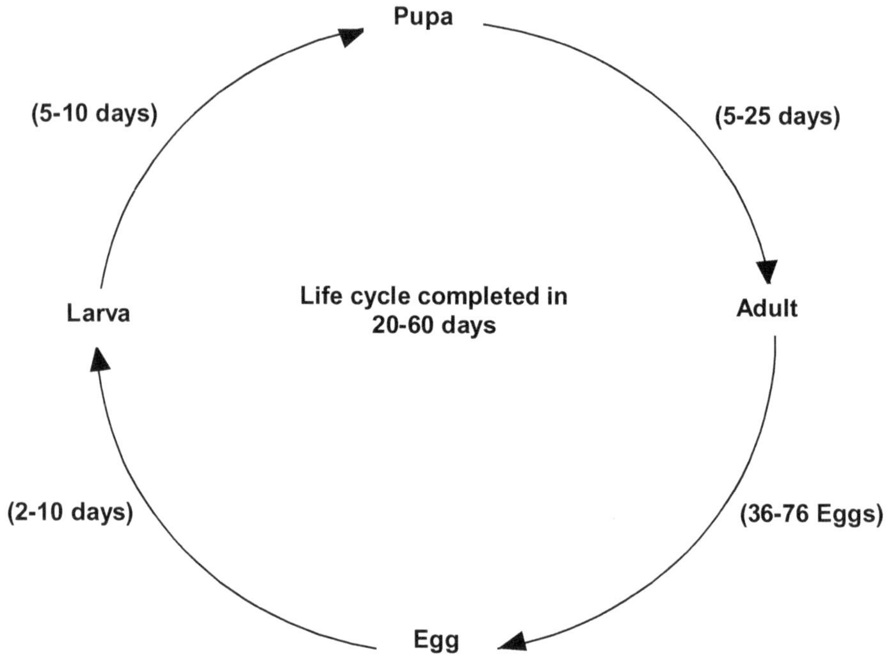

Figure 62: Life Cycle of *Phyllocnistis citrella*.

3. **Pupa**: Full-grown larva pupates in a leaf curl within a silken cocoon. Pupal period is 5 to 25 days.
4. **Adult**: Life cycle is completed in about 20 to 60 days. About 9 to 13 overlapping generations completed per year.

Marks of Identification

1. **Larva**: The full growth caterpillars are 5.1 mm in length, yellowish green in colour and are found inside galleries made in the leaf tissues.
2. **Adult**: Adult is a tiny silvery white moth with wing expanse of 6 mm. Hind wings are pale white while forewings have brownish strips and a prominent black spot near the tip. Both the pairs of wings have fringe of hairs.

Host Crops

Citrus, cinnamon, jasmine, pomelo, bael.

Nature and Symptoms of Damage

☆ Damage is caused mainly by caterpillars by mining into the tender leaves and feeding on inner leaf tissues leaving both the epidermal layers intact.

- ☆ The infested leaves show transparent, zigzag galleries. Attacked leaves later on turn pale, curl up and finally dry up.
- ☆ Damage by this pest may further lead to secondary infections of fungi and bacteria causing "citrus canker".
- ☆ About 34 to 46 per cent leaves are damaged by the pest.

Economic Threshold Level (ETL)

10 per cent affected leaves

Management

(A) Mechanical Measures

1. Prune heavily the infested leaves in winter and burn them.
2. Clipping of first attacked leaves and crushing larvae in galleries, help to minimize the further infestation.

(B) Chemical Measures

1. Pest can be controlled effectively by spraying following insecticides at 15 days interval commencing from the appearance of pest: Dimethoate 0.05 per cent or Quinalphos 0.05 per cent or Chlorpyriphos 0.05 per cent or Malathion 0.07 per cent or Fenvalerate 0.01 per cent.
2. Spraying of Azadirachtin (1500 ppm) 1 per cent (100 ml/10 lit. of water) or Neem oil @ 100 ml/10 litres of water should also be beneficial.
3. Spray NSKE 5 per cent (50 g/L) or neem cake extract 5 per cent or neem oil 3 per cent or imidacloprid 17.8 SL 125 ml per ha.
4. Spray dichlorvos 76 WSC 1.0 L, dimethoate 2.0 L per ha. Use 5-15 L of water per tree/1500-2000 L of water per ha.

(C) PESTS OF PAPAYA

1. Fruit Fly

Also known as: Oriental fruit fly

Scientific Name, Order and Family

Bactrocera dorsalis H.

(Diptera; Tephritidae)

Taxonomical Position/Scientific Classification

Kingdom:	Animalia
Phylum:	Arthropoda
Class:	Insecta
Order:	Diptera

Family:	Tephritidae
Genus:	Bactrocera
SubGenus:	Bactrocera
Species:	***B. dorsalis***

Biology and Life Cycle

1. **Egg**: A single female can lay 150 - 200 eggs (average 50) in about a month. The eggs are laid in clusters of 2 - 15 eggs and these hatches in 2 - 3 days during March and 1 - 1½ days during April.
2. **Larva (Maggot)**: Maggot duration is 6 days in summer and extends up to 19 days with the fall in temperature.
3. **Pupa**: Pupation usually takes place 80 - 160 mm below the soil surface and pupal period ranges from 6 days (summer) - 44 days (winter).
4. **Adult**: The adult fly is light brown with transparent wings. Adult flies are very conspicuous. These are about 7 mm long, with hyaline wings (expanse : 13 - 15 mm), thorax ferrugineous without yellow middle stripe, legs yellow, abdomen conical in shape and dark brown in colour. Preoviposition period is 2 - 5 days.

Marks of Identification

Adult is 5 mm long, small fly with single pair of transparent out stretched wings amd conical shaped yellowish brown abdomen. The brown and grey coloured patches distinguish these flies from others. The maggots are dirty white, legless and 12 mm in length when full grown. Body of maggot goes on tapering at one end. Eggs are white, oval and elongated.

Host Crops

Papaya, mango, guava, pulm, peach, pear, fig, ber, apple, banana, pomegranate, citrus, cucurbits *etc.*

Nature and Symptoms of Damage

On hatching, maggots feed inside on the pulp of fruits. The infested fruits can be identified by the presence of brown resinous juice which oozes out of the punctures. These punctures serve as entry for various bacteria and fungi. As a result, infested fruits start rotting, get distorted and malformed in shape and fall down prematurely. About 50 per cent fruits are damaged by this pest.

- ✰ Maggot bore into semi-ripen fruits with decayed spots and dropping of fruits.
- ✰ Oozing of fluid.
- ✰ Brownish rotten patches on fruits.

Economic Threshold Level (ETL)

10 per cent affected fruits

Management

(A) Cultural Measures

1. Fruits should be harvested before they start ripening.
2. Fallen fruits should be collected and destroyed.
3. Raking soil or ploughing after harvest to destroy hibernated pupae.
4. Mulching beneath the plant helps to kill the pupae inside the soil, if infestation is of severe nature.

(B) Mechanical Measures

1. Monitor the activity of flies with methyl eugenol sex lure traps.
2. Installation of yellow sticky traps.

(C) Chemical Measures

1. Application of spray bait containing 20 ml Malathion + 200 g jaggery + 20 litres of water.
2. Spray Malathion 0.05 per cent or 0.2 per cent Carbaryl during flowering to reduce intensity of infestation.
3. Spray application of imidacloprid 200SL at 0.01 per cent or triazophos 40EC at 0.06 per cent during heavy infestation.
4. Bait spray - combing any one of the insecticides and molasses or jaggery 10 g/l,
   - ✰ Fenthion 100 EC @ 1ml/litre of water,
   - ✰ Malathion 50EC 2 ml/litre of water,
   - ✰ Dimethoate 30 EC 1 ml/litre of water,
   - ✰ Carbaryl 50 WP 4 g/l two rounds at 2 weeks interval before ripening of fruits.
5. Prepare bait with methyl eugenol 1 per cent solution mixed with Malathion 0.1 per cent.
6. Take 10 ml of this mixture per trap and keep them in 25 different places in one hectare

(D) Biological Measures

1. Field release of natural enemies: *Opius compensates* and *Spalangia Philippines.*
2. Parasitoids: *Opius fletcheri, Fopius arisanus, Diachasmimorpha kraussi.*
3. Predator: Ants.

2. Aphid

Also known as: Green peach aphid

Scientific Name, Order and Family

Myzus periscae (Sulzer)

Aphis gossypii (G.)

(Homoptera, Aphididae)

Taxonomical Position/Scientific Classification

Kingdom:	Animalia
Phylum:	Arthropoda
Class:	Insecta
Order:	Hemiptera
Family:	Aphididae
Genus:	Myzus
Species:	***M. persicae***

Biology and Life Cycle

- **Egg:** Eggs are very tiny, shiny-black and are found in the crevices of bud, stems and barks of the plant. Aphids usually do not lay eggs in warm parts of the world.
- **Nymph:** Nymphs (immature stages) are young aphids, they look like the wingless adults but are smaller. They become adults within 7 to 10 days.
- **Adult:** Adults are small, 1 to 4 mm long, soft-bodied insects with two long antennae that resemble horns. Most aphids have two short cornicles (horns) towards the rear of the body.

Host Crops

Papaya, cotton, okra *etc.*

Nature and Symptoms of Damage

- Both nymph and adult suck the cell sap from leaves and there by devitalized the plant.
- The affected plants gradually die away.
- Stunted growth Development of black sooty mould due to the excretion of honeydew.
- The pest act as vector of virus disease causing phenomenal loss. The most common disease transmitted by aphids are papaya mosaic and little leaf.

☆ Mosaic stats as a necrotic dots on leaves, later the lamina become yellowish with blistered patches. In leaf ant disease, the young leaves remain small, curled and wrinkled.

☆ The plants remain dwarfed and fruits do not develop properly.

Management

(A) Cultural Measures

1. If disease is noticed, cut and destroy the affected leaves to prevent the disease from spreading.

(B) Chemical Measures

1. As prophlyatic measure, give regular sprays with Dimethoate @ 0.03 per cent, especially in seed bed and plots as a vector control.

(C) Biological Measures

1. Parasitoids: *Aphidius colemani, Aphelinus* sp.
2. Predators: Fire ant, Robber flies, Big-eyed bug (*Geocoris* sp.), Earwig, Ground beetle, Cecidomyiid fly, Dragon fly, Praying mantis, Lacewing, Ladybird beetle, Spider *etc.*

3. White Fly

Scientific Name, Order and Family

Bemisia tabaci (Gennadius)

(Homoptera, Aleurodidae)

Biology and Life Cycle

☆ **Egg:** The females mostly lay eggs near the veins on the underside of leaves. Each female can lay about 300 eggs in its lifetime. Eggs are small (about 0.25 mm), pear-shaped, and vertically attached to the leaf surface through a pedicel. Newly laid eggs are white and later turn brown.

☆ **Nymph:** Upon hatching, the first instar larva (nymph) moves on the leaf surface to locate a suitable feeding site. Hence, it is commonly known as a "crawler." If then inserts its piercing and sucking mouthpart and begins sucking the plant sap from the phloem. Adults emerge from puparium through a T-shaped slit, leaving behind empty pupal cases or exuviae.

☆ **Adult:** The whitefly adult is a soft-bodied, and moth-like fly. The wings are covered with powdery wax and the body is light yellow in colour. The wings are held over the body like a tent. The adult males are slightly smaller in size than the females. Adults live from one to three weeks.

Marks of Identification

☆ **Egg** - pear shaped, light yellowish

- ☆ **Nymph -** Oval, scale-like, greenish white, Settle down on a succulent part of leaves.
- ☆ **Adult -** White, tiny, scale-like adults.

Nature and Symptoms of Damage

- ☆ Both the adults and nymphs suck the plant sap and reduce the vigor of the plant. In severe infestations, the leaves turn yellow and drop off. When the populations are high they secrete large quantities of honeydew, which favors the growth of sooty mould on leaf surfaces and reduces the photosynthetic efficiency of the plants.
- ☆ Beside causing direct damage, it transmit leaf curl virus disease in papaya.

Management

(A) Cultural Measures

1. Field sanitation.
2. Removal of host plants.

(B) Mechanical Measures

1. Installation of yellow sticky traps.

(C) Chemical Measures

1. Spray application of imidacloprid 200SL at 0.01 per cent or triazophos 40 EC at 0.06 per cent during heavy infestation.
2. Spray neem oil 3 per cent or NSKE 5 per cent.

(D) Biological Measures

1. Release of predators *viz.*, Coccinellid predator, *Cryptolaemus montrouzieri*, *Dicyphus hesperus*, Lacewing, Ladybird beetle, Big-eyed bugs (mirid bug) (Geocorissp).
2. Release of parasitoids *viz.*, *Encarsia haitierrsis* and *E.guadeloupae*, *Encarsia formosa*, *Eretmocerus* spp., *Chrysocharis pentheus*.

4. Mite

Also known as: Red spider mite and two-spotted spider mite

Scientific Name, Order and Family

Tetranychus urticae (Koch)

(Trombidiformes, Tetranychidae)

Taxonomical Position/Scientific Classification

Kingdom:	Animalia
Phylum:	Arthropoda
Class:	Arachnida
Order:	Trombidiformes
Family:	Tetranychidae
Genus:	Tetranychus
Species:	***T. urticae***

Biology and Life Cycle

- ✰ **Egg:** Eggs reddish, spherical and provided with a small filament. Incubation period is 4-6 days, before hatching becomes light orange colour.
- ✰ **Nymph:** Upon hatching, it will pass through a larval stage and two nymphal stages before becoming adult. Developmental stages include six legged larva, protonymph and deutonymph.
- ✰ **Adult:** Adult female elliptical in shape, bright crimson anteriorly and dark pruplish brown posteriorlym. Mites spin a web of silken threads on the leaf. Each developmental stage is followed by a quiescent stage and life cycle completed in 10--14 days.

Marks of Identification

Tetranychus urticae is extremely small, barely visible with the naked eye as reddish or greenish spots on leaves and stems; the adult females measure about 0.4 mm long. The red spider mite, which can be seen in greenhouses and tropical and temperate zones, spins a fine web on and under leaves.

Host Crops

These include most vegetables and food crops – including papaya, peppers, tomatoes, potatoes, beans, maize and strawberries – and ornamental plants such as roses.

Nature and Symptoms of Damage

1. Spider mites usually extract the cell contents from the leaves using their long, needle- like mouthparts. This results in reduced chlorophyll content in the leaves, leading to the formation of white or yellow speckles on the leaves.
2. In severe infestations, leaves will completely desiccate and drop off. The mites also produce webbing on the leaf surfaces in severe conditions.
3. Under high population densities, the mites move to using strands of silk to form a ball-like mass, which will be blown by winds to new leaves or plants, in a process known as "ballooning."

Economic Threshold Level (ETL)

10 no./leaf.

Management

(A) Chemical Measures

1. Dicofol 18.5 EC 2.5ml/lit or wettable sulphur 50 WP @ 2 g/lit.

(B) Biological Measures

1. Predators: Anthocorid bugs (*Orius* spp.), mirid bugs, syrphid/hover flies, green lacewings (*Mallada basalis* and *Chrysoperla* sp.).
2. Predatory mites (*Amblyseius alstoniae, A. womersleyi, A. fallacies* and *Phytoseiulus persimilis*), predatory coccinellids (*Stethorus punctillum*), staphylinid beetle (*Oligota* spp.), predatory cecidomyiid fly (*Anthrocnodax occidentalis*), predatory gall midge (*Feltiella minuta*), Predatory thrips *etc.*

(D) PESTS OF BANANA

1. Rhizome Weevil

Common name/Also known as: Rootstock weevil/Banana weevil/Corm weevil

Scientific Name, Order and Family

Cosmopolites sordidus Germ.

(Coleoptera; Curculionidae)

Taxonomical Position/Scientific Classification

Kingdom:	Animalia
Phylum:	Arthropoda
Class:	Insecta
Order:	Coleoptera
Family:	Curculionidae
Genus:	Cosmopolites
Genus:	***C. sordidus***

Biology and Life Cycle

1. **Egg**: Weevils deposit 15 to 70 white eggs singly in the cavities made in the skin of the corm of banana plant or present on the upper part of rhizome. Hatching period is usually about a week, but sometimes extended upto 4 weeks.
2. **Larva**: Grub development is completed in period of 2-6 weeks. Groub is Apodous, yellowish white with red head.

3. **Pupa**: White in colour, pupation takes place inside larval tunnels made on corm of banana plant. Pupal period is about a week.
4. **Adult**: Weevils live for few months to two years. They can survive upto 6 months without food.

Marks of Identification

Adult is stoutly built, 10 to 13 mm long, reddish brown weevil with long curved snout. Five punctures are present on elytra. Grubs are yellowish with reddish head, apodus, spindle shaped and 8 to 12 mm in length. The adult weevil is sluggish and nocturnal in habit.

Host Crops

Banana (Monophagous).

Nature and Symptoms of Damage

- ✰ The young grub tunnels into the base of suckers, roots and rhizome/corm.
- ✰ Presence of larval tunnels on the entire length of pseudostems due to which plant become wilted.
- ✰ Yellowing and withering of leaves, reduced plant vigour, root destruction, reduced fruit production and are easily blown over by the wind.

Management

(A) Cultural Measures

1. Clean cultivation and destruction of feeding and sheltering places of weevil.
2. At harvest, cut the pseudostems below ground levels.
3. At planting, use healthy planting materials *i.e.* suckers/rhizomes free from infestation of weevil.
4. Trimming the rhizome
5. Deep ploughing of field to expose the pupae to sunlight.

(B) Mechanical Measures

1. Regular removal of pseudostems below the ground level and also the dead leaves.
2. Use cosmolure trap at 5/ha.
3. Use of pheromone trap @ 16 traps/ha for trapping adults.
4. Disc-on-stump traps can be used for trapping weevils.
5. Trap adult weevils with pseudostem cut into small pieces about 1/2 m kept near infested clump at 65/ha.

(C) Chemical Measures

1. Soil incorporation at the time of planting: carbofuran 3 G 10g, phorate 10 G 5 g/plant, lindane 1.3 D @ 20 g/plant.
2. Sucker dipping in triazophos solution (2.5 ml in 1 lit water) for about 20 minutes kills the eggs and grubs.
3. Remove the entire plants after harvest and treat the pit with carbaryl (1g/lit) or chlorpyrifhos (2.5ml/lit).
4. Before planting, the suckers should be dipped in 0.1 per cent Quinalphos emulsion.
5. Apply castor cake 250g or Carbaryl 50g dust or Phorate 10g per pit before planting also prevents infestation.
6. Severe attack dimethoate, Methyl demeton, or Phosphamidon may be sprayed around the collar region.
7. At the time of planting, treat the pits with 5 per cent Chlordane dust @ 60 gm/pit.
8. Suckers/Rhizomes are soaked in 0.1 per cent solution of Chlordane before planting.

(D) Biological Measures

1. Application of bio-control agents, *Beauveria bassiana* and *Metarhizium anisopliae*, causes more than 90 per cent mortality of the weevils.

2. Pseudostem Borer

Also known as: Stem weevil

Scientific Name, Order and Family

Odoiporus longicollis

(Coleoptera; Curculionidae)

Taxonomical Position/Scientific Classification

Kingdom:	Animalia
Phylum:	Arthropoda
Class:	Insecta
Order:	Coleoptera
Family:	Curculionidae
Genus:	Odoiporus
Genus:	***O. longicollis***

Biology and Life Cycle

Eggs thrust within air chamber @ one egg/air chamber in leaf sheath through oviposition slits made by rostrum. Eggs are laid in the psedostem about 1-1.5 m above ground level. Egg period is 4-8 days. Grub apodous, grub period is 30-65 days with five larval instars. Grubs pupate in tunnel towards the periphery in a cocoon made from pieces of fibrous sheath materials. Pupal period is 24-44 days. The total life cycle completes about in 55-60 days.

Marks of Identification

1. **Egg:** Creamy, cylindrical with rounded ends.
2. **Larva:** They are fleshy, yellowish white dark brown head and apodous.
3. **Pupa:** Pupate in fibrous cocoon, pale yellow colour and is exarate.
4. **Adult:** Brownish black and measure 23-39 mm.

Host Crops

Banana

Nature and Symptoms of Damage

- ☆ Both grubs and adults attack all members of genus *Musa*. Infestation of the weevil starts in 5 month old plants.
- ☆ The damage caused by adult weevil is primarily the result of destruction of tissues of corm. Sometimes, this damage may be made more prominent due to secondary infection of micro-organisms.
- ☆ Early symptoms of the infestation are the presence of small pinhead sized holes on the stem, fibrous extrusions from bases of leaf petiole and exudation of a gummy substance from the holes on the pseudostem.
- ☆ The grubs bore into the corm of banana making tunnels. The seedlings are killed due to the attack of boring larvae. The infested plants break out.
- ☆ Central shoot is killed; plants show premature withering; leaves become scarce, fruits become undersized and suckers are killed outright.
- ☆ The estimated yield loss due to this pest is between 10 – 90 per cent depending on the growth stage in which the infestation occurs and it is the highest in 5 months old crop.

Management

(A) Cultural Measures

1. Maintain healthy plantation by periodical removal of dry leaves and suckers.
2. Prune the side suckers every month.
3. After harvesting the bunch, remove the Pseudostem from ground level and destroy them in order to avoid it serving as a breeding site for the pest.
4. Adopt good cultivation practices to improve weevil tolerance.
5. Do not dump infested materials into manure pit.

(B) Mechanical Measures

1. Use Disc-on-stump or longitudinal Pseudostem traps @ 100/ha for trapping weevils.
2. Uproot infested trees, chop into pieces and burn.

(C) Chemical Measures

1. Inject pseudostem with monocrotophos 36 WSC (50 ml + 350 ml water) @ 2 ml at 45 cm height and another at 150 cm height from ground level at monthly intervals from 5th - 8th month. Beyond 8 months (after flowering), this should not be done.
2. Swabbing with chlorpyripos 2.5ml/l + adjuvant 1ml/l on the stem prevents infestation of banana stem weevil.
3. When jelly exudation is noticed, inject 2ml triazophos solution (350 ml in 150 ml water). Two injections per plant at 2 and 4 feet above the ground level till flowering. The injection needle should enter only two or three leaf sheaths and should not touch the central core.
4. Celphos (3 tablets/plant) is recommended for control of egg, larva, pupa and adult population of the insect, application of. After placing the tablet inside the pseudostem, the slit should be plastered with mud.
5. Application of Carbofuran (3 g of granules/stool) is very effective to control the pest. Alternately, application of Carbaryl WP (0.1 per cent) also controls the pest population.

(E) PESTS OF GUAVA

1. Bark Borer

Also known as: Bark eating caterpillar.

Scientific Name, Order and Family

Indarbela quadrinotata (Walker)

I. tetraonis (Moore)

(Lepidoptera; Metarbelidae)

Taxonomical Position/Scientific Classification

Kingdom:	Animalia
Phylum:	Arthropoda
Class:	Insecta
Order:	Lepidoptera
Family:	Metarbelidae
Genus:	Indarbela
Species:	***I. quadrinotata***

Biology and Life Cycle

1. **Egg**: Females lay about 350 spherical eggs in groups of 15 to 25 in cracks and crevices in the bark of tree trunk in the month of May and June. Hatching period is 8 to 10 days.
2. **Larva**: Stout and dirty brown in colour. Larval period is 10 to 11 months.
3. **Pupa**: Pupal period is 21 to 41 days. Pupation takes place inside larval tunnels made in tree trunk or under silken webbings.
4. **Adult**: Stout yellowish –brown moth with brown wavy markings on the forewings. Hindwings is white colour. Males are smaller than the females. Adults are short lived (3 days only). Pest completes one generation in a year.

Marks of Identification

Adult is short, stout, ashy grey coloured moth with wing expanse of about 35 to 50 mm. Male moth has bipectinate antennae at apex. Full grown larvae are cylindrical, 38 to 45 mm long and dirty brown in colour with darkish head region.

Host Crops

Guava, cashew, jackfruit, ber, fig, mango, citrus, rose, jamun, falsa, Pomegranate, litchi.

Nature and Symptoms of Damage

- ✰ The newly hatched larvae feed on bark of the stem or branches during night.
- ✰ In advanced stages, larvae prepare silken galleries or webs on the stem with the help of silken thread secreted by the larva, it's own excreta and fine wood particles. Thereafter remaining below such protective covering, larvae make tunnels in irregular directions.
- ✰ As a result of feeding, sap conducting tissues are damaged which interrupt the translocation of sap within plant body. The severely infested plant shows sickly appearance.
- ✰ Some varieties of guava like "Seedless guava" and "Apple guava" are comparatively more susceptible while "Cattley", "Allahabad safeda", "Strawberry" are tolerant to the attack of bark borer.

Management

(A) Cultural Measures

1. Avoid growing susceptible varieties like Seedless guava, Apple guava *etc.*, and keep the orchard clean.
2. Remove and destroy dead and severely affected branches of the tree.
3. Remove alternate host, silk cotton and other hosts.
4. Clean the webs around the affected portion and inject kerosene oil into the holes and seal with mud.

(B) Mechanical Measures

1. As and when infestation is noticed, kill the caterpillars mechanically by inserting an iron spike into the holes made by these caterpillars.

(C) Chemical Measures

1. As and when pest appears, remove the silken webbings and inject borer solution (2 parts of carbon disulphide + 1 part of creosote oil + 1 part of chloroform) OR ED/CT mixture OR carbon disulphide into larval tunnels.
2. Inject larval holes with quinalphos @ 0.01 per cent or fenvalerate @ 0.05 per cent.
3. Spray with carbaryl @ 0.04 per cent or dichlorovos @ 0.08 per cent on the stem or on affected part.
4. Swab Coal tar + Kerosene @ 1:2 or Carbaryl 50 WP 20 g/l (basal portion of the trunk - 3 feet height) after scraping the loose bark to prevent oviposition by adult beetles.
5. If infestations are severe then apply the copper oxychloride paste on the trunk of the tree.
6. Hook out the grub from the bore hole - apply monocrotophos 36 WSC 10 to 20 ml/hole.
7. One celphos tablet (3 g aluminum phosphide) per hole
8. Apply carbofuran 3G 5 g per hole and plug with mud.
9. Insecticides like 0.1 per cent Dichlorvos or 0.05 per cent Trichlorofan can be injected into larval tunnels using plastic syringe.

2. Oriental Fruit Fly

Scientific Name, Order and Family

Bactrocera diversa

(Diptera; Tephritidae)

Taxonomical Position/Scientific Classification

Kingdom:	Animalia
Phylum:	Arthropoda
Class:	Insecta
Order:	Diptera
Family:	Tephritidae
Genus:	Bactrocera
Species:	***B. diversa***

Biology and Life Cycle

1. **Egg**: Female flies lay 150 to 200 eggs beneath the rind or on soft skin of developing fruits, singly or in cluster of 2 to 15. Hatching period is 2 to 10 days.
2. **Larvae/Maggot**: Maggot pale cream, cylindrical, 5-8 mm in length and have yellowish apodous. Maggot development is completed in period of 6 to 29 days. Maggot pupates is in soil. Pupal period lasts for 8 to 10 days. Pest hibernates as a pupa in soil.
3. **Adult/Fly**: Adult smoky brown with greenish black thorax having yellow marking with transparent wing.

Marks of Identification

- ☆ **Adult -** Brown or dark brown with hyaline wings and yellow legs.

Host Crops

Guava and Papaya.

Nature and Symptoms of Damage

- ☆ Both adults and maggots are capable to cause damage.
- ☆ Females with their needle like ovipositor deposit eggs beneath the rind of developing fruits. The incision made while laying eggs serves as entry for fermenting organisms. The maggots on hatching enter inside the fruit and feed on the pulp.
- ☆ Maggot bore into semi-ripen fruits with decayed spots and dropping of fruits.
- ☆ Oozing of fluid.
- ☆ As the maggots grow inside the brown patch extends. As a result, infested fruits start rotting, get distorted and malformed in shape and fall down prematurely.
- ☆ As a result of maggot damage and deposition of excreta, brown spots are developed at the site of injury.
- ☆ The late maturing varieties of mango and guava are more susceptible.

Economic Threshold Level (ETL)

10 per cent affected fruits.

Management

(A) Cultural Measures

1. Fruits should be harvested before they start ripening.
2. Collect and destroy infested and fallen fruits by burning them in very deep pit.
3. The undersized fruits left on the tree should be picked and destroyed.

4. Sanitation of orchard is the most important measure against fruit flies.
5. Fruit fly infestation can be avoided by harvesting fruits before ripening.
6. Deep ploughing around the trees during winter to expose and kill pupae.
7. Bagging the fruits with paper bags or cloth bag is effective but it can not be practiced on large scale as it is expensive and laborious.

(B) Mechanical Measures

1. Monitor the activity of flies with methyl eugenol sex lure traps (25/ha).

(C) Chemical Measures

1. Spray Malathion 50 EC @ 0.05 per cent or 0.2 per cent Carbaryl during flowering to reduce intensity of infestation.
2. Application of spray bait containing 20 ml Malathion + 200 g jaggery + 20 litres of water.
3. Adult flies may be trapped and killed by poison baiting/spraying with mixture of 20 ml of Malathion or 50 ml of Diazinon + 200 gm of molasses + 2 lit. of water for baiting and 20 lit of water for bait spraying.
   - ☆ Prepare methyl eugenol and malathion 50 EC mixture at 1:1 ratio(take 10 ml mixture/trap)
   - ☆ Bait spray combining molasses or jaggery 10g/l and one of the insecticides,
   - ☆ Malathion 50 EC 2 ml/l
   - ☆ Dimethoate 30 EC 1ml/lit, two rounds at fortnight interval before ripening of fruits

(D) Biological Measures

1. Field release of parasitoids such as *Opius compensates*, *Spalangia philippinensis* and *Diachasmimorpha kraussi*.

(F) PESTS OF POMEGRANATE

1. Anar Butter-fly

Also known as: Fruit borer/Anar Caterpillar

Scientific Name, Order and Family

Deudorix isocrates (Fab.)

(Lepidoptera; Lycaenidae)

Taxonomical Position/Scientific Classification

Kingdom:	Animalia
Phylum:	Arthropoda
Class:	Insecta

Order:	Lepidoptera
Family:	Lycaenidae
Genus:	Deudorix
Species:	***D. isocrates***

Biology and Life Cycle

1. **Egg**: Butterflies lay eggs singly on calyx of flowers or on young fruits. Incubation period is 7 to 10 days.
2. **Larva**: The freshly emerged larva bores into the fruit and feeds on the pulp and seeds and becomes full-grown in about 18 to 47 days.
3. **Pupa**: Full grown larva pupates mostly inside the fruit and rarely outside. Pupal period varies from 8 to 34 days.
4. **Adult**: Life cycle is completed in 1-2 months. Pest breeds throughout the year. Four overlapping generations completed in a year.

Marks of Identification

Adult is glossy bluish violet (male) to brownish violet (female) butterfly with wing expanse of about 50 mm. Larva is short, stout, dark brown in colour with hairs and whitish patches on the body and measures 18-19 mm in length when full grown. Eggs are shiny white in colour, 1-3 mm in size and oval in shape.

Host Crops

Apple, Guava, Ber, Litchi, Tamarind, Pear, Peach, Plum, Sapota, Loquat, Citrus, Mango and other wild fruits.

Nature and Symptoms of Damage

- ☆ The larva on hatching enters into the fruit by boring through the thin and soft rind of fruits and grows inside feeding on internal contents of the fruits. As many as, eight caterpillars may be found feeding in single fruit.
- ☆ Infested fruits are subsequently attacked by fungi and bacteria causing fruits to rot. Affected fruits ultimately fall down and are of no use.
- ☆ Such fruits fetch less price in the market. Pest causes 40 to 90 per cent damage to fruits.

The conspicuous symptoms are:

- ☆ Offensive smell and excreta of caterpillar at the entry hole of fruit.
- ☆ The affected fruits ultimately falling down.
- ☆ It also causes rottening of fruits.

Economic Threshold Level (ETL)

5 eggs/plant

Management

(A) Cultural Measures

1. Remove and destroy all infested and fallen fruits.
2. Clean cultivation as weed plants serve as alternate hosts.
3. Clip off calyx cup immediately after pollination followed by two applications of neem oil @ 3 per cent.

(B) Mechanical Measures

1. Bagging of developing fruits with paper bag or butter paper or plastic bag or cloth bag on small scale when the fruits are up to 5 cm.
2. Use light trap @ 1/ha to monitor the activity of adults.

(C) Chemical Measures

1. Spraying of Malathion 50 EC 0.1 per cent or Dimethoate 30 EC 0.06 per cent, two rounds, one at flower formation and next at fruit set.
2. Flowering stage - spray NSKE 5 per cent or neem formulations 2 ml/1
3. Apply dimethoate 30 EC 1.5 ml/litre of water.
4. Spray deltamethrin 2.8 EC (1.5 ml/litre of water) at fortnightly interval from the stage of flowering to fruit development.
5. Spray Malathion 50 EC 0.1 per cent or methomyl 40 SP @ 1.0 ml/l or azadirachtin 1500 ppm @ 3.0 ml/l at 15 days intervals commencing from initiation of flowering up to the harvesting subjected to the presence of fruit borer.
6. Give four sprays with Carbaryl 50 WP @ 0.2 per cent or Deltamethrin 2.8 EC @ 0.002 per cent at 3 weeks interval starting from appearance of bud or fruit set.

Schedule of Application

1. 1st spray: When buds appear or fruits set.
2. 2nd application: 3 weeks after 1st spray.
3. 3rd application: 3weeks after 2nd spray.
4. 4th application: 3 weeks after 3rd spray.

(D) Biological Measures

1. Release *Trichogramma chilonis* at one lakh/acre.

2. Fruit Fly

Scientific Name, Order and Family

Bactrocera dorsalis H.

(Diptera; Tephritidae)

Taxonomical Position/Scientific Classification

Kingdom:	Animalia
Phylum:	Arthropoda
Class:	Insecta
Order:	Diptera
Family:	Tephritidae
Genus:	Bactrocera
SubGenus:	Bactrocera
Species:	***B. dorsalis***

Biology and Life Cycle

1. **Egg**: A single female can lay 150 - 200 eggs (average 50) in about a month. The eggs are laid in clusters of 2 - 15 eggs and these hatches in 2 - 3 days during March and 1 - 1½ days during April.
2. **Larva/Maggot**: Maggot duration is 6 days in summer and extends up to 19 days with the fall in temperature.
3. **Pupa**: Pupation usually takes place 8-16 cm below the soil surface and pupal period ranges from 6 days in summer and 44 days in winter.
4. **Adult**: The adult fly is light brown with transparent wings. Adult flies are very conspicuous. These are about 7 mm long, with hyaline wings (expanse : 13 - 15 mm), thorax ferrugineous without yellow middle stripe, legs yellow, abdomen conical in shape and dark brown in colour. Preoviposition period is 2 - 5 days.

Marks of Identification

Adult is 5 mm long, small fly with single pair of transparent out stretched wings amd conical shaped yellowish brown abdomen. The brown and grey coloured patches distinguish these flies from others. The maggots are dirty white, legless and 12 mm in length when full grown. Body of maggot goes on tapering at one end. Eggs are white, oval and elongated.

Host Crops

Papaya, mango, guava, pulm, peach, pear, fig, ber, apple, banana, pomegranate, citrus, cucurbits *etc.*

Nature and Symptoms of Damage

On hatching, maggots feed inside on the pulp of fruits. The infested fruits can be identified by the presence of brown resinous juice which oozes out of the punctures. These punctures serve as entry for various bacteria and fungi. As a result, infested fruits start rotting, get distorted and malformed in shape and fall down prematurely.

- ☆ Maggot bore into semi-ripen fruits with decayed spots and dropping of fruits.
- ☆ Oozing of fluid.
- ☆ Brownish rotten patches on fruits.

Economic Threshold Level (ETL)

10 per cent affected fruits

Management

(A) Cultural Measures

1. Fruits should be harvested before they start ripening.
2. Fallen fruits should be collected and destroyed.
3. Raking soil or ploughing after harvest to destroy hibernated pupae.
4. Mulching beneath the plant helps to kill the pupae inside the soil, if infestation is of severe nature.

(B) Mechanical Measures

1. Monitor the activity of flies with methyl eugenol sex lure traps.

(C) Chemical Measures

1. Application of spray bait containing 20 ml malathion + 200 g jaggery + 20 litres of water.
2. Spray Malathion 0.05 per cent or 0.2 per cent Carbaryl during flowering to reduce intensity of infestation.
3. Bait spray - combing any one of the insecticides and molasses or jaggery 10 g/l,
   - ☆ Malathion 50EC 2 ml/litre of water,
   - ☆ Dimethoate 30 EC 1 ml/litre of water,
   - ☆ Carbaryl 50 WP 4 g/l. two rounds at 2 weeks interval before ripening of fruits.
4. Prepare bait with methyl eugenol 1 per cent solution mixed with Malathion 0.1 per cent.
5. Take 10 ml of this mixture per trap and keep them in 25 different places in one hectare
6. Spray Fenthion 100 EC 2 ml/lit or Malathion 50 EC 2ml/lit.

(D) Biological Measures

1. Field release of natural enemies *Opius compensates* and *Spalangia philippinensis.*

(G) PESTS OF BER

1. Fruit Fly

Scientific Name, Order and Family

Carpomyia vesuviana Costa.

(Diptera; Tephritidae)

Taxonomical Position/Scientific Classification

Kingdom:	Animalia
Phylum:	Arthropoda
Class:	Insecta
Order:	Diptera
Family:	Tephritidae
Genus:	Carpomya
Species:	***C. vesuviana***

Biology and Life Cycle

1. **Egg**: female flies make cavities in the skin of fruit and lay 12 to 18 eggs (maximum 22) singly or in small batches of 2 to 4. Eggs are dull, spindle shaped, creamy white and 0.8 to 1.0 mm long. Hatching period 1 to 4 days.
2. **Larva**: Maggot development completed in 7-10 days.
3. **Pupa**: Pupae are barrel shaped, 4-5 mm long, light brown when freshly formed. Pupation takes place in soil. Pupal period 14-30 days. Aestivation is found in pupal stage.
4. **Adult**: Adult is smaller than housefly. Adult is small fly with black spots on the thorax and dark spots on the wings.Three to four generations are completed in a year.

Marks of Identification

Adult is 6 mm long, brownish yellow coloured black spotted fly with banded wings. Wings are hyaline, transparent with four yellowish cross bands. Maggots are creamy white, legless and 6mm in length when full grown.

Host Crops

Ber (Monophagous).

Nature and Symptoms of Damage

- ✰ The damage is caused by maggots which feed on the pulp inside the ber fruits.
- ✰ Female flies lay eggs inside the rind of ripening fruits by puncturing them with their needle like ovipositor. The development of fruit around the

punctures is arrested. The maggots on hatching feed on fleshy and juicy pulp.

- ☆ The infested fruits turn dark brown, rot and smell offensively.
- ☆ As many as, 18 maggots have been found within a single fruit. One maggot is sufficient to destroy the entire fruit.
- ☆ Fleshy varieties of the ber are more seriously damaged than the less fleshy.

Economic Threshold Level (ETL)

3 per cent fruit infestation.

Management

(A) Cultural Measures

1. Collect and destroy fallen and infested fruits by dumping in a pit and covering with a thick layer of soil.
2. Remove all wild ber bushes from the vicinity of grafted ber orchards.
3. Rake the soil around the trees during summer to expose larvae/pupae to feed by natural enemies. Raked soil may be mixed with Chloropyriphos 1.5 per cent dust to enhance the killing.
4. Cultivate fruit fly resistant varieties such as Safeda Ilaichi, Chinese, sanaur-1, Mirchia, Tikadi and Umran.

(B) Chemical Measures

1. Spray Dimethoate 30 EC @ 1 lt/ha during Feb-March and as per need, repeat the spray after 15 days.
2. Spray Carbaryl 0.1 per cent, four times at an interval of 21 days.
3. Poison baiting with 50 ml of diazinon or 20 ml of Malathion + 200 gm of jaggary or molasses + 2 lits of water has been found effective.

(C) Biological Measures

1. Conserve parasitoids *Opius compensates* and *Spalangia philippinensis.*

(H) PESTS OF LITCHI

1. Litchi Bug

Scientific Name, Order and Family

Chrysocoris stolii

(Hemiptera; Pentatomidae)

Taxonomical Position/Scientific Classification

Kingdom:	Animalia
Phylum:	Arthropoda
Class:	Insecta
Order:	Hemiptera
Family:	Pentatomidae
Genus:	Chrysocoris
Species:	*C. stolii*

Biology and Life Cycle

1. **Egg**: female lay 32 spherical and creamy white eggs singly or in 2-3 batches. Hatching per iod is 5-7 days.
2. **Nymph**: nymph is bright in colur. Nymph having 5 moults. Nymphal period is completed in 19-29 days with 5 moults.
3. **Adult**: Red on emergence and gradually become greenish. There are 6 distinct spots arrange in rows on thorax.

Host Crops

Litchi, Banana, Lantana and berries.

Nature and Symptoms of Damage

Both adults and nymph suck mostly on tender plant parts such as growing buds, leaf petioles, fruit stalks and tender branches of litchi tree. Excessive feeding causes drying of growing buds and tender shoots and ultimately fruit drop. The bugs when feed on the developing fruit, it causes the fruits to fall a couple of days later.

Management

(A) Cultural Measures

1. This pest is combated by shaking the trees in winter, collecting and dropping them into kerosene.
2. The eggs of pest are in group and visible which can be removed and destroyed.

(B) Chemical Measures

1. If chemicals are used, the timing of sprays is critical because the bugs vary in their susceptibility to insecticide at different times of the year, depending on body fat content and its nature. Many of these bugs may be controlled with dimethoate and fenthion.
2. At serious attack, spraying is done with Dimethoate 30 EC @ 2 lt/ha.

(C) Biological Measures

1. There are natural enemies which parasitize 70 to 90 per cent of eggs laid late in the season. The adults are attacked by several fungi, birds and red ants may also be used as biological means of control.

2. Leaf Roller

Scientific Name, Order and Family

Platypeplus aprobola (Meyer)

(Lepidoptera; Tortricidae)

Taxonomical Position/Scientific Classification

Kingdom:	Animalia
Phylum:	Arthropoda
Class:	Insecta
Order:	Lepidoptera
Family:	Tortricidae
Genus:	Platypeplus
Species:	***P. aprobola***

Biology and Life Cycle

The female moth lays eggs under the surface of newly emerged tender leaves which hatch within 2-8 days. The last instar larvae pupate in larval clip, a small portion of the leaf on the margin, both anteriorly and posteriorly and conceal themselves by bending and sealing the clipped piece of the leaf.

The breeding season of the leaf roller on litchi leaves is from August to February when new leaf flush is available and restricted breeding takes place during off-season (March to July) on alternate hosts such as kath-jamun (*Eugenia jambolana*) and chhota amaltas (*Cassia tora*) growing around litchi orchards. It may attack flower also.

Host Crops

Litchi and plants belonging to genus *Hibiscus* act as main host plant.

Nature and Symptoms of Damage

- ☆ The incidence of leaf roller is reported during July to February.
- ☆ The symptoms of leaf injury by the larvae are manifested through rolling of tender leaves and feeding inside. As a result of larval injuries, the infested twigs distort and wither.
- ☆ Litchi trees whose foliage is attacked by the larvae, very poor flowering or in case of younger tree, no flowering is seen in the season. Thus, the crop yield gets reduced considerably.

Management

(A) Cultural Measures

1. Plants belonging to genus Hibiscus act as main host plant therefore, such plants should not be grown near litchi orchards.
2. Collect and destroy the rooled leaves that contain larvae manually during light infestations as soon as rolling are observed on young plants.

(B) Chemical Measures

1. Spray Dursban 20 EC (chloropyriphos) @ 2m/litre as soon as the attack is seen particularly in August-September.
2. If necessary, carbaryl 2g/l can be applied when 20 per cent of leaf flushes are infested to minimize damage to young trees or at critical periods of leaf growth in older trees.

3. Mite

Common name: Leaf curl mite or Erionose mite

Scientific Name, Order and Family

Aceria litchi Channa.

(Acarina/Prostigmata, Eriophyidae)

Taxonomical Position/Scientific Classification

Kingdom:	Animalia
Phylum:	Arthropoda
Class:	Arachnida
Order:	Prostigmata
Family:	Eriophyidae
Genus:	Aceria
Species:	***A. litchi***

Biology and Life Cycle

1. **Egg**: Female lay very small eggs singly on the under surface of leaves. Hatching period is 2-3 days.
2. **Nymph**: Nymphal period is completed in 8-12 days.
3. **Adult**: Smaller in size. The adults start multiplying from the end of March and the peak activity is noticed around July. The life cycle is completed in 13 days under favourable conditions.

Marks of Identification

The body is warm like, 100-135 µm long, 30 µm wide and yellowish to reddish in colour. The Gnatosoma is some-what downcured. The predorsal shield has definite longitudinal central lines and is granular laterally; the scapular tubercles are 16 µm apart, set on the rear margin; the scapular setae are 17 µm long, projecting caudal. The coxae are strongly tuberculate; the setae are ahead of the junction of coxae. The genitalia are 15 µm wide and 10 µm long. The empodium is 5-rayed.

Host Crops

Litchi only.

Nature and Symptoms of Damage

- ☆ Both **nymphs and adults** damage the leaves, inflorescence and young developing fruits.
- ☆ They puncture and lacerate the tissues of the leaves with their stout rostrum and suck the cell sap. As a result of its infestation, undersurfaces of the infested leaves show abnormal growth of epidermal cells in the form of hair like velvety growth of chocolate brown colour.
- ☆ The infestation generally begins from the lower portion of the trees and gradually extends upwards. In addition to leaves, the mite also infests the newly formed leaf bud; inflorescence and epicarp of the newly formed fruit.
- ☆ Young plants and plantlets in nurseries are very susceptible and could even be killed if leaf drop is excessive.
- ☆ Dispersal of the erineum mite from one orchard to the other usually takes place through old infested litchi leaves or by winds by direct contact between trees or carried around by orchard workers, wind and bees.

Economic Threshold Level (ETL)

10/Leaf

Management

(A) Preventive Measures

Litchi mite control measures must be preventive. Once the mite is established, it is almost impossible to eradicate, hence depending upon infestation it is recommended that:

1. Layers should be prepared only from non-infested plants.
2. Layer saplings may be sprayed with 0.05 per cent dimethoate when they leave the nursery. Prior to planting out, the operation should be repeated twice at 10-14 day intervals.
3. The leaves should be checked regularly for symptoms over summer and autumn. All trees in an orchard are not to be flushed or infested at the same time. Therefore, branches infested with the mite should be cut off and burnt. Velvety growth on mite affected leaves Insect Pest Management.
4. After harvesting in June, infested branches must be removed.

5. In September-October, trees must be treated just prior to vegetative flushing with 0.05 per cent dimethoate either alone or in combination with 0.12 per cent dicofol. Spraying should be repeated two weeks later and monthly thereafter until new growth is free of all symptoms of infestation. Infested leaves should be gathered and burnt or buried deeply into the ground.
6. In December-January, just before flush/flower buds, the affected shoots must be removed and spraying of 0.15 per cent kelethane may be done. In the month of Feb., two sprays, one each before and after flowering have been found useful.

(B) Cultural Measures

1. Prune as much of the infested foliage from the tree as possible and destroy it. The mite is easily spread on nursery plants especially marcots taken from infested trees.
2. Use only clean, mite-free planting material. Where possible obtain your planting material from mite-free orchards and dip the trees in miticide before planting out. Wind and bees can also spread these mites.

(C) Chemical Measures

1. Spray wettable sulphur 4g litre of water after harvest of fruit in July and again in January.
2. Spray Rogor (Dimethoate 35 EC) @ 2ml/L before and after flowering.
3. Spray with Dicophol 18.5 EC @ 2.5 ml/lt. or Avamectin 1.8 EC @ 0.5 ml/lt. of water.
4. Spray *Pongamia* oil @ 2 ml/lt. of water mixed with acaricide.

(I) PESTS OF TAMARIND

1. Mealy Bug

Scientific Name, Order and Family

Perissopneumon tamarinda

(Homoptera; Pseudococcidae)

Biology and Life Cycle

The females descend the trees and enter into the soil for laying eggs. On hatching, the nymphs crawl up to the aerial portion of the tree and settle on succulent and tender foliage for feeding. The body is covered with white waxy fluffy secretions which is characteristic to these mealy bugs.

Host Crops

Tamarind

Nature and Symptoms of Damage

- Both the nymph and adult stages suck the cell sap on growing twigs and developing pods.
- It results in premature shedding of flower buds and immature pods.
- Besides the fruit setting capacity is affected, the size of the fruit is much reduced and the shape malformed.
- The honey dews excreted by the mealy bugs retard the growth of fruit development.

Management

(A) Cultural Measures

1. Plough around the trees to expose and kill the nymphs.
2. Prune the trees periodically.
3. Remove infested twigs and destroy the nymphal and adult stages in the initial stage of attack.
4. Destroy nests of ants around the trees.
5. Tie polythene sticky strapes of 30cm width round the tree trunks to prevent nymphs from ascending to the ground.
6. Banding the trunk with carbaryl, keeps off the attendantants and allows natural enemies to clean up the mealy bug infestation.

(B) Chemical Measures

1. Drench the infested twigs with 2 per cent neem oil suspension.
2. Spray dichlorvos (0.2 per cent) mixed with fish oil rosin soap (25g/lit) and treat the soil with 1.3 per cent diazinon dust at 500g/tree.
3. In case of severe infestation, spray with 0.1 per centdiazinon or Monocrotophos, Malathion 0.05 per cent ordimethoate 0.03 per cent.

(J) PESTS OF CASHEW

1. Tea Mosquito Bug

Also known as: Kajji bug/Plant lice

Scientific Name, Order and Family

Helopeltis antonii (Signoret)

(Hemiptera; Miridae)

Taxonomical Position/Scientific Classification

Kingdom:	Animalia
Phylum:	Arthropoda
Class:	Insecta
Order:	Hemiptera
Family:	Miridae
Genus:	Helopeltis
Species:	***H. antonii***

Biology and Life Cycle

1. **Egg**: The bugs lay about 30 to 50 eggs singly deep inside the tissues of tender shoots or floral branches or axis of inflorescence. Hatching period lasts for 6 to 7 days.
2. **Nymph**: Nymphal development comprising of five instars occupies 15 days. The build of population synchronises with the emergence of new flushes after cessation of monsoon shower.
3. **Adult**: Pest infestation starts in October and reaches to peak in January and pest disappears with the onset of monsoon.

Marks of Identification

Adult is small, active, reddish brown bug with a black head, red thorax and black and white abdomen. A pin head like knobbed process arises from mid dorsal aspect of the thorax. Adults measure about 10 mm in length. Freshly formed nymph is ant like and yellowish in colour initially and later becoming orange red. Eggs are creamy white, kidney shaped and 1.5 to 2 mm long.

Host Crops

Cashew, neem, cocao, tea, coffee, mulberry, avocado, tamarind, guava, apple, grape.

Nature and Symptoms of Damage

- Both nymphs and adults suck cell sap from leaves, tender shoots, floral branches, developing nuts and apples.
- Injury made by the suctorial type of mouth parts causes tender shoot to exude the resinous gummy substances. The tissues around the point of entry of stylets become necrotised and brown or black scabs/lesions formed presumably due to the action of phytotoxin present in the saliva of insect injected into the plant tissue at the time of feeding. The adjoining lesions unite and finally affected shoots/panicles dry up.
- Brownish – black necrotic patches develop on foliage.
- Elongate streaks and patches develop on shoots.
- Corky scab formation on fruits.

Management

(A) Cultural Measures

1. Collect and destroy the damaged plant parts.
2. Removal of stalks containing eggs while plucking.

(B) Mechanical Measures

1. Monitoring the infestation level in the field.

(C) Chemical Measures

1. Three sequential applications of insecticides are recommended for controlling this pest.
   - 1st spray with 0.05 per cent Monocrotophos at the time of emergence of new/vegetative flushes.
   - 2nd spray with 0.05 per cent Quinolphos at the time of initiation of panicle.
   - 3rd spray with 0.1 per cent Carbaryl or 0.05 per cent Phosalone at the time of fruit setting of the crop.
2. Application of Quinalphos 25 EC @ 750 ml/ha or chlorpyrifos 20 EC @ 750 ml/ha or fenthion 80 EC @ 200 ml/ha or quinalphos 25 EC + dichlorvos 76 EC @ 750+250 ml/ha. Spraying may be undertaken during early mornings or evenings when these bugs are active.

(D) Biological Measures

1. Encouraging the egg parasitoid, *Erythmelus helopeltidis* population to build up.

2. Stem and Root Borer

Scientific Name, Order and Family

Plocaederus ferrugineus (Linn.)

(Lepidoptera; Cerambycidae)

Taxonomical Position/Scientific Classification

Kingdom:	Animalia
Phylum:	Arthropoda
Class:	Insecta
Order:	Coleoptera
Family:	Cerambycidae
Genus:	Plocaederus
Genus:	***P. ferrugineus***

Biology and Life Cycle

1. **Egg**: Female beetle lays eggs singly inside the cracks and crevices present on the exposed portions of roots of cashew or inside loose bark on tree trunk upto one meter height from the ground level. Incubation period is 4 to 6 days.
2. **Larva**: Grub development is completed in 6 to 7 months.
3. **Pupa**: Pupation takes place inside larval tunnel made in the root or collar region of the plant. The pupal period varies from 20 to 60 days.
4. **Adult**: Pest completes one generation in a year.

Marks of Identification

Adult is medium sized, 50 mm long, reddish brown longicorn beetle, the head and thorax of which are dark brown or almost black. Full grown grubs are whitish, fleshy and 75 to 100 mm long.

Host Crops

Cashewnut (Monophagous).

Nature and Symptoms of Damage

- ☆ The grubs on hatching bore into fresh tissues of the bark and feed on the subepidermal and sapwood tissues making tunnels in irregular directions.
- ☆ Due to the injury to living cell, a resinous material oozes out which on exposure to air gets hardened as gummy substance and the same resembles as **"gummosis"**.
- ☆ When the vascular tissues are damaged, the ascent of plant sap is arrested. As a result, the leaves turn yellow and are shed subsequently.
- ☆ The both young and older plants are infested. Young plants are killed outright and the older trees become weak initially and the death appears in due coarse of time.

The series of symptoms noticed are:

- ☆ Presence of small holes at collar region.
- ☆ Gummosis.
- ☆ Extrusion of frass through holes.
- ☆ Yellowing and shedding of leaves.
- ☆ Drying up of twigs.
- ☆ Death of the plant.

Management

(A) Cultural Measures

1. Destroy infested plant parts.

2. The dead trees and those which are beyond the stage of recovery may be cut and destroyed by burning as they serve as reservoir for the multiplication of this pest.

(B) Mechanical Measures

1. Destruction of grubs by peeling the bark mechanically.

(C) Chemical Measures

1. Drench the basal trunk and root region with 1.15 lit. Lindane 20 EC in 250 litres of water per hactare to kill the grubs.
2. Swab Coal tar + Kerosene @ 1:2 or Carbaryl 50 WP 20 g/l on the basal portion of the trunk (3 feet height) After scraping the loose bark to prevent oviposition by adult beetles.
3. Spray carbaryl @ 0.1 per cent or Dichlorvos 0.05 per cent.
4. Inject Carbon disulphide or ED/CT mixture into the larval tunnels and plaster with mud.
5. Padding with monocrotophos 36 WSC 10 ml in 2.5 cm/tree soaked in absorbent cotton.
6. If infestations are severe, then apply the copper oxychloride paste on the trunk of the tree.
7. Hook out the grub from the bore hole and apply one celphos tablet (3 g) per hole.
8. Apply carbofuran 3G 5 g per hole and plug with mud.

(K) PESTS OF SAPOTA

1. Leaf Weber

Also known as: Chiku moth

Scientific Name, Order and Family

Nephopteryx eugraphella (Ragonot)

(Lepidoptera; Pyralidae)

Taxonomical Position/Scientific Classification

Kingdom:	Animalia
Phylum:	Arthropoda
Class:	Insecta
Order:	Lepidoptera
Family:	Pyralidae
Genus:	Nephopteryx
Species:	***N. eugraphella***

Biology and Life Cycle

1. **Egg**: Female moth lays upto 374 eggs singly or in clusters of 2 to 9 on tender leaves and shoots. Egg stage lasts for 2 to 5 days.
2. **Larva**: Larval period is 13 to 26 days.
3. **Pupa**: Pupation takes place in the leaf web itself. Pupal period varies from 8 to 13 days.
4. **Adult**: One generation is completed in 26 to 38 days. There are 7 to 9 generations per year.

Marks of Identification

Adult is slender, gray coloured moth having black forewings with yellow spots on the basal half and black transverse lines on the remaining half. Adult has wing expanse of 18 mm. Caterpillars are pinkish with three dorso-lateral pinkish brown stripes on each side. Full-grown caterpillar is 25 mm long. Eggs are oval in shape and pale yellowish white in colour.

Host Crops

Sapota, Bakul and cured tobacco.

Nature and Symptoms of Damage

- ✰ Caterpillars web together a bunch of leaves with silken threads and excreta and feed within on chlorophyll leaving behind a fine network of veins.
- ✰ It also attacks tender fruits and flower buds throughout the year by boring holes in them.
- ✰ Leaves dried and hanging from the webbed shoots.
- ✰ Caterpillar bores into flower buds and tender fruits - withered and shed.
- ✰ Pest is active during June- July.

Management

(A) Cultural Measures

1. Remove and destroy the infested plant parts (leaf webs) along with larvae.
2. Collect and remove the dried clusters of leaf web.

(B) Chemical Measures

1. Insecticides: Spray phosalone 35 EC 2 ml/lit or phosphamidon 40 SL 2 ml/lit or neem seed kernel extract 5 per cent.
2. Spray application of cypermethrin 0.025 per cent affords protection.
3. Spray the plant with 0.05 per cent Quinolphos.

2. Fruit Fly

Scientific Name, Order and Family

Bactrocera dorsalis H.

(Diptera; Tephritidae)

Taxonomical Position/Scientific Classification

Kingdom:	Animalia
Phylum:	Arthropoda
Class:	Insecta
Order:	Diptera
Family:	Tephritidae
Genus:	Bactrocera
SubGenus:	Bactrocera
Species:	***B. dorsalis***

Biology and Life Cycle

1. **Egg**: A single female can lay 150 - 200 eggs (average 50). The eggs are laid in clusters of 2 - 15 eggs and these hatches in 2 - 3 days during March and 1 - 1½ days during April.
2. **Larva/Maggot**: Maggot duration is 6 days in summer and extends up to 19 days with the fall in temperature.
3. **Pupa**: Pupation usually takes place 80 - 160 mm below the soil surface and pupal period ranges from 6 days (summer) - 44 days (winter).
4. **Adult**: The adult fly is light brown with transparent wings. Adult flies are very conspicuous. These are about 7 mm long, with hyaline wings (expanse : 13 - 15 mm), thorax ferrugineous without yellow middle stripe, legs yellow, abdomen conical in shape and dark brown in colour. Preoviposition period is 2 - 5 days.

Marks of Identification

Adult is 5 mm long, small fly with single pair of transparent out stretched wings amd conical shaped yellowish brown abdomen. The brown and grey coloured patches distinguish these flies from others. The maggots are dirty white, legless and 12 mm in length when full grown. Body of maggot goes on tapering at one end. Eggs are white, oval and elongated.

Host Crops

Papaya, mango, guava, pulm, peach, pear, fig, ber, apple, banana, pomegranate, citrus, cucurbits *etc.*

Nature and Symptoms of Damage

On hatching, maggots feed inside on the pulp of fruits. The infested fruits can be identified by the presence of brown resinous juice which oozes out of the punctures. These punctures serve as entry for various bacteria and fungi. As a result, infested fruits start rotting, get distorted and malformed in shape and fall down prematurely.

- ✰ Maggot bore into semi-ripen fruits with decayed spots and dropping of fruits.
- ✰ Oozing of fluid.
- ✰ Brownish rotten patches on fruits.

Economic Threshold Level (ETL)

10 per cent affected fruits.

Management

(A) Cultural Measures

1. Fruits should be harvested before they start ripening.
2. Fallen fruits should be collected and destroyed.
3. Raking soil or ploughing after harvest to destroy hibernated pupae.
4. Mulching beneath the plant helps to kill the pupae inside the soil, if infestation is of severe nature.

(B) Mechanical Measures

1. Monitor the activity of flies with methyl eugenol sex lure traps.

(C) Chemical Measures

1. Application of spray bait containing 20 ml Malathion + 200 gms jaggery + 20 litres of water.
2. Spray Malathion 0.05 per cent or 0.2 per cent Carbaryl during flowering to reduce intensity of infestation.
3. Bait spray - combing any one of the insecticides and molasses or jaggery 10 g/l,
   - ✰ Malathion 50EC 2 ml/litre of water,
   - ✰ Dimethoate 30 EC 1 ml/litre of water,
   - ✰ Carbaryl 50 WP 4 g/l. two rounds at 2 weeks interval before ripening of fruits.
4. Prepare bait with methyl eugenol 1 per cent solution mixed with Malathion 0.1 per cent.

5. Take 10 ml of this mixture per trap and keep them in 25 different places in one hectare.
6. Spray Fenthion 100 EC 2 ml/lit or Malathion 50 EC 2ml/lit.

(D) Biological Measures

1. Field release of natural enemies *Opius compensates* and *Spalangia Philippines.*

Pest Management of Stored Grains

On the basis of nature of damage caused by different species of beetles, weevils and moths, the insect pests of stored grains can be classified into two main groups as follows:

1. Primary Pests

The primery pests are those, which are capable of causing damage to the whole or sound grains. Larvae of these insects are confined within the seed during development. *e.g.*, Rice weevil, granary weevil, pulse beetle.

They are of two types:

a) **Internal feeders**: In these, the larvae feed entirely within the kernals or grain. This group includes Rice weevil, Granary weevil, Lesser grain borer, Pulse Beetle and Angoumois grain moth.

b) **External feeders**: In this, larvae feed from outside of the grain. This group included Khapra beetle, Indian meal moth, Fig moth and Rice moth.

2. Secondary Pests

These do not damage or attack sound grains but feed on damaged, broken or milled grains. Larvae are free living. These pests are Red rust flour beetle, Saw toothed grain beetle, Long headed flour beetle and Flat grain beetle.

1. Pulse Beetle

Also known as: Cowpea weevil.

Scientific Name, Order and Family

Callosobruchus chinensis L.

(Coleoptera; Bruchidae)

Taxonomical Position/Scientific Classification

Kingdom:	Animalia
Phylum:	Arthropoda
Class:	Insecta
Order:	Coleoptera
Family:	Bruchidae
Genus:	Callosobruchus
Species:	***C. chinensis***

Biology and Life Cycle

1. **Eggs**: Eggs are whitish, elongated and stuck on the grains or on pods and sometimes on the surface of the container. Fecundity is about 100 eggs per female. Incubation period is 3-6 days.
2. **Larvae/Grub**: Grubs are scarabeiform or eruciform, plump and with short legs and yellowish in colour. First instar larvae bear functional legs and a pair of thoracic plates to facilitate boring into the seeds. They feed on the inner contents of the grain and may damage several grains during

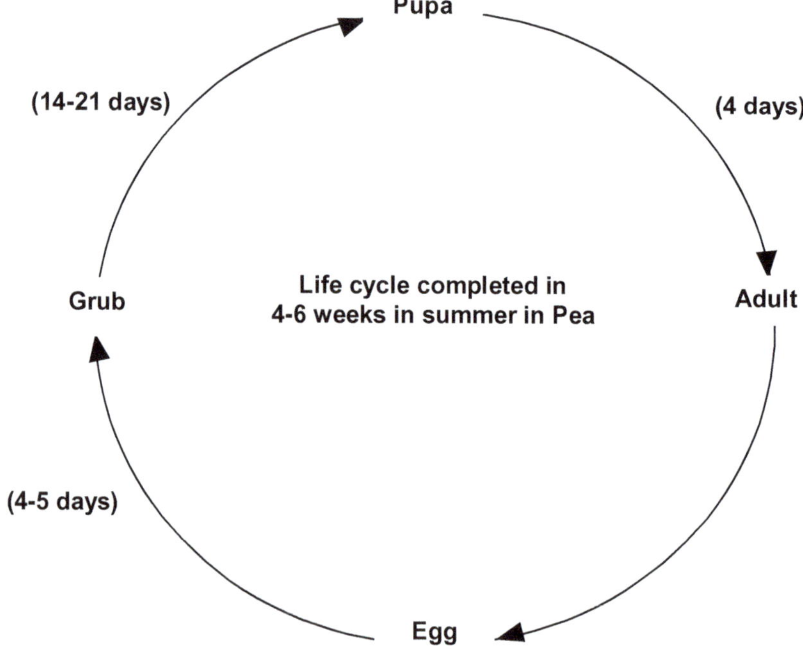

Figure 63: Life Cycle of *Callosobruchus chinensis*.

development. A fully developed larva is about 5.5 – 6.5 mm in length, cylindrical in shape, fleshly and strongly wrinkled. It is perfectly white except in the region of its mouth where the mouth parts are brown. It remains lying on the grain in a curved state. Larval period may vary between 12-20 days.

3. **Pupa**: Pupation takes place inside the grain and pupa is dark brown in colour. Occasionally pupation may take place outside the grain in a cocoon made of excretory matter. The pupal stage lasts for about 4 days during summer season but in winter season this stage lasts for a considerably longer period.
4. **Adult**: Adult beetle is 3-4 mm long, female being larger, brownish in colour, broader at shoulders and rounded posteriorly. There are dark patches on elytra and thorax. Adults show sexual dimorphism. Males possess deeply emarginated or indented eyes and prominently serrate antennae, while in female these characters are not distinctly marked. In females tip of abdomen is exposed while in males it is covered by elytra. They are active beetles and readily fly when disturbed. The adult emerges after cutting a circular hole in the seed coat. The time taken to complete the life cycle varies according to temperature and the nature of grains on which the pest breed. In soya-beans it is completed in 41 days, in lentil it takes 34-41 days, while in arhar it is completed in 25-34 days and there may be 6-7 overlapping generations in a year.

Marks of Identification

The adult male is about 2-3 mm in length and 1.5-2.5 mm in width. The female adult is slightly bigger than the male, measuring 3-4 mm in length and 2-3 mm width. The beetle, when viewed from above, presents a heart shaped appearance like spots in the middle of the body and a posterior spot at the junction of head and thorax the dorsal surface.

The insect tapers a little towards the head which is hypognathus in normal resting position. Its colour is usually pale or dark brown and it has a conspicuous swollen abdomen more apparent in female loaded with eggs. In male, the elytra cover almost the entire abdomen but in female it is not so due to the distended abdomen.

Host Crops

It is a major pest of pulses like moong, gram, bean, masoor and urd, cowpea, soybean, gram, pigeon pea, lablab *etc.*

Nature and Symptoms of Damage

- ✰ It is a very serious stored pest, causing enormous damage to almost all kind of pulse grains.
- ✰ Damage to the pulse grain is mainly caused by the developing larvae. Just after hatching young larvae bores into the grain, feed upon the contents of the grain making them almost hollow and empty.

- ☆ Adults cut out circular emergence holes while coming out from the seed.
- ☆ Pulse beetle is a primary pest of whole pulses. The flying adults or white egg spots on the seed are indicative of their presence. These beetles do not attack split pulses which are traded in India as "Dals".
- ☆ Pest infestation starts in the field on the ripening crop and is continued further in storage.

Economic Threshold Level (ETL)

Appearance of live insects.

Management

Also given at last of Chapter.

(A) Cultural Measures

1. Intercropping maize with cowpeas, and not harvesting crops late significantly reduced the infestation by *C. chinensis*.
2. Good store hygiene plays an important role in limiting infestation by these species. The removal of infested residues from last season's harvest is essential, as is general hygiene.
3. Solarization (or drying and heating) can be used to control infestations of *C. chinensis* without affecting seed germination.

(B) Chemical Measures

1. Fumigation treatment with phosphine.

(C) Biological Measures

Biological control has not been widely used against *Callosobruchus* species, although natural populations of *C. chinensis* are often subject to high levels of parasitism, and there have been suggestions that biological control programmes of *C. hinensis* may be viable.

1. The parasitic wasps *Dinarmus basalis, Lariophagus distinguendus* and *Anisopteromalus calandrae* have been associated with a number of *Callosobruchus* species. It is probable that several other wasps known to parasitize a wide range of *Coleoptera,* such as *Theocolax elegans,* will also use *Callosobruchus* species as hosts.

2. Rice Weevil

Scientific Name, Order and Family

Sitophilus oryzae Linn.

(Coleoptera; Curculionidae)

Taxonomical Position/Scientific Classification

Kingdom:	Animalia
Phylum:	Arthropoda
Class:	Insecta
Order:	Coleoptera
Family:	Curculionidae
Genus:	Sitophilus
Species:	***S. oryzae***

Biology and Life Cycle

1. **Eggs**: After copulation, adult female bores a hole in the grain with the help of its powerful jaws and deposit a single egg in the grain cavity. Each egg is tiny, white, oval structure measuring 0.7 mm long and 0.3 mm broad. The eggs are elastic enough to fit in the hole made by the mother, who thereafter covers the egg with gelatinous fluid. A female may lay as many as 300 to 550 eggs in 4-5 months. Under optimum conditions (during August- September) the eggs hatches into larvae in 3-4 days but during winter the hatching may take 6-9 days.
2. **Larvae**: The tiny, white, fleshy, legless grubs with yellowish- brown head and biting jaws bores down into the grain, feeding on its starchy content and hallowing it out leaving the shell intact. Grubs are always found inside the kernels and never outside. The grub stage lasts for 19-34 days.

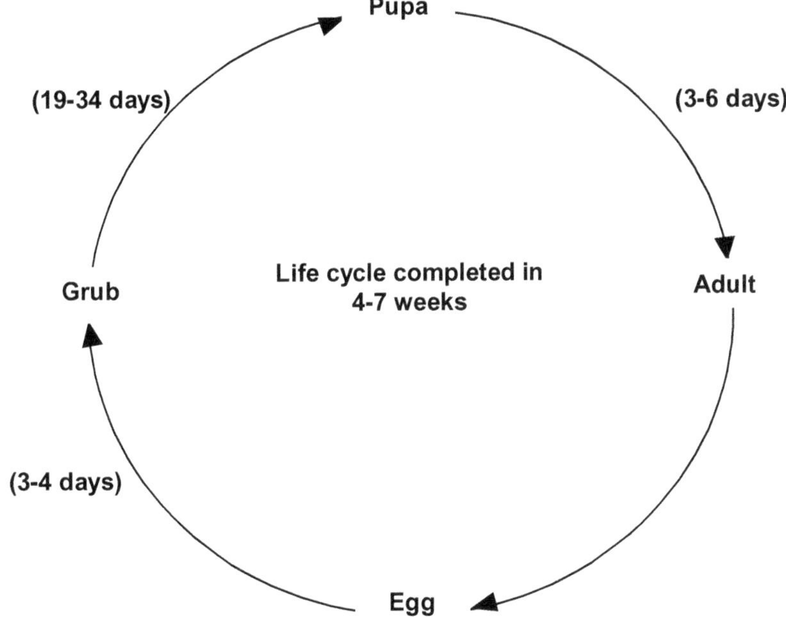

Figure 64: Life Cycle of *Sitophilus oryzae*.

3. **Pupa**: A fully matured grub makes a pupal cell inside the grain and pupates. The pupal stage lasts for 3-6 days (July-September) but in unfavourable conditions (winter and summer months) it may extent up to 20 days.
4. **Adult**: Adult is a small, 3 to 4 mm long reddish brown to dark brown or almost black weevil with the head having the shape of a long slender snout. The wings have four light reddish or yellowish spots. The adult is able to fly. The adults formed after pupation bores its way out of the grains. Immediately after the emergence, the adult weevils are ready for breeding. The duration of life cycle of rice weevil and number of generations completed in a year depends upon the weather conditions, like temperature and humidity. Adults survive for 2 to 5 months. In Indian conditions, 5-7 generations of this weevil is generally completed in a year.

Marks of Identification

Adult

1. Dark brown in colour.
2. 2.0 to 3.5 mm long.
3. Long elongate snout characteristic of weevils.
4. Generally have four lighter coloured areas on the back, two on each wing cover which are light red to yellow in colour.
5. Pronotum is pitted with deep, round punctures.
6. Adults do not fly readily.
7. Bumping outside of bagged grain or walking on surface of bulk grain will cause adults to appear at the surface a few minutes later. This is one method that can be used to detect an infestation.

Host Crops

Wheat, corn, rice, maize, jowar, barley, bajra *etc.*

Nature and Symptoms of Damage

1. Both grubs and adults feed voraciously on the grain so much that the grain becomes unfit not only for consumption but also for seed purpose. Apart from rice, it also feeds upon wheat, com, jaw, barley *etc.*
2. The larvae are more destructive as they feed voraciously on the content of the grain but leave the shell of the grain intact. Adults may also, feed upon flour (milled cereals) but the larvae cannot develop in it unless the material is caked.
3. In case of severe infestation, the grain becomes a mass of broken vegetable matter. The adults eat a small amount of grain making shallow holes with rugged edges into it but the amount of damage thus caused is negligible.

4. The young tiny grubs bore into the grain kernels and feed on starchy content. The severely damaged grains are hollowed out, leaving only the outer shell intact. Pest infests the grains both in storage and in field.
5. It is universally regarded a serious pest of cereals.
6. The damage caused by them may extent up to 50 per cent of the total grain stored in a particular go down.

Economic Threshold Level (ETL)

Appearance of live insects.

Management

Given at last of Chapter.

3. Rice Grain Moth

Scientific Name, Order and Family

Corcyra cephalonica Stainton.

(Lepidoptera; Galleridae)

Taxonomical Position/Scientific Classification

Kingdom:	Animalia
Phylum:	Arthropoda
Class:	Insecta
Order:	Lepidoptera
Family:	Pyralidae
Genus:	Corcyra
Species:	***C. cephalonica***

Biology and Life Cycle

1. **Eggs**: Moths are short lived but realise a fecundity of 150—200 eggs per female within a few days after emergence. Eggs are laid anywhere, on the grains, among grains, on the containers or on any surface near the grains, either singly or in clusters. Eggs are whitish, oval in shape, 0.5 mm long and having an incubation period of 4-5 day. Eggs take about 2-3 days to hatch.
2. **Larvae**: The larvae are generally creamish – white except for the head capsule and the prothoracic tergite, which are brown. There are well-developed prolegs on abdominal segments 3-6 and 10. A fully matured larva measures 15 mm. There is considerable variation in the number of larval instars; however, males generally have 7 and females have 8. The last-instar larvae pupate within the food. Larval period is 25-35 days in summer and may be extended in winter. Optimum conditions for larval development of *C. cephalonica* are 30 – 32.5°C and 70 per cent RH.

3. **Pupa**: Pupation takes place inside an extremely tough, opaque whitish cocoon that is surrounded by webbed grains. Pupal period is about 10 days but may extend to 40-50 days to tide over winter moths. The anterior portion of the cocoon has a line of weakness through which the adult emerges.
4. **Adult**: The adults are small. The hind-wings are pale-buff, and the fore-wings are mid-brown or greyish-brown with thin vague lines of darker brown colour along the wing veins. The males are smaller than the females. The adults emerge through the anterior end of the cocoon, where there is a line of weakness. The sex ratio is 1:1. The adult moth is nocturnal and is most active at nightfall.

Marks of Identification

Adult is medium sized, pale grayish brown moth with a wing expanse of about 25 mm. Forewings are spotless, uniformly pale buff-brown in colour. The young larva is creamy white with prominent, broad, yellowish head. The prothoracic shield is very distinct and light yellow in colour. The full-grown larva measures about 25 mm in length.

Host Crops

It is particularly a pest of stored rice, jowar and maize. It is also found on other cereals, dals, pulses, oilseeds, nuts and dry fruits. It may also be seen attacking milled spices, flour, suji.

Nature and Symptoms of Damage

- ☆ The larvae alone damage the grains of rice and maize.
- ☆ The young larvae wander about for food and readily feed on the broken parts of the grain but as they grow older, they become capable of feeding on entire grains also.
- ☆ The larvae bore into the grains and pass their life feeding on starchy matter which reduces marketing quality of the grains.
- ☆ They web silken shelter which cover grain, frass *etc.* so densely that it is difficult to detect their presence inside.
- ☆ Grains are rendered unfit for human consumption.

Economic Threshold Level (ETL)

Appearance of adult moths

Management

Given at last of Chapter.

4. Red Rust Flour Beetle

Scientific Name, Order and Family

Tribolium castaneum (Herbat)

(Coleoptera, Tenebrionidae)

Taxonomical Position/Scientific Classification

Kingdom:	Animalia
Phylum:	Arthropoda
Class:	Insecta
Order:	Coleoptera
Family:	Tenebrionidae
Genus:	Tribolium
Species:	***T. castaneum***

Biology and Life Cycle

1. **Eggs**: A female beetle lay about 400-500 eggs. Eggs are laid singly in flour and dust of the grains. They soon become covered with small particles of dust and flour as they are moist and sticky when freshly laid. The eggs are minute, slender, and cylindrical in shape, rounded at both ends and of whitish colour. The incubation period varies from 5 to 12 days, depending on temperature. A temperature of about 27 °C has been observed to be the most favourable for its development.
2. **Larvae**: The freshly hatched grub is small, worm like, slender, cylindrical and wiry in appearance. The body segments have a number of fine hairs, the terminal segment being in addition furnished with a pair of spine like appendages. The full grown grub is about 3/16 inch long and is pale yellowish in colour. The larval period varies from 27 to 29 days but it

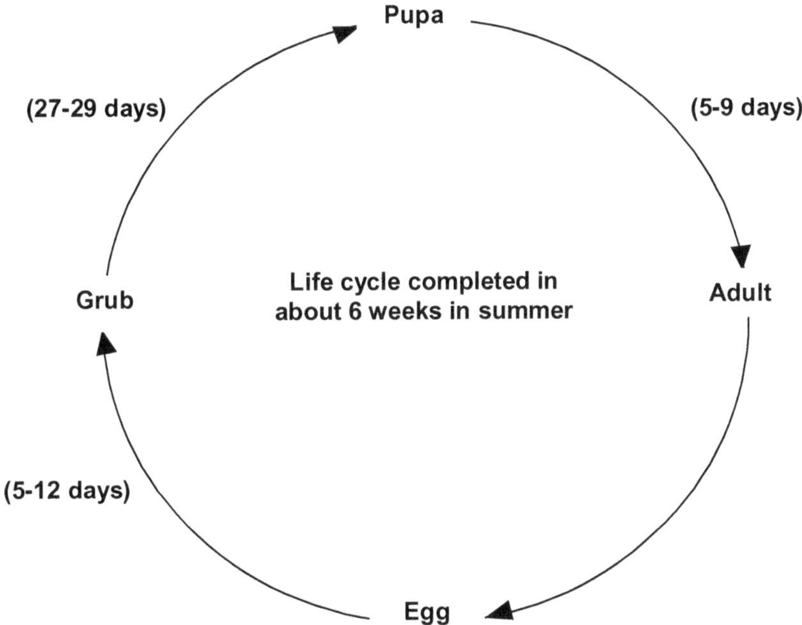

Figure 65: Life cycle of *Tribolium castaneum*.

may get prolonged up to 90 days, according to the food available and the prevailing temperature.

3. **Pupa**: Pupation takes place generally on the surface of the food. The pupa is naked. At first it is white but gradually becomes yellowish, the dorsal surface having hairs and processes resembling those of the grub. Pupal stage lasts for 5-9 days.
4. **Adult**: The total life cycle from egg to the emergence of adult takes about six weeks during August and September, but is greatly prolonged in the cold weather. There may be four to seven generations in one year.

Marks of Identification

- ✰ Although small beetles, about 1/8 of an inch long, the adults are long-lived and may live for more than three years.
- ✰ The red flour beetle is reddish-brown in color and its antennae end in a three-segmented club. Whereas the confused flour beetle is the same color but its antennae end is gradually club-like, the "club" consisting of four segments.
- ✰ The head of the red flour beetle is visible from above, does not have a beak and the thorax has slightly curved sides. The confused flour beetle is similar, but the sides of the thorax are more parallel.
- ✰ The body of the adult beetle is flattened reddish-brown in colour and measures 3-4 mm in length. The head, thorax and abdomen are distinct. The antennae are well developed, the last few segments being abruptly much larger than the preceding ones.

Host Crops

It is a cosmopolitan insect found throughout the world. In India it is considered as serious stored grain pest, well distributed all over the country. This pest is especially common in granaries, mills, 'warehouses *etc.*

Nature and Symptoms of Damage

- ✰ This is a serious pest of prepared cereal products such as atta, maida and suji and is found in abundance in flour mills. This pest is found infesting all stored products like seeds, grains, flour, dry fruits, nuts, oil cakes *etc.*
- ✰ In case of heavy infestation, flour or maida turns greyish-yellow or develops red taints which subsequently becomes mouldy and emits offensive pungent smell.
- ✰ It also infests dry museum specimens and stuffed animals. *Tribolium castaneum* does not cause as much damage as Sitophilus, Rhizopertha and others in whose company it is mostly found.
- ✰ Neither the larva nor the adult can generally damage sound grains, but they feed on those grains only which have already been damaged by other insect pests.

Economic Threshold Level (ETL)

Appearance of adult beetles.

Management

Given at last of Chapter.

5. Khapra Beetle

Also known as: Wheat beetle/Cabinet beetle

Scientific Name, Order and Family

Trogoderma granarium Everts.

(Coleoptera; Dermistidae)

Taxonomical Position/Scientific Classification

Kingdom:	Animalia
Phylum:	Arthropoda
Class:	Insecta
Order:	Coleoptera
Family:	Dermestidae
Genus:	Trogoderma
Species:	***T. granarium***

Biology and Life Cycle

1. **Eggs**: After 2-3 days of emergence, copulation takes place between adult males and females. The female start laying eggs after five days of mating. The eggs are generally laid in crevices in go downs or in grain heaps. The average number of eggs laid by a single female per day is 25 which last for 5-7 days. A female lays about 125 eggs in its life. The eggs are white and cylindrical in shape. They hatch in about a week or two depending upon the temperature and humidity. In humid atmosphere the incubation period is 5-7 days.
2. **Larvae**: The larvae are yellowish brown in colour, with the body covered by bundles of long, reddish- brown, movable and erectile hair. A fully grown larva measures 4.5 mm in length. The larval period varies considerably. Under favourable conditions the male larva moults four times and the larval duration lasts for 30-50 days. In case of females it is slightly longer. During the unfavourable conditions the number of moults may be 8-10 and the larval duration may prolong for a period of 200 days to four years. During winter months or in absence of food the larvae remain inactive and live in cracks and crevices or any other concealed places.
3. **Pupa**: At the last ecdysis, the larval skin splits, but the pupa remains within this skin for the whole of its life. The pupa is of the exarate type; male

smaller than female, average lengths being 3.5 mm and 5 mm, respectively. Pupal stage lasts for 6-16 days.

4. **Adult**: After pupal period, adults weevils emerge out which becomes sexually matured in 2-3 days. Adults are 2 – 3.5 mm, oval, hairy, with elytra unmarked or light markings. The antennal club has three to eight segments, joined symmetrically. The adults are short-lived, mated females living 4-7 days, unmated females 20-30 days and males 7-12 days; The total development takes 4-6 weeks at 95°F which is the optimal temperature for this pest, at 70°F the life cycle needs 220 days to be completed.

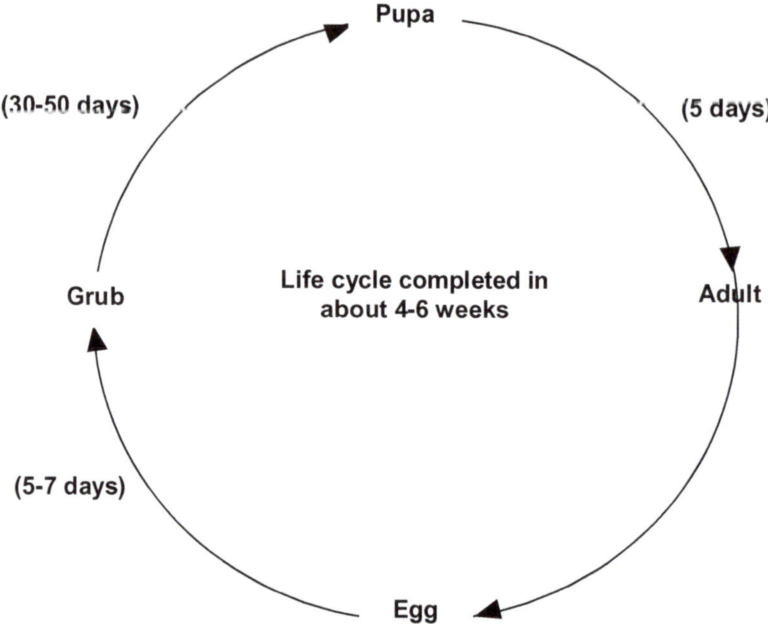

Figure 66: Life Cycle of *Trogoderma granarium*.

Marks of Identification

1. **Eggs**: Initially milky-white, later pale-yellowish; typically cylindrical, 0.7 mm long and 0.25 mm broad; one end rounded, the other more pointed and bearing a number of spine-like projections, broader at the base and tapering distally.
2. **Larvae:** The larvae at hatching are approximately 1.6 to 1.8 mm long, more than half of this length consisting of a tail made up of hairs on the last abdominal segment. Larvae are uniformly yellowish white, except head and body hairs are brown. As the larvae increase in size, their body color changes to a golden or reddish brown, more body hairs develop, and the tail becomes proportionally shorter. Mature larvae are approximately 6 mm long and 1.5 mm wide.

Larvae bear characteristic body hairs: (1) simple hairs in which the shaft bears many small, stiff, upwardly directed processes, and (2) barbed hairs with a constricted shaft in which the apex is a barbed head as long as the preceding 4-segmented-like constrictions (Hadaway 1955, Anonymous 1981).

3. **Adults:** The adults are oblong-oval beetles, approximately 1.6 to 3.0 mm long and 0.9 to 1.7 mm wide. Males are brown to black with indistinct reddish brown markings on their elytra. Females are slightly larger than males and lighter in color. The head is small and deflexed with short 11-segmented antennae. The antennae have a club of three to five segments, which fit into a groove in the side of the pronotum. The adults are covered with hairs (Buss and Fasulo, 2006).

Host Crops

All cereals especially wheat, cereal products, oil cakes, peanuts, jowar, oat and rice.

T. granarium is a general storage pest which occurs mainly on cereals and cereal products, oilseeds (especially groundnuts and oilcakes), pulses and pulse products, as well as on compound animal feeding stuffs. The occurrence on other products, as on empty sacks and gums, *etc.*, is probably accidental by cross infestation.

Nature and Symptoms of Damage

- It is considered to be a serious pest of stored wheat grains. It also, attack rice, oat, maize, jaw, pulses, oil seeds and their products, malting, copra, dry fruits *etc.*
- The damage to the grain is caused by the larvae/grub while the adults are harmless and do not feed.
- These pests are most active from July-October during which they are capable of causing heaviest damage to the stored grains.
- The infestation occurs mainly in superficial layers of grain as the pest is not able to penetrate deep into the grain.
- The destruction of the embryo end of the grain is the major damage caused by this pest but during the heavy infestations complete grain is destroyed.
- However, in case of heavy infestations, it may destroy the entire lot.

Economic Threshold Level (ETL)

Appearance of adult beetles.

Management

Given at last of Chapter.

6. Lesser Grain Borer

Also known as: Australian wheat weevil.

Scientific Name, Order and Family

Rhizopertha dominica Fab.

(Coleoptera; Bostrychidae)

Taxonomical Position/Scientific Classification

Kingdom:	Animalia
Phylum:	Arthropoda
Class:	Insecta
Order:	Coleoptera
Family:	Bostrichidae
Genus:	Rhyzopertha
Species:	***R. dominica***

Biology and Life Cycle

1. **Eggs**: After copulation the mother beetle lays 300-500 eggs singly or in clusters either near the embryo-end of the grain or the eggs is simply dropped down in between the grains or in the powdery starchy materials lying outside grains. Eggs are pear shaped, glistening white when freshly laid, but they become pinkish opaque as the larvae develop inside the egg shell. The duration of egg stage is 5-6 days during summer, 7-11 days during autumn and much longer in winter.
2. **Larvae**: The newly hatched larva is quite active and is campo-deiform in shape. It burrows at once into the grain or crawls actively about, feeding on the loose starchy material. The grubs complete their development either within the grains or in the flour where they undergo larval moults. There are three larval instars. The fully grown larva is dirty white with a light brown head and curved abdomen. It bears three pairs of legs with swollen anterior end. The whole body is covered with tiny hairs. The average larval period is about 40 days.
3. **Pupa**: The prepupal and pupal periods last for about a week and after that it changes into adult form.
4. **Adult**: The adults eat their way out of the grain. The total life cycle from egg up to the emergence of the adult is about 6-8 weeks. There are generally five generations in a year. Often the beetle has been found breeding in flour and the whole life cycle may be completed outside the grain.

Marks of Identification

- ✰ Rhizopertha dominica can be distinguished from other store pests by its cylindrical shape and small size. It measures about 3 mm in length.

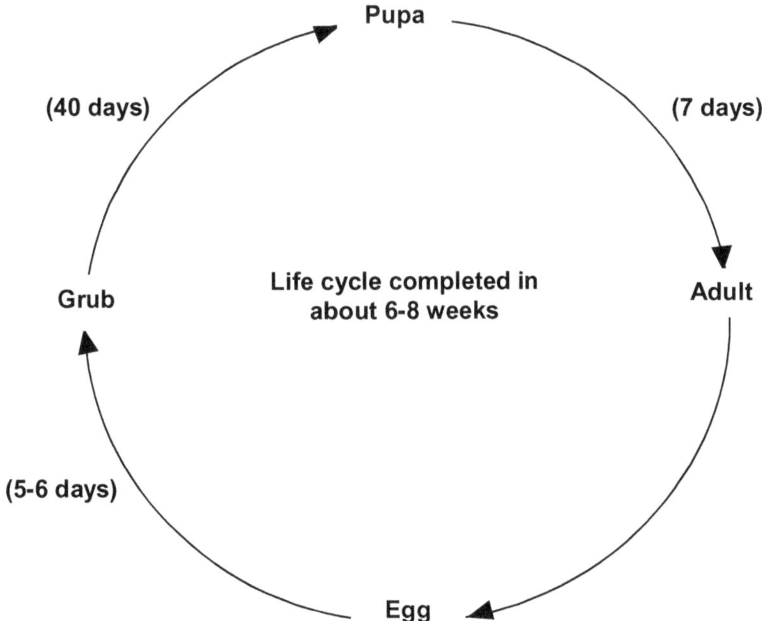

Figure 67: Life Cycle of *Rhizopertha dominica*.

Adult beetle is polished dark brown or black in colour with a somewhat roughened surface. Its head is inserted into a hood-like triangular structure lying under the thorax. Head possesses powerful jaws with which it causes serious damage to the grain and occasionally any part of the wooden structure in the store or warehouse to tide over an unfavourable period.

- Grub is dirty white, cylindrical, light brown head, body clothed with tiny hairs.

Host Crops

Cereal grains of all types, cereal products, pulses and root crops.

Nature and Symptoms of Damage

- *R. dominica* originally belongs to a family of wood borer but later on it developed affinity with cereals and in course of time became the major pest of nearly all types of cereals, grocery and dried fruits.
- Both grubs and adults cause damage within warm climate, attacking variety of grains like wheat, rice, jowar, bajra, maize, pulses, paddy *etc.*
- They completely hollow but the grain kernels and only the bran coat is left. Adults chew a great deal of more grain than they need for their consumption. The milled foodgrain products are converted into tangled mass of excreta.
- This pest has lower power of dispersal and it also does not infest the grains before harvest.

☆ Pest has less importance as a primary pest of stored cereals than that of rice weevil.

Economic Threshold Level (ETL)

Appearance of adult moths

Management

Given at last of Chapter.

Control Measures of Store Grain Pests

The measures adopted for the control of stored grain pests are of two types:

1) **Preventive measures:** Measures which are employed to prevent pests attacking stored foodgrains and grain products.
2) **Curative or Remedial measures:** Measures which are employed to control the pests when the grain infestation has already taken place.

1) Preventive Measures

For the protection of grains against the attack of pests, the maxim that prevention is better than cure is applicable the most. Much damage to the stored grains can be avoided by adopting preventive measures.

i) Storage Facility

Development of infestation in storage is guided largely by the type of facility used for the storage of grain. Darkness and dampness would provide better condition for insect multiplication. A facility that is airtight, when filled to capacity helps in keeping the infestation controlled through depletion of oxygen. A facility that can be made sufficiently air-tight for carrying out fumigation of the whole shed is better suited for keeping grain free of infestation. From the point of view of prevention as well as control of insects, therefore, an air-tight facility must be preferred.

ii) Warehousing Facilities

The simple steps that can help to maintain grain in storage for long durations are:

a) Thorough cleaning of the facility.
b) White washing where necessary before use.
c) Placing dunnage between the grain and the floor.
d) Building uniform sized rectangular stacks when grain is in sacks.
e) Filling the facility to capacity when grain is in bulk.
f) Regular inspections.
g) Maintaining hygienic conditions during storage.

iii) Moisture Content of Grain

Food grains with moisture content lower than 11 per cent are relatively resistant to insect attacks. Much of the grain to be stored over long duration in India is harvested in March- June and this invariably is very dry. Substantial quantities of wheat and gram coming into storage have moisture content 8 to 10 per cent. Only a few insects like *Rhizopertha dominica* and *Trogoderma granaria* infest such grain. Low moisture content is not conducive to infection by microorganism and infestation by mites. So, it is essential that before grains are put into storage, the moisture content should preferably be less than 8 per cent. At the same time, care may be taken to prevent absorption of moisture from atmosphere in storage grains.

iv) Sun Drying

This was the only technique followed in the country till recently and is still followed in respect of large quantities of grain. Moist grains are spread in the open exposed to the sun for varying periods depending upon the initial moisture and intensity of heat. Floors used for drying are either plastered with mud or bitumen or in some cases cemented. One tonne of grain takes about 10 sq.m. of space. The requirements of labour and the duration of operation depend on the moisture content of grains meant for storage. Recently, heated as well as unheated air and some chemicals are also used for drying grain.

v) Mixing of Inert Dusts

It has been adopted in India for centuries. Inert dusts such as clay, ash *etc.* helps in destruction of part of the waxy layer that covers the insect cuticle, thus destroying its water proofing properties. The dusts used either abrase or absorb the waxy layers resulting in dessication and death of insects. Inert dusts have now lost popularity because of the demand for clean grain. Addition of dusts would also add into costs of storage and transport.

vi) Use of Insect Barrier

Use of insect proof and polythene sacks have been investigated from the point of view of providing a barrier to insects. Grains stored in sacks can be covered with polythene. Only effective barrier for insects in case of grain today is either the steel or cement concrete air-tight bin. Polythene is not very effective barrier.

vii) Elimination of Sources of Infestation

The healthy grains kept in insect free receptacles or godowns remain free from pests. It is therefore imperative that the godowns and the bags used for storage and the carriages used for transport are insect free. This can be achieved by thorough cleaning of godowns or receptacles, plugging of all the holes, cracks and crevices and by spraying malathion 0.5 per cent emulsion on the walls, the ceiling and the floor.

As far as possible new bags should be used for storage. If old bags are to be used, they need to be made insect free by dipping them in 0.1 per cent malathion emulsion for 10 minutes and then drying to avoid cross infestation.

Optimum Doses of Pesticides Recommended by the Storage and Research Division of Ministry of Food

Sl.No.	*Name of Pesticide*	*Dose of Pesticide*	*Method of Application*	*Against*	*Remarks*
1.	Malathion 50 per cent EC	In ratio of 1:100 with water at the rate of 3 litres/100 sq.m. For flying insects at the rate of 1 litre/300 cu.m.	Spray on grain bags, walls, floors *etc.* Should not be sprayed on foodgrains directly.	As prophylactic treatment against insect pests	Fortnightly spraying
2.	Ethylene Dibromide (EDB)	3 ml/quintal for small storage 22 gms/cu.m. for large storage.	The ampules are inserted in storage structures after breaking and making the structure airtight.	As fumigant against insect pests (except for oilseeds/ flour)	7 days exposure period
3.	ED/CT mixture	30 to 40 kg/100 cu.m. in large scale storage. For small storage 55 ml. per qtl.	Fumigation in airtight condition on bags/in bulk.	As fumigant against insect pests (except for oilseeds/ flour)	7 days exposure period

viii) Seed Treatment

The grains meant for seed can be protected by mixing insecticidal dust with it. Malathion 5 D, if mixed @ 250 gms per quintal of seed is effective. Insecticides mixed with seed offer better and longer protection than when they are used for surface treatment.

ix) Use of Improved Receptacles

The grains can be best protected by using improved insect proof receptacles. The metal bins are now available for storing small quantities of grain ranging from 1 to 6 tonnes. The grains stored in bins can be easily fumigated. For storing larger quantities, it is advisable to construct improved godowns.

2) Curative Measures

Most practicable and useful curative method is the fumigation. Fumigation may be defined as the treatment of commodity or a space with a gaseous material to kill the insect pest present. They are highly volatile and able to penetrate deep and kill insect within a large mass of food stuff. In case of small scale storage, first sieve the grains and remove different stages of pests. But in large scale storage, it is not possible to sieve and clean the grains and hence direct fumigation has to be carried out. Fumigation is possible only under air tight conditions. For small scale fumigation, metal bins or 'kothis' can be made air-tight. For large scale fumigation, however, the dump method is used.

Dump Method or Gas Proof Cover Fumigation

The grains are enveloped in air proof cover. The covers for dump method are prepared from balloon fabrics or rubberized cloth or polythene sheet of varying sizes. The standard size of cover being 20′ x 15′ x 15′ which can accommodate about 500 bags at a time. Put the cover over the grains to be fumigated and the sides touching the ground are covered with dry earth to prevent the leakage. The fumigant should be introduced (poured) from the opening at the top which should subsequently be closed. After the expiry of exposure time, the cover should be gradually opened from the sides so that the operator will not be exposed to the fumes for a longer period. The bags should not be disturbed for at least 24 hours after removing covers to allow the enclosed gases to escape. After this, the grains can safely be used. The commonly used fumigants for grain storage are given in the table along with the rate, time of exposure and the precautions to be taken while using particular fumigant.

Appendices

Appendix I: Major Pests of Crops

Crop/Pest Name	Scientific Name	Order	Family
Rice			
1. Gall midge	*Orseolia oryzae* (Wood-Mason)	Diptera	Cecidomyidae
2. Yellow stem borer	*Scirphophaga incertulas* (Walker)	Lepidoptera	Pyralidae
3. Green Leaf hopper (GLH)	*Nephotetrix nigropictus* (Stal) and *N. virescens* (Distant)	Homoptera	Cicadellidae
4. Brown Leaf hopper (BPH)	*Nilaparvata lugens* (Stal)	Homoptera	Delphacidae
5. White-backed plant hopper	*Sogatella furcifera* (Horvath)	Homoptera	Delphacidae
6. Leaf folder/Leaf roller	*Cnaphalocrocis medinalis* (Guence)	Lepidoptera	Pyralidae
7. Case worm	*Nymphula depunctalis* (Guence)	Lepidoptera	Pyralidae
8. Army worm	*Mythimna saparata* (Walker)	Lepidoptera	Noctuidae
9. Swarming caterpillar	*Spodoptera mauritia*	Lepidoptera	Noctuidae
10. Gundhi bug/Ear head bug	*Leptocorisa acuta* (Thunberg)	Hemiptera	Coreidae
Wheat			
1. Pink Stem borer	*Sesamia inferens* (Walker)	Lepidoptera	Noctuidae
2. Termites/White Ant	*Odentotermis obesus* (Rambur) *Microtermes obesus* (Holmgren)	Isoptera	Termitidae
Sorghum and Maize			
1. Stem borer/Maize Stalk borer	*Chilo partellus* (Zell)	Lepidoptera	Pyralidae
2. Shoot fly	*Atherigona soccata (Rondani)*	Diptera	Muscidae
Urd and Moong			
1. Pod borer	*Maruca vitrata* (Geyer)	Lepidoptera	Crambidae
2. Red hairy caterpillar	*Amsacta moorei* (Butler)	Lepidoptera	Arctiidae
Soybean			
1. Girdle beetle	*Oberea brevis*	Coleoptera	Cerambycidae
2. Stem fly	*Melanagromyza sojae*	Diptera	Agromyzidae
3. Red Hairy caterpillar	*Amsacta moorei* (Butler) *A. albistriga* (Walker)	Lepidoptera	Arctiidae
Pigeonpea			
1. Pod borer/Spiny pod borer	*Etiella zincknella* (Treitschke)	Lepidoptera	Pyralidae
2. Pod fly	*Melanagromyza obtuse* (Malloch)	Diptera	Agromyzidae
3. Plume moth/Pod caterpillar	*Exelastis atomosa* (Walsingham)	Lepidoptera	Pterophoridae
4. Pod bug	*Clavigralla gibbosa* (Spinola)	Hemiptera	Coreidae
Gram/Chickpea			
1. Pod borer	*Helicoverpa armigera* (Hubner)	Lepidoptera	Noctuidae
2. Cut worm	*Agrotis ypsilon* (Hufnagel)	Lepidoptera	Noctuidae

Contd...

Appendix I–*Contd...*

Crop/Pest Name	*Scientific Name*	*Order*	*Family*
Pea			
1. Aphid/Green aphid	*Acyrthosiphon pisum* (Mordvilko)	Hemiptera	Aphididae
2. Pod borer	*Etiella zincknella* (Treitschke)	Lepidoptera	Pyralidae
Lathyrus			
1. Thrips	*Caliothrips indicus*	Thysanoptera	Thripidae
Groundnut			
1. Aphids	*Aphis craccivora* (Koch)	Homoptera	Aphididae
2. Leaf minor	*Stomoperyx nertaria* (Meyrick) *Aproaerema modicella* (Deventer)	Lepidoptera	Gelechiidae
3. White grub/Root grub	*Holotrachia conseguina* (Blanchard)	Coleoptera	Scarabaeidae
4. Red Hairy caterpillar	*Amsacta moorei* (Butter)	Lepidoptera	Arctiidae
	A. albistriga (Walk)		
Sesamum			
1. Gall fly	*Asphondylia sesame*	Diptera	Cecidomyidae
2. Howk moth/Dead head moth/Sphinx Moth	*Acherontia styx* (Westwood)	Lepidoptera	Sphingidae
3. Leaf webber	*Antigastra catalaunalis* (Duponchel)	Lepidoptera	Pyralidae
Linseed			
1. Bud fly/Gall Midge	*Dasineura lini* (Barnes)	Diptera	Cecidomyiidae
2. Caterpillar	*Spodoptera exigua* (Hubner)	Lepidoptera	Noctuidae
3. Thrips	*Thrips lini*	Thysanoptera	Thripidae
4. Jassid	*Empoasea kerri* var. *motti*	Hemiptera	Cicadellidae
Safflower			
1. Aphid	*Uroleucon carthami* *Dactynotus carthami*	Homoptera	Aphididae
2. Bud fly/Capsule fly	*Acanthiophilus helianthi*	Diptera	Tephritidae
Sunflower and Niger			
1. Bihar hair caterpillar	*Spilosoma obliqua* (Walker)	Lepidoptera	Arctiidae
Mustard			
1. Aphid	*Lipaphis erysimi* (Kalt)	Homoptera	Aphididae
2. Sawfly	*Athalia lugens proxima* (Klug)	Hymenoptera	Tenthridinidae
3. Painted bug	*Bagrada hilaris*, *B. cruciferarum* (Krik.)	Hemiptera	Pentatomidae
Caster			
1. Semilooper	*Achaea janata* (Linn.)	Lepidoptera	Noctuidae
2. Capsule borer	*Conogethes punctiferalis* (Guenee)	Lepidoptera	Pyralidae
3. Tussock hairy caterpillar	*Notolophus posticus*	Lepidoptera	Lymantriidae

Contd...

Appendix I–*Contd...*

Crop/Pest Name	*Scientific Name*	*Order*	*Family*
Cotton			
1. Pink Bollworm	*Pectinophora gossypiella* (Saunders)	Lepidoptera	Gelechidae
2. Spotted Bollworm	*Earias vitella, E. insulana* (Boisduval)	Lepidoptera	Noctuidae
3. American Bollworm	*Helicoverpa armigera* (Hubner)	Lepidoptera	Noctuidae
4. Jassid/Leaf hopper	*Amrasca bigutulla* (Ishida)	Homoptera	Cicadellidae
5. Red Cotton Bug	*Dysdercus koenigii* (Fabricius)	Hemiptera	Pyrrhocoridae
6. White Fly	*Bemisia tabaci* (Gennadius)	Homoptera	Aleurodidae
7. Aphid	*Aphis gossypii* (Glover)	Homoptera	Aphididae
Sunhemp			
1. Hairy caterpillar	*Utetheisa pulchella* (Linn.)	Lepidoptera	Arctiidae
Mesta			
1. Spiral borer	*Agrilus acutus (Thunb.)*	Coleoptera	Buprestidae
Sugarcane			
1. Top shoot borer/Top borer	*Scirpophaga novella* (Fabricius)	Lepidoptera	Pyralidae
2. Stem/Shoot/Internode borer	*Chilo sacchariphagus indicus* (Kapur)	Lepidoptera	Pyralidae
3. Leaf hopper/Pyrilla	*Pyrilla purpusilla* (Walker)	Homoptera	Fulgoridae
4. White Fly	*Aleurolobus barodensis* (Signoret)	Homoptera	Aleurodidae
5. Mealy bug	*Saccharicoccus sacchari* (Cockll)	Homoptera	Pseudoco-ccidae
Potato			
1. Tuber Moth	*Phthorimaea operculella* (Zeller)	Lepidoptera	Gelechiidae
2. Aphid/Green peach aphid	*Myzus persicae*	Hemiptera	Aphididae
3. Cut Worm/Greasy cutworm	*Agrotis ypsilon* (Hufnagel)	Lepidoptera	Noctuidae
Okra/Bhindi			
1. Shoot and Fruit borer	*Earias vitella* (Fab.) *E. insulana* (Boisd.)	Lepidoptera	Noctuidae
2. Jassid/Leaf hopper	*Amrasca biguttula biguttula* (Ishida)	Homoptera	Cicadellidae
3. White fly	*Bemisia tabaci* (Gennadius)	Homoptera	Aleurodidae
Brinjal			
1. Shoot and fruit borer	*Leucinodes orbonalis* (Guenee)	Lepidoptera	Crambidae
2. Stem borer	*Euzophera perticella* (Rag.)	Lepidoptera	Pyralidae
3. Mite	*Tetranychus urticae* *Tetranychus telaris* (L.)	Acarina	Tetranychidae

Contd...

Appendix I–*Contd...*

Crop/Pest Name	*Scientific Name*	*Order*	*Family*
Chilli			
1. Thrips	*Scirtothrips dorsalis* (Hood) *Thrips tabaci* (Lind.)	Thysanoptera	Thripidae
2. Fruit borer/Pod borer	*Helicoverpa armigera* (Hubn) *Spodoptera litura* (Fabricius)	Lepidoptera	Noctuidae
Tomato			
1. Fruit borer	*Helicoverpa armigera* (Hubn)	Lepidoptera	Noctuidae
2. Leaf minor	*Liriomyza trifolii*	Diptera	Agromyzidae
3. Stem borer	*Euzophera perticella*	Lepidoptera	Pyralidae
Onion and Garlic			
1. Onion Thrips	*Thrips tabaci (Lindermann)*	Thysanoptera	Thripidae
2. Tobacco caterpillar	*Spodoptera litura* (Fabricius)	Lepidoptera	Noctuidae
3. Onion fly	*Delia (Hylemya) antique*	Diptera	Anthomyiidae
Sweet Potato			
1. Weevil	*Cylas formicarius* (Fabricius)	Coleoptera	Curculionidae
Ginger			
1. Shoot borer/Rhizome borer	*Dichocrocis punctiferalis* (Guenee)	Lepidoptera	Pyralidae
2. Fly maggot/Rhizome fly	*Mimegralla coeruleifrons*	Diptera	Micropisidae
3. Rhizome scale	*Aspidiella hartii*	Hemiptera	Coccidae
Coriander			
1. Aphid	*Hyadaphis coriandri* (Das)	Hemiptera	Aphididae
2. White fly	*Bemisia tabaci* (Gennadius)	Hemiptera	Aleurodidae
3. Flower sting bug	*Nezara viridula*	Hemiptera	Pentatomidae
Cucurbits			
1. Red Pumpkin Beetle	*Raphidopalpa foveicollis* (Lucas) *Aulacophora foveicollis* (Lucas)	Coleoptera	Chrysomelidae
2. Fruit fly	*Dacus cucurbitae* (Cog.)	Diptera	Tephritidae
3. Vine borer/Squash Vine borer	*Melitta curcurbitae* (Harris)	Lepidoptera	Sesiidae
Cruciferous vegetables			
1. Cabbage Semilooper	*Trichoplusia ni*	Lepidoptera	Noctuidae
2. Diamond back moth	*Plutella xylostella* (Linnaeus)	Lepidoptera	Plutellidae
3. Leaf webber	*Crocidolomia binotalis* (Zeller)	Lepidoptera	Pyralidae
4. Cabbage borer/Head borer	*Hellula undalis* (Fabricius)	Lepidoptera	Pyralidae
Mango			
1. Inflorescence midge	*Erosomyia indica*	Diptera	Cecidomyiidae
2. Shoot gall psylla	*Apsylla cistella*	Homoptera	Psyllidae
3. Leaf hopper	*Amritodus atkinsoni* (Lethiery)	Homoptera	Cicadellidae

Contd...

Appendix I–*Contd...*

Crop/Pest Name	*Scientific Name*	*Order*	*Family*
4. Mealy bug	*Drosicha mangiferae* (Green)	Homoptera	Pseudococcidae/ Margarodidae
5. Stone weevil	*Sternochaetus mangiferae* (Fab.)	Coleoptera	Curculionidae
6. Stem borer/Tree borer	*Batocera rufomaculata* (DeG.) *Batocera rubus* (Linn.)	Coleoptera	Cerambycidae
Citrus			
1. Lemon Butterfly/ Citrus Caterpillar	*Papilio demoleus* (Linn.)	Lepidoptera	Papilionidae
2. Citrus Psyllia	*Diaphorina citri* (Kuwayama)	Homoptera	Psyllidae
3. Fruit sucking moth	*Ophideres conjuncta, Eudocima* (Othreis) *fullonica* (Linn.), *E. materna*	Lepidoptera	Noctuidae
4. White fly	*Dialeurodes citri*	Homoptera	Aleurodidae
5. Leaf miner	*Phyllocnistic citrella* (Stainton)	Lepidoptera	Phyllocnistidae
Papaya			
1. Fruit Fly	*Bactrocera* (Dacus) *dorsalis*	Diptera	Tephritidae
2. Aphid	*Myzus periscae, Aphis gossypii*	Homoptera	Aphididae
3. White fly	*Bemisia tabaci* (Gennadius)	Homoptera	Aleurodidae
4. Mite	*Tetranychus urticae* (Koch)	Trombidiformes	Tetranychidae
Banana			
1. Rhizome/Rootstock weevil	*Cosmopolites sordidus* (Germar)	Coleoptera	Curculionidae
2. Pseudostem borer// Stem weevil	*Odoiporus longicollis*	Coleoptera	Curculionidae
Guava			
1. Bark borer/Bark eating caterpillars	*Indarbela quadrinotata* (Walker) *Indarbela tetraonis* (Moore.)	Lepidoptera	Metarbelidae
2. Oriental Fruit fly	*Bactrocera diversus, B. zonatus*	Diptera	Tephritidae
Pomegranate			
1. Anar butterfly/Fruit borer/ Anar caterpillar	*Deudorix isocrates* *Virachola isocrates*	Lepidoptera	Lycaenidae
2. Fruit fly	*Bactrocera* (Dacus) *dorsalis*	Diptera	Tephritidae
Ber			
1. Fruit fly	*Carpomyia vesuviana* (Costa)	Diptera	Tephritidae
Litchi			
1. Litchi bug	*Chrysocoris stolii*	Hemiptera	Pentatomidae
2. Leaf roller	*Platypeplus aprobola* (Meyer)	Lepidoptera	Tortricidae
3. White fly			
4. Black hussain fly	*Simulinum* sp.		
5. Mite/Leaf curl mite	*Aceria litchi* (Channa/Keifer.)	Acarina	Eriophyidae

Contd...

Appendix I–*Contd...*

Crop/Pest Name	*Scientific Name*	*Order*	*Family*
Tamarind			
1. Fruit borer	*Aphomia gularis*	Lepidoptera	Pyralidae
2. Mealy bug	*Perissopneumon tamarinda*	Homoptera	Pseudoco-ccidae
Cashewnut			
1. Tea mosquito bug	*Helopeltis antonii* (Signoret)	Hemiptera	Miridae
2. Stem and root borer	*Plocaederus ferrugineus* (Linn.)	Lepidoptera	Cerambycidae
Sapota/Chiku			
1. Leaf webber/Chiku moth	*Nephopteryx eugraphella* (Regonot)	Lepidoptera	Pyralidae
2. Fruit fly	*Bactrocera* (Dacus) *dorsalis*	Diptera	Trypetidae
Stored Grain Pests			
1. Pulse beetle	*Callosobruchus chinensis* (Linn.)	Coleoptera	Bruchidae
2. Rice Weevil	*Sitophillus oryzae* (Linn.)	Coleoptera	Curculionidae
3. Rice grain moth	*Corcyra cephalonica* (Stainton)	Lepidoptera	Galleriidae
4. Red rust flour beetle	*Tribolium castaneum* (Herbat)	Coleoptera	Tenebrionidae
5. Khapra beetle/Wheat beetle	*Trogoderma granarium* (Everts)	Coleoptera	Dermestidae
6. Lesser Grain borer/ Australian wheat weevil	*Rhizopertha dominica* (Fab.)	Coleoptera	Bostrichidae
Polyphagus Pest			
1. White grub	*Leucopholis lepidophora*	Coleoptera	Scarabaeidae
2. Locust	*Schistocera gregaria*	Orthoptera	Acrididae
3. Surface grasshopper	*Chrotogonus trachypterus*	Orthoptera	Acrididae
4. Termite	*Odontotermes obesus*	Isoptera	Termitidae
5. Red hairy caterpillar	*Amsacta moorei*	Lepidoptera	Arctiidae
6. Bihar hair caterpillar	*Spilosoma obliqua*	Lepidoptera	Arctiidae
7. Greasy cutworm	*Agrotis ypsilon*	Lepidoptera	Noctuidae

Appendix II: Damaging Symptoms of Major Crop Pest

Crop/Pest Name	*Scientific Name*	*Damage Symptoms*
Rice		
1. Gall midge	*Orseolia oryzae*	☆ Maggots enter into growing shoot and lacerate the leaf tissue. ☆ Silver shoot/onion leaf in rice.
2. Yellow stem borer	*Scirphophaga incertulas*	☆ Larvae bore into central shoots of paddy seedlings and tillers and cause dead hearts.
3. Green Leaf hopper	*Nephotetrix nigropictus*	☆ Nymphs and adults suck the leaf sap and infested leaves turn brown.
4. Brown Leaf hopper	*Nilaparvata lugens*	☆ Nymphs and adults congregate at the base of the stem just above water level. ☆ Heavy infestation causes yellowing and drying of plant, this condition known as "Hopper burn".
5. White-backed plant hopper	*Sogatella furcifera*	☆ Heavy infestation causes drying of plant, also showing "Hopper burn".
6. Leaf folder/Leaf roller	*Cnaphalocrocis medinalis*	☆ Larvae fold leaf marginally and feds the folded leaves by scraping chlorophyll.
7. Case worm	*Nymphula depunctalis*	☆ Leaf cases hanging from paddy leaf and cut leaf bits floating in water.
8. Army worm	*Mythimna saparata*	☆ Cut the earhead and appearance of field as if grazed by cattle.
9. Swarming caterpillar	*Spodoptera mauritia*	☆ Caterpillars damage paddy crop by cutting off leaf tips, leaf margins, leaves and even the plants at the base.
10. Gundhi bug/Ear head bug	*Leptocorisa acuta*	☆ Individual rice grains become white and chaffy. ☆ Buggy odour in rice fields during milky stage.
Wheat		
1. Pink Stem borer	*Sesamia inferens*	☆ Caterpillars bores into the stem and feed upon the tissues of the central shoot. The infected plant produces 'dead heart'.
2. Termites/White Ant	*Odentotermis obesus* *Microtermes obesus*	☆ Termites feed on the roots and stem parts of the plants. ☆ Damaged plants dry up completely and are easily pulled out.

Contd...

Appendix II–*Contd...*

Crop/Pest Name	Scientific Name	Damage Symptoms
Sorghum and Maize		
1. Stem borer	*Chilo partellus*	☆ Dead heart in early stage of crop. ☆ Pin holes on the whorl of newly opened leaves.
2. Shoot fly	*Atherigona soccata*	☆ Maggots cause dead heart in young crop with more side tillers.
Urd and Moong		
1. Pod borer	*Maruca vitrata*	☆ Webbing of leaves, flowers and pods and feeds by resting inside these webs. ☆ Larva feeds on developing pods.
2. Red hairy caterpillar	*Amsacta moorei*	☆ Early instars as skeletoniser. ☆ Late instars as defoliator, leaving only midribs.
Soybean		
1. Girdle beetle	*Oberea brevis*	☆ The grub tunnel into the stem upto bottom and may be filled with excreta. ☆ The tunneled portion wilts and dies.
2. Stem fly	*Melanagromyza sojae*	☆ Mining of leaves is observed and larva bores into leaf petiole and tunnels into tender stem. ☆ The tunneled portion above the stem, wilts and die.
3. Red Hairy caterpillar	*Amsacta moorei*	☆ Early instars as skeletoniser. ☆ Late instars as defoliator, leaving only midribs.
Pigeonpea		
1. Pod borer/Spiny pod borer	*Etiella zincknella*	☆ Larvae enter the pod and feed on the seed. ☆ Dropping of flowers.
2. Pod fly	*Melanagromyza obtuse*	☆ Maggots feed on the seeds as internal borer. ☆ Pin head sized bore hole near the septum and shriveling the developing seeds. ☆ Half eaten seeds can be seen inside the pod.
3. Plume moth/Pod caterpillar	*Exelastis atomosa*	☆ Larva found inside the irregular bore hole in pod. ☆ Also feed on flowers, buds and on developing grains.
4. Pod bug	*Clavigralla gibbosa*	☆ Nymphs and adults suck the sap from tender shoots and seeds and these seeds become shriveled and lose germination.

Contd...

Appendix II–*Contd...*

Crop/Pest Name	*Scientific Name*	*Damage Symptoms*
Gram/Chickpea		
1. Pod borer	*Helicoverpa armigera*	☆ Young larvae feeds by scraping green tissue on the leaves and older larvae feeds on buds, flowers and pods. ☆ Circular bore holes on pods plugging by the head of a larva.
2. Cut worm	*Agrotis ypsilon*	☆ Cutting of tender stem and defoliates the leaves.
Pea		
1. Aphid/Green aphid	*Acyrthosiphon pisum*	☆ Aphids extract sap from the terminal leaves and stem. ☆ Their feeding can result in deformation, wilting, or death of the host.
2. Pod borer	*Etiella zincknella*	☆ Larvae enter the pod and feed on the seed. ☆ Dropping of flowers.
Lathyrus		
1. Thrips	*Caliothrips indicus*	☆ Nymphs and adults feed on tender leaves and stunting of plants can be seen in case of severe infestation.
Groundnut		
1. Aphids	*Aphis craccivora*	☆ Yellowing and distorted terminal shoots in groundnut. ☆ Nymphs and adults congregate on growing tips and foliage and desap plants resulting in chlorotic patches and curling of leaves.
2. Leaf minor	*Stomoperyx nertaria* *Aproaerema modicella*	☆ Larvae mines into the leaves. ☆ White blotches are produced. ☆ Webbing and drying of terminal leaf lets in groundnut.
3. White grub	*Holotrachia conseguina*	☆ Wilting of groundnut plants in patches with damaged roots.
4. Red Hairy caterpillar	*Amsacta moorei*	☆ Early instars as skeletoniser. ☆ Late instars as defoliator, leaving only midribs.
Sesamum		
1. Gall fly	*Asphondylia sesame*	☆ Maggots feed on inner contents by boring into capsules. ☆ Gall like swelling on the pod.

Contd...

Appendix II–*Contd...*

Crop/Pest Name	*Scientific Name*	*Damage Symptoms*
2. Howk moth	*Acherontia styx*	☆ Larvae defoliate the plants.
3. Leaf webber	*Antigastra catalaunalis*	☆ Larva webs the top leaves and feed by remaining inside and even it damages the capsules by boring into it.
Linseed		
1. Bud fly/Gall Midge	*Dasineura lini*	☆ Maggots, which feed on the flower buds and prevent their proper opening. Consequently the seed does not set properly. ☆ Due to their feeding, galls are produced and there is no pod formation.
2. Caterpillar	*Spodoptera exigua*	☆ Young larvae feed gregariously and skeletonise foliage.
3. Thrips	*Thrips lini*	☆ Nymphs and adults feed on tender leaves and stunting of plants can be seen in case of severe infestation.
4. Jassid	*Empoasea kerri* var. *motti*	☆ Yellowing and cupping of leaves is observed due to sucking of sap by adults and nymphs.
Safflower		
1. Aphid	*Uroleucon carthami* *Dactynotus carthami*	☆ Nymphs and adults found in group and suck the sap from leaves and capsules.
2. Bud fly/Capsule fly	*Acanthiophilus helianthi*	☆ Newly hatched larvae feed on the floral parts including thalamus and soft parts of the capsules. ☆ Affected buds show small bore holes. ☆ The infested buds rotten with foul smelling ooze coming out of the apices.
Sunflower and Niger		
1. Bihar hair caterpillar	*Spilosoma obliqua*	☆ Larvae defoliate the plants.
Mustard		
1. Aphid	*Lipaphis erysimi*	☆ Curly leaf tip with poor pod set.
2. Sawfly	*Athalia lugens proxima*	☆ Defoliation of leaf by a poly pod larva, encircling the leaf edge by caudal end. ☆ Margin of tender leaves are cut.
3. Painted bug	*Bagrada hilari*	☆ Due to feeding by bugs, the vigour of the crop is reduced.

Contd...

Appendix II–*Contd...*

Crop/Pest Name	*Scientific Name*	*Damage Symptoms*
Caster		
1. Semilooper	*Achaea janata*	☆ Larvae feed gregariously and defoliate the plants. ☆ In severe infestation, only midribs are left.
2. Capsule borer	*Conogethes punctiferalis*	☆ Larvae bore into terminal shoots and capsules. ☆ Capsules that are attacked are webbed together and galleries are made.
3. Tussock hairy caterpillar	*Notolophus posticus*	☆ The newly emerged larvae scraped leaves and feeding on the paranchymatus tissue of the leaves. ☆ Later on, the caterpillar feed voraciously on the castor leaves leaving behind only the midrib and stalk.
Cotton:		
1. Pink Bollworm	*Pectinophora gossypiella*	☆ Larvae feed on flower buds, flowers and seed. ☆ Feeds on seeds by moving from one locule to another in boll. ☆ Pin head size hole on bolls. ☆ Rosette flowers.
2. Spotted Bollworm	*Earias vitella, E. insulana*	☆ Larvae attack shoots which results in wither and they drop. ☆ Holes on damaged bolls are plugged with excreta. ☆ Small and circular hole with faecal matter on the fruiting bodies. ☆ Flared squares. ☆ Drooping and drying of shoots.
3. American Bollworm	*Helicoverpa armigera*	☆ Larvae feed voraciously on leaves, square and flowers. ☆ Larger and irregular hole.
4. Jassid/Leaf hopper	*Amrasca biguttula biguttula*	☆ Nymphs and adults suck sap causing yellowing, curling and reddening of leaves. ☆ Drying and falling of leaves is observed.
5. Red Cotton Bug	*Dysdercus koenigii*	☆ Nymphs and adults suck sap from developing green bolls affecting the quality of lint.

Contd...

Appendix II–*Contd...*

Crop/Pest Name	Scientific Name	Damage Symptoms
6. White Fly	*Bemisia tabaci*	☆ Nymphs and adults suck sap leading to premature defoliation, uniform bronzing and also development of sooty mould due to honeydew secretion. ☆ Shedding of squares and bolls in severe infestation ☆ It is a vector of leaf curl virus
7. Aphid	*Aphis gossypii*	☆ Crinking and curling of leaves is observed due to aphid colonies sucking the sap by staying on the under surface of leaves. ☆ Leaves appear shiny and sticky due to honeydew secretion.
Sunhemp		
1. Hairy caterpillar	*Utetheisa pulchella*	☆ The just hatched larva either defoliates (vegetative stage) or bores into pods (in heading stage). ☆ The larvae feed on sun hemp and defoliate the plant.
Mesta		
1. Spiral borer	*Agrilus acutus*	☆ Larvae start feeding upon the woody tissues, making a spiral around the stem beneath the bark. ☆ A portion of the infested region swells up considerably to form an elongated gall. ☆ The fibre obtained from the infected plants become useless.
Sugarcane		
1. Top shoot borer/Top borer	*Tryporza novella*	☆ Shoot holes are noticed on leaves once they open. ☆ Bunchy top appearance in growing up sugarcane due to formation of side shoots.
2. Stem/Early shoot/ Internode borer	*Chilo sacchariphagus indicus*	☆ Dead heart in young sugarcane crop by cutting of the growing shoot/Internode.
3. Leaf hopper/Pyrilla	*Pyrilla purpusilla*	☆ Nymphs and adults suck sap from leaves. ☆ Plant infested loose vigour, become shrunken, sucrose content and quality of juice is affected. ☆ Honeydew secretion leads to development of sooty mould.
4. White Fly	*Aleurolobus barodensis*	☆ Nymphs and adults suck sap from leaves causing characteristic yellow streaks. ☆ Leaves remain dry and plants become stunted in severe infestation.
5. Mealy bug	*Saccharicoccus sacchari*	☆ Nymphs and adults stick to the nodes under leaf sheath. ☆ They suck the juice from the cane and reduces the vigour of the plant.

Contd...

Appendix II–*Contd...*

Crop/Pest Name	Scientific Name	Damage Symptoms
Potato		
1. Tuber Moth	*Phthorimaea operculella*	☆ Mines on leaves and bore holes on tender shoot and tubers in potato.
2. Aphid/Green peach aphid	*Myzus persicae*	☆ Nymphs and adults suck sap from leaves.
3. Cut Worm/Greasy cutworm	*Agrotis ypsilon*	☆ Caterpillar feed on young leaves and also cut tender seedlings and stems at ground level.
Okra/Bhindi		
1. Shoot and Fruit borer	*Earias vitella, E. insulana*	☆ Caterpillars bore into tender shoots and feed on the inner content as a result of which affected shoots wither and die and due to which side shoots may arise. ☆ Buds and fruits are damaged and dropping of fruits and buds is observed.
2. Jassid/Leaf hopper	*Amrasca biguttula biguttula*	☆ Nymphs and adults suck sap causing yellowing, curling and reddening of leaves.
3. White fly	*Bemisia tabaci*	☆ Nymphs and adults suck sap from under surface of leaves and excrete honeydew. ☆ It is a vector of Yellow mosaic virus.
Brinjal		
1. Shoot and fruit borer	*Leucinodes orbonalis*	☆ Larva bores into tender shoots resulting in drying of tip. ☆ Also bores into developing fruits and bore hole plugged with excreta in brinjal fruit can be observed.
2. Stem borer	*Euzophera perticella*	☆ Caterpillars bore into stem near the ground level and damaged plant wilt.
3. Mite	*Tetranychus urticae* *Tetranychus telaris*	☆ Adults and nymphs suck sap. ☆ White specks on leaves.
Chilli		
1. Thrips	*Thrips tabaci*	☆ Nymphs and adults lacerate the leaf tissue. ☆ Inward curling of leaves, stunted plant growth and dropping of flower buds and fruits.
2. Fruit borer/Pod borer	*Helicoverpa armigera*	☆ Caterpillars bore into fruit and dropping of fruits and may also feed on seeds.

Contd...

Appendix II–*Contd...*

Crop/Pest Name	*Scientific Name*	*Damage Symptoms*
Tomato		
1. Fruit borer	*Helicoverpa armigera*	☆ Caterpillars bore into fruits.
2. Leaf minor	*Liriomyza trifolii*	☆ Larvae mines into the leaves and in cases of severe damage leaves dry up.
3. Stem borer	*Euzophera perticella*	☆ Caterpillars bore into stem near the ground level and damaged plant wilt.
Onion and Garlic		
1. Onion Thrips	*Thrips tabaci*	☆ Nymphs and adults feed on tender leaves and stunting of plants can be seen in case of severe infestation. ☆ Leaf tip drying in onion.
2. Tobacco caterpillar	*Spodoptera litura*	☆ Larvae defoliates the plant leaves.
3. Onion fly	*Delia (Hylemya) antique*	☆ Maggots bore on bulb and stem.
Sweet Potato		
1. Weevil	*Cylas formicarius*	☆ Grubs bore into vines resulting in wilting of branches and also bore into tubers ☆ Adults nibble the vines and tubers. ☆ Bore holes on tuber in field and storage.
Ginger		
1. Shoot borer/Rhizome borer	*Dichocrocis punctiferalis*	☆ The larvae bore into pseudostems and feed on internal tissues resulting in yellowing and drying of leaves of infested pseudostems.
2. Fly maggot/Rhizome fly	*Mimegralla coeruleifrons*	☆ Maggots on hatching bore into the rhizomes of ginger and damage them. ☆ The yellowing of plants and rotting of rhizomes takes place due to severe infestation of pest. ☆ 'Dead hearts' are also seen due to primary injury of the maggots.
3. Rhizome scale	*Aspidiella hartii*	☆ Scale feed on plant sap in field or on rhizome.
Coriander		
1. Aphid	*Hyadaphis coriandri)*	☆ Nymph and adult suck the cell sap from leaves, flowers and fruits and excuding copious quality of honeydew which favours the growth of shooty mould. ☆ Yellowing, curling and subsequent drying of leaves takes place, resulting in poor and shriveled seed formation.

Contd...

Appendix II–*Contd...*

Crop/Pest Name	*Scientific Name*	*Damage Symptoms*
2. White fly	*Bemisia tabaci*	☆ Nymphs and adults suck sap from under surface of leaves and excrete honeydew. ☆ It is a vector of yellow mosaic virus.
3. Flower sting bug	*Nezara viridula*	☆ Stink bugs feed on developing pods, the seeds do not develop. ☆ Stink bugs inflict mechanical injury to the seed as well as transmit the yeast-spot disease organism. ☆ In damaged seeds, germination is adversely affected.
Cucurbits		
1. Red Pumpkin Beetle	*Raphidopalpa foveicollis*	☆ Irregular holes on the cucurbit leaves.
2. Fruit fly	*Dacus cucurbitae*	☆ Maggots feed on the pulp and seeds inside. ☆ The fed fruits may fall premature. ☆ Infested fruits become soft and rotten.
3. Vine borer/Squash Vine borer	*Melitta curcurbitae*	☆ Stems can be girdled by larvae, which prevent water and nutrients from circulating in the plant, plants marked by a hole with yellow granular or sawdust-like frass exuding from it. ☆ Injured vines often decay and become wet and shiny.
Cruciferous vegetables		
1. Cabbage Semilooper	*Trichoplusia ni*	☆ Caterpillars act as defoliator.
2. Diamond back moth	*Plutella xylostella*	☆ Mining and skeletanization of cabbage leaves. ☆ Larvae also feed on cabbage.
3. Leaf webber	*Crocidolomia binotalis*	☆ Caterpillars webs the green leaves with silken thread and skeletonizes the webbed leaves. ☆ Larvae feed on web leaves by remaining inside.
4. Cabbage borer/Head borer	*Hellula undalis*	☆ Larvae bore into stem, stalk and cabbage and consume its inner content.
Mango		
1. Inflorescence midge	*Erosomyia indica*	☆ Maggot affects the crop at floral bud burst, fruit set and tender leaves particularly encircling the inflorescence. ☆ The inflorescence show stunting growth and ultimately dies before fruit set.

Contd...

Appendix II–*Contd...*

Crop/Pest Name	*Scientific Name*	*Damage Symptoms*
2. Shoot gall psylla	*Apsylla cistella*	☆ Growing buds of shoots are attacked which results in formation of galls on leaf axil.
3. Leaf hopper	*Amritodus atkinsoni*	☆ Drying of inflorescence with large scale withering and shredding of mango flowers.
4. Mealy bug	*Drosicha mangiferae*	☆ Large numbers of bugs are found on stalks and fruits where they affect the quality of fruit and if infestation is found on leaves, the leaves dry up.
5. Stone weevil	*Sternochaetus mangiferae*	☆ 'T' shaped marking on marble sized mango fruits. ☆ Grubs tunnel through pulp, endocarp and finally reach the cotyledon and feed on them. ☆ Germination is affected in cotyledons.
6. Stem borer/Tree borer	*Batocera rufomaculata* *Batocera rubus*	☆ Bore holes on mango stem with protrusion of frass and drying of bunches.
Citrus		
1. Lemon Butterfly	*Papilio demoleus*	☆ Defoliation in citrus by caterpillars.
2. Citrus Psyllia	*Diaphorina citri*	☆ Drying and withering of terminal shoot in citrus with the presence of large number of ants.
3. Fruit sucking moth	*Ophideres conjuncta*	☆ Adults suck the juice from ripening fruits resulting in dropping. ☆ Shedding of citrus fruits with feeding punctures.
4. Leaf miner	*Phyllocnistic citrella*	☆ Zig zag leaf mines in citrus leaves resulting in wilting and curling of leaves.
Papaya		
1. Fruit Fly	*Bactrocera* (Dacus) *dorsalis*	☆ Maggots feed on the pulp of the fruit and the affected portion rot and the fruit drops off.
2. Aphid	*Myzus periscae,* *Aphis gossypii*	☆ Nymphs and adults suck sap from leaves and later on leaves become curled and crinkled and unfit for curing.
3. White fly	*Bemisia tabaci*	☆ Nymphs and adults suck sap from under surface of leaves and excrete honeydew. ☆ It is a vector of Yellow mosaic virus.
4. Mite	*Tetranychus urticae*	☆ Eruptions can be observed on leaves and under severe attack, the leaves curl.

Contd...

Appendix II–*Contd...*

Crop/Pest Name	*Scientific Name*	*Damage Symptoms*
Banana		
1. Rhizome/Rootstock weevil	*Cosmopolites sordidus*	☆ Grubs tunnel within corm and feed due to which presence of bore holes are observed. ☆ Plants wilt because of grubs boring into rhizome.
Guava		
1. Bark borer	*Indarbela quadrinotata*	☆ Feeding of caterpillar on bark and also the presence of small chips of wood and excreta with sticky gum adhered to the branches and tunneled into the stem.
2. Oriental Fruit fly	*Bactrocera diversus*	☆ Maggots feed on the pulp of the fruit and the affected portion rot and the fruit drops off.
Pomegranate		
1. Anar butterfly	*Deudorix isocrates*	☆ Bore hole on pomegranate fruits surrounded by faecal material and feeds on inner content *i.e.* pulp and seeds.
2. Fruit fly	*Bactrocera (Dacus) dorsalis*	☆ Maggots feed on the pulp of the fruit and the affected portion rot and the fruit drops off.
Ber		
1. Fruit fly	*Carpomyia vesuviana*	☆ Maggots feed on fresh pulp and fruits get deformed, turn dark brown and rot.
Litchi		
1. Mite/Leaf curl mite	*Aceria litchi*	☆ Eruptions can be observed on leaves and under severe attack, the leaves curl.
Cashewnut		
1. Tea mosquito bug	*Helopeltis antonii*	☆ Nymphs and adults suck sap from leaves, petioles and tender shoots causing brown blackish patches. ☆ Corky out growth or scab formation.
2. Stem and root borer	*Plocaederus ferrugineus*	☆ Grubs enter into stem by making holes resulting in yellowing and gradual wilting of tree and the bored holes are filled with frass and excreta.

Contd...

Appendix II–*Contd...*

Crop/Pest Name	*Scientific Name*	*Damage Symptoms*
Sapota/Chiku		
1. Leaf webber/Chiku moth	*Nephopteryx eugraphella*	☆ Larvae web the leaves and feeds on the chlorophyll by remaining inside the web.
2. Fruit fly	*Bactrocera* (Dacus) *dorsalis*	☆ Maggots feed on the pulp of the fruit and the affected portion rot and the fruit drops off.
Stored Grain Pests		
1. Pulse beetle	*Callosobruchus chinensis*	☆ Grubs feed on the inner content of pulse seed and pupate inside the seed.
2. Rice Weevil	*Sitophillus oryzae*	☆ Grub feeds inside and pupates inside and emerges by making a irregular hole through the grain.
3. Rice grain moth	*Corcyra cephalonica*	☆ Larvae web together the grains and feed within inside. ☆ Also attacks broken grains and also milled products.
4. Red rust flour beetle	*Tribolium castaneum*	☆ Grubs and adults feed from outside and they are serious pests of processed foods like flour and grains.
5. Khapra beetle/Wheat beetle	*Trogoderma granarium*	☆ Grubs mainly attack stored wheat and other grains.
6. Lesser Grain borer	*Rhizopertha dominica*	☆ Grubs feed on the inner content of grains like wheat and rice.

Appendix III: ETL of Crop Pest

Crop/Pest Name	*Scientific Name*	*ETL*
Rice		
1. Gall midge	*Orseolia oryzae*	1 gall/m^2 or 5 per cent affected tillers or 5-10 per cent silver shoots in non-endemic areas
2. Yellow/Stem borer	*Scirphophaga incertulas*	5 per cent Dead hearts or white ear
3. Green Leaf hopper	*Nephotetrix nigropictus*	2 insects/hill in tungro endemic area or 20-30 insects/hill in tungro areas
4. Brown Leaf hopper	*Nilaparvata lugens*	5-10 insects/hill
5. White-backed plant hopper	*Sogatella furcifera*	5-10 insects/hill
6. Leaf folder/Leaf roller	*Cnaphalocrocis medinalis*	1 damaged leaf/hill or 10 per cent folded leaves
7. Case worm	*Nymphula depunctalis*	1-2 cases/hill
8. Army worm	*Mythimna saparata*	1 caterpillar/hill
9. Swarming caterpillar	*Spodoptera mauritia*	1 caterpillar/hill
10. Gundhi bug/Ear head bug	*Leptocorisa acuta*	1-2 bugs/m^2
Wheat		
1. Pink Stem borer	*Sesamia inferens*	10 per cent dead hearts
Sorghum and Maize		
1. Stem borer	*Chilo partellus*	5-10 per cent infestation
2. Shoot fly	*Atherigona soccata*	15 per cent dead hearts
Urd and Moong		
1. Pod borer	*Maruca vitrata*	5 webs/plant
2. Red hairy caterpillar	*Amsacta moorei*	8 egg masses/100 meter
Soybean		
1. Girdle beetle	*Obereopsis brevis*	4 beetles/m or 5 per cent incidence
2. Stem fly	*Melanagromyza sojae*	5 per cent infestation
3. Red Hairy caterpillar	*Amsacta moorei*	5 larvae/meter row
Pigeonpea		
1. Pod borer/Spiny pod borer	*Etiella zincknella*	2-3 eggs/plant or 1 mature larvae/ 10 plants
2. Pod fly	*Melanagromyza obtuse*	Presence of 5 per cent oviposition on pod basis
3. Plume moth/Pod caterpillar	*Exelastis atomosa*	5 larvae per plant
4. Pod bug	*Clavigralla gibbosa*	1 egg mass/plant
Gram/Chickpea		
1. Pod borer	*Helicoverpa armigera*	3 eggs or 2 small larvae/plant
2. Cut worm	*Agrotis ypsilon*	5 per cent plant mortality

Contd...

Appendix III–*Contd...*

Crop/Pest Name	Scientific Name	ETL
Pea		
1. Aphid/Green aphid	*Acyrthosiphon pisum*	40 to 50 aphids per 25 cm (10 inch) or shorter stem
2. Pod borer	*Etiella zincknella*	10 per cent affected parts
Lathyrus		
1. Thrips	*Caliothrips indicus*	25 per cent of the damaged leaves or 5 thrips per terminal shoot.
Groundnut		
1. Aphids	*Aphis craccivora*	5-10 aphids/terminal at seedling stage
2. Leaf minor	*Stomoperyx nertaria* *Aproaerema modicella*	5 mines/plant at 30 days of crop age, 10 mines/plant up to 45 days of emergence
3. White grub	*Holotrachia conseguina*	5 plants/m^2
4. Red Hairy caterpillar	*Amsacta moorei*	8 egg masses/100 meter
Sesamum		
1. Gall fly	*Asphondylia sesame*	2-5 galls/plant or 10 per cent damage
2. Howk moth	*Acherontia styx*	1-2 larvae/plant
3. Leaf webber	*Antigastra catalaunalis*	2-5 webs/plant
Linseed		
1. Bud fly/Gall Midge	*Dasineura lini*	5 per cent bud infestation or 2 insects per 10 flax
2. Caterpillar	*Spodoptera exigua*	5-8 larvae (worms)/25 plants
3. Thrips	*Thrips lini*	8-10 Adults or Nymphs per Leaf
4. Jassid	*Empoasea kerri* var. *motti*	25 per cent damaged leaves 5-10 per cent nymphs/adults per plants
Safflower		
1. Aphid	*Uroleucon carthami* *Dactynotus carthami*	50-70 nymphs or adults/plant (5cm apical twig)
Sunflower and Niger		
1. Bihar hair caterpillar	*Spilosoma obliqua*	At 20-25 per cent leaf defoliation
Mustard		
1. Aphid	*Lipaphis erysimi*	50-60 aphids/10 cm terminal portion of central shoot
2. Painted bug	*Bagrada hilari*	One bug/10 ft.^{2} on plants
Cotton		
1. Pink Bollworm	*Pectinophora gossypiella*	5-10 per cent boll infestation
2. Spotted Bollworm	*Earias vitella, E. insulana*	5-10 per cent boll infestation
3. American Bollworm	*Helicoverpa armigera*	1 larvae/plant or 5-10 per cent boll infestation

Contd...

Appendix III–*Contd...*

Crop/Pest Name	Scientific Name	ETL
4. Jassid/Leaf hopper	*Amrasca bigutulla*	1-2 nymphs/leaf or 50 per cent yellowing and curling of the plants
5. White Fly	*Bemisia tabaci*	8-10 adults/leaf or 20 nymphs/leaf or 10 thrips/leaf
6. Aphid	*Aphis gossypii*	15-20 per cent infestation or appearance of honeydew on 50 per cent plants
Sugarcane		
1. Top shoot borer/Top borer	*Tryporza novella*	5 per cent damage of terminal portion
2. Stem/Shoot/Internode borer	*Chilo sacchariphagus indicus*	15-20 per cent dead hearts
3. Leaf hopper/Pyrilla	*Pyrilla purpusilla*	3-5 nymphs or adults/leaf
4. White Fly	*Aleurolobus barodensis*	10 nymphs/Leaf
Potato		
1. Tuber Moth	*Phthorimaea operculella*	15-20 moths/trap consecutive for 3 nights
2. Aphid/Green peach aphid	*Myzus persicae*	20 aphids/100 leaves
3. Cut Worm/Greasy cutworm	*Agrotis ypsilon*	3 per cent Attacked Plants
Brinjal		
1. Shoot and fruit borer	*Leucinodes orbonalis*	1-5 per cent shoot/fruit infestation
2. Mite	*Tetranychus urticae* *Tetranychus telaris*	10/Leaf or 30 per cent of plants infested
Chilli		
1. Thrips	*Thrips tabaci*	6 thrips/leaf or 10 per cent affected plants
2. Fruit borer/Pod borer	*Helicoverpa armigera*	1 egg or 1 larvae/plant or 1 damaged fruit
Tomato		
1. Fruit borer	*Helicoverpa armigera*	1 egg or 1 larvae/plant or 1-5 per cent fruit damage
2. Leaf minor	*Liriomyza trifolii*	2-5 miner/leaf
Onion and Garlic		
1. Onion Thrips	*Thrips tabaci*	5 thrips/leaf
2. Tobacco caterpillar	*Spodoptera litura*	1-5 per cent incidence
3. Onion fly	*Delia* (Hylemya) *antique*	1 maggot/hill
Ginger		
1. Shoot borer/Rhizome borer	*Dichocrocis punctiferalis*	1 egg mass/m^2
Coriander		
1. Aphid	*Hyadaphis coriandri*)	1 aphid/plant
2. Flower sting bug	*Nezara viridula*	2 per cent bug damage
Cucurbits		
1. Red Pumpkin Beetle	*Raphidopalpa foveicollis*	1/10 Plants (at seedling stage) or 1/Plants (at crop stage)

Contd...

Appendix III–*Contd...*

Crop/Pest Name	*Scientific Name*	*ETL*
Cruciferous vegetables		
1. Cabbage Semilooper	*Trichoplusia ni*	20 larvae/l0 plants or 5 per cent defoliation
2. Diamond back moth	*Plutella xylostella*	1-5 per cent incidence
3. Cabbage borer/Head borer	*Hellula undalis*	15-25 per cent of plants infested
Mango		
1. Inflorescence midge	*Erosomyia indica*	10 spots/twig or inflorescence
2. Leaf hopper	*Amritodus atkinsoni*	5 adults/panicle
3. Mealy bug	*Drosicha mangiferae*	On Appearance
4. Stem borer/Tree borer	*Batocera rufomaculata* *Batocera rubus*	Appearance of pest
Citrus		
1. Lemon Butterfly/Caterpillar	*Papilio demoleus*	20-30 per cent foliar damage
2. Citrus Psyllia	*Diaphorina citri*	6/Leaf
3. Leaf miner	*Phyllocnistic citrella*	10 per cent affected leaves
Papaya		
1. Fruit Fly	*Bactrocera* (Dacus) *dorsalis*	10 per cent affected fruits
2. Mite	*Tetranychus urticae*	10/Leaf
Guava		
1. Oriental Fruit fly	*Bactrocera diversus*	10 per cent affected fruits
Pomegranate		
1. Anar butterfly	*Deudorix isocrates*	5 eggs/plant
2. Fruit fly	*Bactrocera* (Dacus) *dorsalis*	10 per cent affected fruits
Ber		
1. Fruit fly	*Carpomyia vesuviana*	1-2 per cent incidence

Appendix IV: Insecticides and their Trade Names

Insecticides	*Trade Name*
Organochlorines	
Dicofol 18.5 EC (miticide)	*Kelthane*
Endosulfan 35EC*	*Thiodan*
Organophosphates:	
Acephate 75 per cent SP	*Asataf, Orthene and Starthene*
Chlorpyrifos 20EC	*Dursban*
Dimethoate 30EC	*Rogor*
Dichlorvos 76EC	*Nuvan, Vapona*
Ethion50EC	*Fosmite*
Fenitrothion 50EC*	*Sumithion*
Monocrotophos 36SL*	*Nuvacron*
Malathion 50EC	*Cythion, Hilthion*
Methyl parathion 50EC*	*Metacid*
Oxydemeton-methyl25EC	*Metasystox*
Phorate 10 G	*Thimet*
Phosphamidon 40SL	*Dimecron*
Phosalone35EC	*Zolone*
Profenophos 50EC	*Curacron*
Triazophos 40EC and 20EC	*Hostathion*
Quinalphos 25EC	*Ekalux*
Carbamates	
Aldicarb 10CG	*Temik*
Carbufuran 3G	*Furadan*
Carbaryl 75WP	*Sevin*
Methomyl 40SP	*Lannate*
Propoxur 1 per cent Aerosol	*Baygon*
Thiodicarb 75WP	*Larvin*
Synthetic pyrethroids	
Betacyfluthrin 2.45 SC	*Bulldock*
Bifenthrin 10EC	*Tal star*
Cypermethrin 10EC and 25 EC	*Cymbush, Bilcyp*
DeltamethrinEtofenprox 10EC	*Decis 2.8EC, K-Obiol 2.5 WP*
Fenpropathrin 10EC	*Meothrin*
Fenvaleratre 20EC and 0.4 per cent DP	*Parafen*
Alphamethrin 10EC	*Alphaguard*

Contd...

Appendix IV–*Contd...*

Insecticides	*Trade Name*
Fumigant	
Aluminium phosphide*	*Celphos*
New insecticides group:	
Neonicotinoids	
Imidacloprid	*Confidor 17.8 SL*
Thiamethoxam	*Cruiser, Actara*
Acetmiprid 20 SP	*Pride*
Pyyrole insecticides	
Fipronil5 SC	*Regent*
Avermectins	
Emamectin benzoate	*Proclaim*
Spinosyns	
Spinosad 45 per cent SC and 2.5 per cent SC	*Tracer, Naturalyte*
Chitin synthesis inhibitors	
Diflubenzuron 25WP	*Dimilin*
Biopesticides	
Bacillus thuringiensis (Liquid and WP)	*Dipel, Delfin, Halt, Spicturin, Biolep*
Verticillium lecanii	*Vertilec*
Beauveria bassiana	*Larvocel, Boverin*
Hirsutella thompsoni	*Mycar*
Metarrhizium anisopliae	*Blomax*
Miscellaneous	
Cartap hydrochloride 4 per cent	*Padan,Caldan*

* Pesticide restricted for use in India.

Level of Hazard of Insecticides

Level of Hazard		*Colour of Triangle Contains*
1. Extremely hazards	-	Bright Red
2. Very hazards	-	Bright Yellow
3. Moderately hazards	-	Bright Blue
4. Relative hazards (low)	-	Bright Green

References

Arora, R. and Dhaliwal, G.S. 1999. The Insects: Diversity, Habits and Management. Kalyani Publishers, New Delhi India.

Atwal, A.S. 1975. Agricultural Pests of India and South East Asia. Kalyani Publishers, New Delhi India.

David, B.V. 1992. Pest Management and Pesticides: Indian Scenario. Namrutha Publications, Chennai, India.

Dhaliwal, G.S. and Arora, R. 1994. Trends in Agricultural Insect Pest Management. Commonwealth Publishers, New Delhi, India.

Dhaliwal, G.S. and Heinrichs, E.A. 1998. Critical Issues in Insect Pest Management. Commonwealth Publishers, New Delhi, India.

Dhaliwal, G.S. and Arora, R. 2006. Integrated Pest Management: Concepts and Approaches. Kalyani Publishers, New Delhi India.

Hameed, S.F. and Singh, S.P. 1998. Handbook of Pest Management. Kalyani Publishers, New Delhi, India.

Ignacimuthu, S. And Jayaraj, S. 2005. Sustainable Insect Pest Management. Narosa Publishing House, New Delhi, India.

Ignacimuthu, S. And Jayaraj, S. 2003. Biological Control of Insect Pests. Phoenix Publishing House, New Delhi, India.

Ignacimuthu, S. And Jayaraj, S. 2006. Biodiversity and Insect Pest Management. Narosa Publishing House, New Delhi, India.

Khare, B.P. 1994. Stored Grain Pests and their Management. Kalyani Publishers, New Delhi India.

Mathur, Y.K. and Upadhyay, K.D. 2000. A Text Book of Entomology. Aman Publishing House, Meerut, India.

Panwar, V.P.S. 1995. Agricultural Insect Pests of Crops and their Control. Kalyani Publishers, New Delhi, India.

Pradhan, S. 1983. Agricultural Entomology and Pest Control. Indian Council of Agricultural Research, New Delhi, India.

ICAR, 2010. Hand book of Agriculture. Indian Council of Agricultural Research, New Delhi, India.

Prakash, A., Rao, J., Pasalu, I.C. and Mathur, S.N. 1987. Rice Storage and Insect Pest Management. B.R. Publishing Corp., New Delhi, India.

Prashad, D. and Puri, S.N. 2002. Crop Pest and Disease Management: Challenges for the Millennium. Jyoti Publishers, New Delhi, India.

Rangarajan, A.V., Chelliah, S. and Jayaraj, S. And Jayaraj. 1985. Pest Management in Field Crops and Stored Products. Tamil Nadu Agricultural University, Coimbtore, India.

Shivankar, V.J. and Singh, S. 2005. Insect Pests of Citrus and their Management. Kalyani Publishers, New Delhi, India.

Singh, A., Sharma, O.P. and Garg, D.K. 2006. Integrated Pest Management Vol (2). CBS Publishers and Distributors, New Delhi, India.

Kumar, A. and Sharma, J.P. 2007. Agriculture Update. International Book Distributing Co., Luchnow, India.

Shrivastava, K.P. 1996. A Textbook of Applied Entomology. Kalyani Publishers, New Delhi, India.

Verma, L.R. Verma, A.K. and Goutham, D.C. 2004. Pest Management in Horticultural Crops: Principles and Practices. Asiatech Publ., New Delhi, India.

Pratiyogita Darpan. 2010. Agriculture Science. Published by Pratiyogita Darpan Group, New Delhi, India.

Sasikumar, K. and Saravankumar, V. 2010. Agriculture Made Easy. Jain Brothers Publications, New Delhi, India.

Index

B

C

D

E

F

G

H

I

J

K

L

S

T

U

www.ingramcontent.com/pod-product-compliance
Ingram Content Group UK Ltd.
Pitfield, Milton Keynes, MK11 3LW, UK
UKHW021445280726
14060UKWH00001BA/258

9 789386 07101